FORGING THE ANVIL

FORGING THE ANVIL

Combat Units in the US, British, and German Infantries of World War II

G. Stephen Lauer

LYNNE RIENNER PUBLISHERS

BOULDER
LONDON

Published in the United States of America in 2022 by
Lynne Rienner Publishers, Inc.
1800 30th Street, Suite 314, Boulder, Colorado 80301
www.rienner.com

and in the United Kingdom by
Lynne Rienner Publishers, Inc.
Gray's Inn House, 127 Clerkenwell Road, London EC1 5DB
www.eurospanbookstore.com/rienner

Library of Congress Cataloging-in-Publication Data
Names: Lauer, G. Stephen, 1952–2020 author.
Title: Forging the anvil : combat units in the US, British, and German infantries of World War II / G. Stephen Lauer.
Description: Boulder : Lynne Rienner Publishers, Inc., [2022] | Includes bibliographical references and index. | Summary: "A landmark comparative study of the key factors that influenced the creation of World War II infantries in the United States, Germany, and Great Britain and sustained them in the crucible of close combat"—
Provided by publisher.
Identifiers: LCCN 2021025743 (print) | LCCN 2021025744 (ebook) | ISBN 9781626379589 (hardback) | ISBN 9781955055031 (ebook)
Subjects: LCSH: Infantry—History—20th century. | Great Britain. Army. Infantry—History—World War, 1939–1945. | United States. Army—Infantry—History—World War, 1939–1945. | Germany. Heer—History—World War, 1939–1945.
Classification: LCC UD15 .L38 2022 (print) | LCC UD15 (ebook) | DDC 356/.1—dc23
LC record available at https://lccn.loc.gov/2021025743
LC ebook record available at https://lccn.loc.gov/2021025744

British Cataloguing in Publication Data
A Cataloguing in Publication record for this book is available from the British Library.

Printed and bound in the United States of America

The paper used in this publication meets the requirements of the American National Standard for Permanence of Paper for Printed Library Materials Z39.48-1992.

5 4 3 2 1

I dedicate this work to the memory of my mother,
Katy (1929–2006), without whose constant example of love,
spirituality, faith in her children, and encouragement
none of the best things in my life would have been possible

Contents

Tables and Figures

Tables

Figures

Foreword

THIS STUDY IS THE CULMINATION OF THE LATE G. STEPHEN Lauer's lifelong engagement with the question of what makes a good infantryman. The seed of this question was most likely planted when Steve was commissioned into the US Marine Corps infantry upon graduation from college and the Reserve Officer Training Corps, if not sooner. Unlike some who wind up in an infantry unit by chance, Steve was the type of marine who wanted to be in the infantry, and he wanted to lead infantry marines. These goals were achieved—along with many others—throughout a long career serving the great nation that is the United States.

Unfortunately, Steve passed away before achieving yet another goal, the publication of this book. The honor of publishing this important study thus fell to me, a close friend, colleague, and intended coauthor of another work (one that will now never see the light of day). As someone who spent countless hours discussing the topic, the research process, and Steve's intended argument, I was determined that the book would be published despite Steve's untimely death.

Writing this book was a joy in Steve's life. He was a natural writer, with a mature style that impressed me from the first time he asked me to read a draft chapter. I was privileged to read many chapters and sections, though able to offer only the most modest comments and suggestions in exchange for that privilege. In the process of preparing the manuscript for publication, I have now had the chance to read it in its entirety and am more certain than ever that it deserves to be published and its arguments presented as a contribution to the field.

Steve's transnational survey of the effectiveness of the combat infantry in the US, German, and British armies is unique. As a German-speaker, Steve was able to read many firsthand accounts by German soldiers, and he spent countless hours in German libraries and archives. As a result, the German infantryman is looked at objectively and juxtaposed to his British and US counterparts. I am convinced that this study is one of the first in the English language to assess impartially the German infantryman of World War II, and for that it is breaking with the trope that dominates the field. The argument here is clear and straightforward, and readers will have to engage the work on its merits, for the author is no longer with us to debate his choices.

I was not the only one who contributed to this work. Kevin Benson and Christoph Rass were both intimately engaged with the project as well. More than a simple thanks is due to Marie-Claire Antoine, Steve's editor at Lynne Rienner Publishers. If it wasn't for her hard work, dedication, guidance, and support, this book never would have been published. Finally, thanks go to Lynne Rienner herself for supporting this project and agreeing to take it on.

—*Christopher Marsh*
Director of Research and Analysis,
Joint Special Operations University

Acknowledgments

MANY HAVE BEEN INSTRUMENTAL IN HELPING ME TO forge this book, and it is my hope to remember their many kindnesses and extraordinary patience as I asked for their thoughts and advice. Their encouragement and critiques were invaluable and can never be repaid.

The opportunity to travel to Europe and to work with scholars such as David French, professor emeritus of history at University College, London, was inspiring. He was ever kind in sharing his unparalleled knowledge of the British Army in World War II, generously offering his time and advice, and he patiently bore my questions and emails as the research went forward. The staff of the Imperial War Museum was superb, especially in aiding my search for oral histories and their photograph collections. The archivists at the Liddell Hart Center for Military Archives at King's College, London, gave me access to truly distinctive material. Stephen Fritz of Eastern Tennessee State University provided his unique perspective on the German soldier, as well as on the use of source material in the form of wartime personal letters and memoirs. Charles Carlton, professor emeritus of history at the University of North Carolina, graciously provided his thoughts on the work in progress. Perhaps most impactful of all, my friend Christoph Rass, professor at the University of Osnabrück, provided remarkable insights into the sociology and demographics of the German soldier of World War II. His counsel and friendship remain one of my finest experiences during my research on the German Army and my excursion through the Bundes-Archiv Millitär-Archiv in Freiburg and the many other sites that make up the superb German archival system. Also, Colonel Karl-Heinz Frieser, head of the Department of World Wars I and II at the Military History Research Institute of the Bundeswehr in Potsdam, and Rolf-Dieter Müller graciously offered many hours of their time and that of their faculty to further my research.

I cannot thank enough the entire staff of the marvelous British National Archives at Kew, London. Their knowledge and expertise in the archival resources in their possession are without equal in any archive I visited. At the Bundesarchiv-Militärarchiv in Freiburg, Helga Weibel and Andrea Meier bore my questions with inestimable patience and forbearance as they guided me through the German archive system. Their knowledge of the material was unsurpassed.

Uwe Fehn, Veronika Belleman, Olivia Belleman, and Stefanie Fehn became wonderful friends during my visits to the archive in Freiburg. Their hospitality, generosity, and encouragement allowed me to get to know the best of Germany. Stefanie Fehn was instrumental in finding works that were no longer in print but invaluable for my research. The extraordinary experience of living in Germany would not have been possible without them.

Timothy Nenninger at the National Archives in College Park, Maryland, was profoundly influential in guiding me through the labyrinth of US Army and Marine Corps archival documents, and he possessed the uncanny ability to find the right sources for the interwar years. I am also indebted to the US Army Military History Institute at Carlisle, Pennsylvania, and the archivists at the Alfred M. Gray Research Center at the Marine Corps University in Quantico, Virginia. The men and women of the Combined Arms Center Library at Fort Leavenworth, Kansas, were simply marvelous in support of my research, especially while I was resident teaching at the college there.

As an assistant professor at the US Army School of Advanced Military Studies, I have been blessed with the comradeship of a powerfully intellectual group of professors known affectionately as the "heretics of the hayloft," due to our residence above the former stables that houses our school. My daily interactions with the staff and faculty of the school created a truly unique combination of discourse and discovery by historians and social scientists that never ceases to drive excellence and promote thinking away from the ordinary and the academically mundane. Their encouragement and willingness to review portions of this work with endless patience as well as challenge were without a doubt responsible for much of the final book.

Most important is my love and thanks for my beautiful wife, Kim, whose undying patience, love, and encouragement helped me through the long days to the end of this part of our journey together. I could not have done this without her.

—G. Stephen Lauer

1

Forging the Anvil

NOVEMBER 11, 1918, THE "WAR TO END ALL WARS"—the Great War—did not. The bloodletting of war's end in 1918, the visual and visceral narrative of the cost of four years of stalemated war in the trenches, the destruction of so many young men, the millions of infantrymen, considered the flower and hope of their generation, dimmed the honor and raised horror at the cost in lives.

The largest war in mankind's history followed—World War II. Conscripted citizens once more formed the basic raw material of war—obligation in service as citizen-soldiers. Millions filled the fighting armies, navies, and air forces. All the technological marvels of the mid–twentieth century, all the industrial, management, and military advances born of the crucible of the Great War, drove a narrative focused on technology and machines—more machines on land, in the air, at sea. Industry needed greater numbers of citizens to build, feed, and fix the technological products of the great industrial nations.

The infantry remained the one indispensable fighting element, the least machine-oriented, required by all armies to hold the ground seized by the machines of land, air, and sea, the force to occupy the enemy country and capital. The infantry became the anvil against which the machines came to bear upon and destroy. The purpose of this book is to define how three combatant nations of World War II forged the anvil—the infantry—upon whose legs victory, the occupation of the enemy's land and capital, and the breaking of their will depended.

A narrative arose during and after World War II of a perception among the infantrymen of the nations opposed to Germany that the average German infantrymen displayed a greater individual and small-unit capability in close combat. The perception manifested itself in the words of historians, and contemporary general officers who observed the performance of German infantry, and other combat soldiers, and noted a qualitative difference in combat performance. John Ellis concluded that German victories in the defensive stages of the war were "often as not infantry victories in which a combination of fixed defenses, a masterly deployment of artillery and assault guns, and a positive genius for the well-timed *local* counter-attack time and again prevented the enemy from completely shattering or rolling up the German front until they were at the very gates of Berlin itself."[1] Field Marshal Harold Alexander, commanding the Allied Fifteenth Army Group in Italy, noted that "the enemy is quicker than we are: quicker at regrouping his forces, quicker at thinning out on a defensive front to provide troops to close gaps at decisive points, quicker in effecting reliefs, quicker at mounting attacks and counter-attacks, and above all quicker at reaching decisions on the battlefield. By comparison our methods are often slow and cumbersome, and this applies to all our troops, both British and American."[2] General Omar Bradley's aide, Major Chester B. Hansen, noted Bradley's exasperation in late 1944, before the German Ardennes offensive, at the resistance of the German infantry. "If we were fighting a reasonable people they would have surrendered a long time ago, but these people are not reasonable." And further: "The German has proved unexpectedly resistant, however, he dies only with great difficulty. . . . It is little wonder, therefore, that we find them fighting our advance savagely, causing us to kill them in great numbers."[3] Unable to visualize another motive for such tenacity in defeat, Hansen attributed it in defeat to propaganda.

Postwar, with the onset of the Cold War and the prospect of fighting the Soviet Red Army, the US Army employed former chief of the German General Staff Franz Halder to create a collection of German wartime operational experiences. The collection reached over 200,000 pages between 1946 and 1961. These operational studies, building upon a narrative of a "clean" German Army, professionals untainted by Nazism, reinforced the perception.[4] Upon completion of the studies, General Halder received the US Meritorious Civilian Service Award for "a lasting contribution to the tactical and strategic thinking of the United States Armed Forces."[5]

A narrative also exists that this notion, a perception of German individual and small-unit combat excellence by their enemies, has been told many times, with works either acknowledging the perception as true, or vehemently denying it. Why write again about something accepted as truth, at least by some historians? Or something believed to be untrue and unprovable? What is new is the availability of statistically relevant wartime demographic information for comparison across the Anglo-American and German infantry. New German demographic data and soldier narrative studies provide information to allow such a comparative.

This work hypothesizes that if the perception and its accompanying narrative existed, an evaluation of evidence in the classification, selection, training, and assignment of citizens to the infantry, and efforts to build and sustain a cohesive social bond within the small units, should demonstrate a bias toward supporting the perception. The starting place for this examination was the method for entering citizens into the army and then into the infantry, and we find here a significant discontinuity between the Anglo-American armies and the German.

The British Army and the US Army classified their conscripted citizens and volunteers for selection and assignment with criteria based upon psychologically and scientifically informed methodologies. There were three principle factors: first, the possession of a civilian job skill or experience that translated with little or no additional training into a military position; and second, the results of individual mental, mechanical, and other test scores indicating the level of intelligence or trainability. Where there was no civilian skill identified, high test scores marked an individual for assignment to jobs assumed to demand greater intelligence and technical aptitude. The final element in selection was the medical examination to determine whether a citizen was physically fit, as one robust or healthy enough for military service in some form. In the US forces, all soldiers fell into one of two physicality categories, general service or limited service, with the majority accepted at induction for general service. For the British Army there was a more complex and stratified physical determination, and three-quarters of the soldiers found suitable for induction qualified for active service at home or abroad.[6]

These methods ensured that, in the largest majority of cases, men with proven civilian skills, high test scores, or both, did not receive assignment to the infantry. The armed forces of both the United

Kingdom and the United States perceived that the technical nature of the war required the appointment of higher-category men to the army's more complex specialties, favoring their navies and air forces.[7] The US Army Ground Forces[8] found a fundamental correlation between intelligence and physical health, in which men of higher intelligence also tended statistically to be the most physically healthy or robust.[9] This correlation is the primary criterion for the comparative methodology used here to determine the effect of measures to classify, select, and train inducted and volunteer citizens for the infantry. The British Army noted this relationship between physicality and intelligence, observing that "it was known that the less fit men tended to be somewhat less intelligent."[10] On this basis, the quality of the Anglo-American infantry suffered when men without a demonstrable civilian skill, with test scores identifying a lower capability in terms of intelligence or trainability, and by correlation, less physically robust and mentally alert, were disproportionately assigned to small units tasked with the greatest likelihood of physical, mental, and moral stress—those destined for frontline combat.[11]

The German Army classification and selection process did not employ the "modern" psychological or scientifically based testing of the US and British Armies. Instead, the German Army applied a traditional methodology, seeking to identify the most physically and mentally robust individuals for service in combat units, through reliance on a medical examination and multiple-perspective interview. A standardized general intelligence test was not part of the German methodology, which instead relied on the level and nature of individual education achieved through the German education system to judge this aspect of a citizen's qualifications. Unlike the Anglo-American forces, the German Army found no requirement for programs to bring illiterate citizens to a minimum level of literacy for purposes of wartime service.[12] Psychological testing, performed from 1926 to 1942, was for officer candidates, pilots, drivers, and technicians, not for the conscripted or volunteer combat soldier.[13]

Citizens judged by the examining physician to possess the higher requisite physical robustness to withstand the rigors of the combat environment received the designation as combat-capable, or *kriegsverwendungsfähig* (k.v.). Army-specific military specialties of the front line, or *fechtende truppe,* such as the infantry, armor, artillery, engineers, and a few others, also required the k.v. designation, as did pilots and aircrew, naval combatant vessel officers, and crewmen.[14]

The German induction and classification process combined the multiple-perspective interview with the judgment of the examining physician for its primary selection decision. Physicality, the presumption of a mental and physical capability to withstand the stress of combat, was the principal criterion for selection for combat service. The combined interview by a local board consisting of representatives of police, labor, mayoral, and education officials provided a community perspective on the character of each inductee.[15] Given the fundamental correlation between physicality and intelligence, average German infantry quality increased when German citizens possessing high physical robustness and its corollary intelligence or trainability were disproportionately assigned to small units tasked with the greatest likelihood of physical, mental, and moral stress—those destined for frontline combat.[16]

The methods through which Germany, the United Kingdom, and the United States classified, selected, assigned, and trained their citizen-soldiers for duty as infantrymen, and how they built and sustained the social fabric of their small unit, their cohesive behavior in combat, are the two themes forming the heart of this investigation. It is the story of how these nations took their conscripted citizens and turned them into soldiers, those destined for the most frightful mental and physical duty, the face-to-face combat of the infantryman during World War II.

Each nation had a tradition of citizen service. The Anglo-American model was one allowing the conscription of citizens only in wartime and then only in defense of the nation. Geography as an island and a continent separated by seas and oceans from threats defined the form and function of Anglo-American understanding of the obligation of a citizen in wartime, and the moral and legal authority of the nation to compel his service in conscription. Ocean barriers and navies gave the Anglo-Americans time to prepare an army without the need for conscription until their major wars of the nineteenth and twentieth centuries, as in the American Civil War, the Boer Wars, and World War I. Prussia-Germany held a different understanding based on its geographic heritage. The lack of natural barriers to invasion created a state tradition obligating some native subjects to military service from the earliest days of Brandenburg-Prussia. By the time of the great wars of the twentieth century, citizens now, rather than subjects, held an obligation to military service in greater Germany and embedded as a societal narrative. The Prussian-German armies substituted for natural barriers because there was no time to raise an army when threatened. The army was the barrier.

Cohesion and the Primary Group

Once selected for assignment to the infantry, the second factor affecting the presence of the perception was the concept of building and sustaining the social space, the cohesive behavior of the individual within his small unit—the primary group—the infantry squad of eight to fourteen men. "Combat is the end toward which all the manifold activities of the Army are oriented, however indirectly. Organized combat is also the activity by which an Army is most differentiated from other social organizations. The role of the combat soldier may well be considered the most important single role for the understanding of the Army."[17] Psychiatrists and historians struggle for an adequate definition of the term *cohesion.* Most often it describes the actions and expectations of individuals within the social sphere of the small unit or "primary group." A principal conclusion of psychiatrists after the war was that the social bond was of principal importance for the lessening of psychiatric breakdown and the sustainment of combat effectiveness. "This bonding maintained, he never faces combat alone."[18]

Notions of comradeship and friendship and esprit de corps illuminated in concept the expectation of cohesive behavior in combat. Carl von Clausewitz listed this as the "moral" constituent of battle and placed it among the key factors of war, from which "one might say that the physical seem little more than the wooden hilt, while the moral factors are the precious metal, the real weapon, the finely-honed blade."[19] The expansion of armies into the millions moving across the stage in the wars of the twentieth century, combined with the introduction of smokeless gunpowder, massed shrapnel artillery, machine guns, trenches to enhance the defense over the offense, all made uncertain the ability of soldiers to advance under the phenomenon of the "empty battlefield." Notions of élan, the primacy of a bond between soldiers, the moral factors needed to advance under these lethal conditions to achieve offensive success, dominated the writings and doctrinal concerns of all armies.

S. L. A. Marshall noted the importance of the bond to the well-being of the soldier. "I hold it to be one of the simplest truths of war that the thing which enables an infantry soldier to keep going with his weapons is the near presence or the presumed presence of a comrade. The warmth which derives from human companionship is . . . essential. . . . It is that way with any fighting man. He is sustained by his fellows primarily and by his weapons secondarily."[20] Building and sustaining this bond appeared essential for success and survival in battle. The collective training, unit organization, leadership, punishment, and recognition, as well as the medical treatment of wounded soldiers, and replacement policies, all combined as factors in the soldier's view of his own organization and his opponent across no-man's land. The result of the selection process and the experience of battle has its focus here in the intent of the armies to create and sustain cohesive behavior of the individuals in their small units.

US and British soldier-authors sought to portray through World Wars I and II this concept of comradeship and obligation in literature and poetry as they observed and experienced this aspect of infantry combat. J. Glenn Gray noted of the US infantry the power of the "determination not to let down his comrades" as the key factor allowing the soldier to continue amid the horrific sensations of battle. Gray expressed the idea of the cost expected of the infantry and their willingness to move forward in the face of the sacrifice of life as one both tragic and noble. "In the German language, men never die in battle. They *fall.* The term is exact for the expression of self-sacrifice when it is motivated by the feeling of comradeship. I may fall, but I do not die, for that which is real in me goes forward and lives on in the comrades for whom I gave up my physical life."[21]

The poetry of the British soldier of World Wars I and II expressed this sense of sacrifice and loss in comradeship in death for these generations of infantry in John McRae's *In Flanders Fields* (1918) and in Alun Lewis's *Lines on a Tudor Mansion* (1942).[22] These poems gave the sense of both the obligation of the survivors to continue to fight in the face of loss and the recognition that it was the young upon whom this burden fell.

Stephen Fritz expressed the universality of the combat experience across the young men of these three nations in noting of the German infantryman, the *Landser,* "amid the despair and cynicism, that affection for those enduring the same horrors created a sense of unity and pride, an intensity of feeling that rose to a level rarely achieved by mere friendship. Loyalty, mutual obligation, a willingness to sacrifice, pride, a sense of duty, even love—these constituted comradeship for the *Landser.*"[23] The young men of each nation sought, amid the horrific sensations of death and wounding, in face-to-face battle with other young men, an enduring sense of worth, "when, through military reverses or the fatiguing and often horrible experiences of combat, the original purpose becomes obscured."[24] It is this sense of cohesion toward which the infantry policies of the various nations here find expression, either as a certain, stated goal, or as an implied and expected outcome. *Cohesion* is a term reflective of the battlefield, more art than science; combat-oriented, and centered on group, mission, and task, its intent is the satisfaction of the primary social needs of the individual for family, respect, and the "sympathy and mutual identification for which 'we' is the natural expression."[25]

The seminal work from World War II on the nature of collective cohesive behavior was Edward A. Shils and Morris Janowitz's *Cohesion and Disintegration in the Wehrmacht in World War II.* Their work on the concept of the "primary group" remains a principal reference for studies of such behavior in combat. They hypothesized that "the extraordinary tenacity of the German Army" was found in the effective influence of the primary group:

> It appears that a soldier's ability to resist is a function of the capacity of his immediate primary group (his squad or section) to avoid social disintegration. When the individual's immediate group, and its supporting formations, met his basic organic needs, offered him the affection and esteem from both officers and comrades, supplied him with a sense of power and adequately regulated his relations with authority, the element of self-concern in battle, which would lead to disruption of the effective functioning of his primary group was minimized.[26]
>
> He was likely to go on fighting, provided he had the necessary weapons, as long as the group possessed leadership with which he could identify himself and as long as he gave affection to and received affection from the other members of his squad and platoon. In other words, as long as he felt himself to be a member of his primary group bound by the expectations and demands of its other members, his soldierly achievement was likely to be good.[27]

The characteristics of the rifleman form the essence of this comparison of the armies of Germany, the United Kingdom, and the United States. Riflemen, then and today, must cross no-man's land to engage an enemy. During World War II, the primary location of the rifleman was within the rifle squads resident in the typical or line infantry division. Elite combat divisions, such as the airborne, serve here to illustrate key distinctions between these atypical infantry divisions. Elites in the Ango-American tradition used different and often more strenuous entry requirements in physicality and intelligence than the line infantry. Anglo-American armored or German Panzer divisions also housed mechanized or motorized infantry. The rifleman received support from other infantry specialists manning crew-served weapons such as machine guns, mortars, and anti-tank weapons, all supporting his task to expose himself to reach and defeat his foe.[28]

The British Army in December 1944 deployed twenty-one infantry, five armored, and two airborne divisions. The US Army in June 1944 mustered sixty-seven infantry, sixteen armored, one cavalry, and five airborne divisions. The German Army on the Eastern Front alone in July 1943 marshaled 151 infantry divisions and nineteen Panzer divisions, excluding three *Schutzstaffel* (Shield Squadron [SS]) Panzer-grenadier divisions. For all armies and nations, the largest combat experience for the conscripted citizen serving in the infantry lay in the line infantry division.[29]

The numbers of riflemen, and infantrymen in general, were quite small in comparison to the total strength of the armies. Fewer than 230,000 riflemen served during 1944 in the eighty-nine ground combat divisions in the US Army, winnowed from 8 million citizen-soldiers serving at that time. Of 2.5 million soldiers in the British Army, there were 81,000 riflemen serving in 1944 in twenty-eight ground combat divisions. In 1941, near 400,000 riflemen served in the ninety-nine infantry divisions, nineteen Panzer divisions, and eleven motorized divisions in the German Eastern Army, or *Ostheer,* within an attacking force of over 3 million soldiers. The numbers of German infantry plummeted as the war progressed, dwindling to an average manning level of at or near 50 percent, or 250,000 riflemen in 151 infantry divisions on the Eastern Front in 1943.[30] Of the 19 million soldiers in these three armies serving during 1943–1944, 561,000 riflemen, or 3 percent of the total for all armies, suffered up to 80 percent of the combat casualties.

Comparing Soldiers

There are few works making direct comparisons of US, British, and German soldiers or infantry. A US historian wrote that a successful work of comparative history "does justice to culture and ideology as well as structure, draws attention to the most significant causal variables, and shows the peculiarities of each case without making one of them the exception to a general pattern represented by the others."[31] Along with Germany, the United Kingdom and the United States share an essential Western narrative of Judeo-Christian beliefs, a history of war as allies and enemies, as well as historical antecedents in economics, law, empire, and revolution. A common Western heritage converges to demonstrate the common legacy of the cultures and national experiences of the soldiers of these nations. During the interwar years, the Great Depression was another point of shared experience in the searing remembrance of poverty and hunger among many citizens. One factor was not shared by the nations—Adolf Hitler—and this shadows any wartime comparison.

The overlay of twelve years of governance of the German state by the National Socialist German Workers Party (NSDAP), the Nazi Party under Adolf Hitler, and its legacy of aggressive war and human extermination, colors any comparison of these soldiers. This fact has the capacity to overwhelm otherwise resonant features of a prior shared Western culture and society and military service. The Nazis murdered millions of human beings.[32] The German Army was the principal means by which the Hitlerite regime imposed its expansionist political and racial goals through war and conquest.[33] Without the offensive mastery of the German Army from 1939 through 1942, and its defensive tenacity from 1943 through 1945, the extension and consolidation of Nazi influence across Europe, the European Soviet Union, and North Africa was not possible. The capture and imprisonment of millions targeted for Nazi policies of extermination occurred following the march of the German Army. The tactical prowess of the German Army provided the space to make Nazi murder policies workable.[34]

This inquiry is not a definitive review of the role of the German armed forces, the *Wehrmacht,* in the Holocaust, or Nazi racial extermination policies, all well-documented and thoroughly researched. The moral judgment against the Hitlerite regime is settled. The intent here is to examine, and compare to their Western adversaries, the manpower policies of the German Army, the *Heer,* within the larger organization of the *Wehrmacht,*[35] and the impact of *Heer* manpower selection policies on the cohesion of the small combat infantry unit.

This work confines its analysis to the drafted and volunteer German male citizen who, upon being selected for the combat infantry, carried the primary burden for close combat in the German Army. The form and type of the principal Nazi criminal formations intended for the destruction of innocents and prisoners of war are not discussed in detail here, including the selection policies of Heinrich Himmler's *Schutzstaffel* or *Gestapo* units of any type, the murder units of the *SS-Einsatzgruppen,* the *SS-Sonderkommando,* or the *Waffen-SS,* the Nazi political-military combat component. Nor are the murder units associated to the *Wehrmacht,* or the *Heer,* such as the police battalions, the field police and gendarmes, the army's security divisions, nor those manning the prisoner-of-war transit camps (*Armee-Gefangenen-Sammelstellen*). Not all who wore German field gray uniforms were combat soldiers.

If it can be demonstrated that the characteristics of the German infantry soldier, and the development of his cohesive behavior in battle, was compatible with the expected battle norms of other Western soldiers, does not this seem to justify a conclusion of moral equivalence between the German and his US and British counterparts? "This objection rests on a common tendency to confuse an explanation of causes with a justification or acceptance of results. What use one makes of a historical explanation is a question separate from the explanation itself."[36] The distinction made between the German combat soldier and those not selected or qualified for assignment to fighting units is not made to condone, or to excuse, or to diminish the horror of the aims of the Nazi state.

It is not deniable that the German generals acquiesced to these aims, and the army facilitated them through its tactical and operational methods. The distinction between men selected for infantry combat, and those not, enables a comparison of infantrymen and infantry small units across the Western nations selected for study. This qualification in the comparison does not hide or shadow any responsibility borne by individual infantrymen of the regular combat divisions who participated, in or out of battle, in the killing of innocents or prisoners of war.

To cite the differences between the infantry of these armies does not elevate the moral purpose to which the German Army was directed. It is these differences that highlight an understanding of the perceptions of that infantry in the eyes and experience of the Anglo-American infantry.

Narratives and Perceptions

British perceptions of the quality of the infantry force and the results of the selection process both during and following the war were remarkable for their candor. In the interwar years, commentary noted the reduction in both the quality and numbers of infantry and highlighted the possible effects in a future war. During the war, reporting by general officers and other commanding officers related their continuing discontent with the manner of selection and assignment to the infantry, noting the experience of this selection on morale and combat performance. After the war, the official British Army reports in the works titled *Morale, Manpower Problems, Personnel Selection,* and *Training in the Army* all highlighted the lessons learned in regard to the reduction in the quality of the infantry force, its training, and combat effectiveness as a result.

It is the contention here that the meta-narrative, the master plot,[37] for the expectations of the British infantry arose from the experience of the mass infantry army of World War I. The character of that master plot was to place the physically best and most intelligent young Britons into the deadliest space—the infantry. The quality of the conscripted army allowed for the general expression of the initiative, alertness, and leadership from the ranks to sustain the army toward victory. There was no mass Royal Air Force to siphon away the leaders available to the army in the infantry of the Great War. The expectation for the infantry regiment, its traditions of bringing into the mass of soldiers a spirit of bonded, cohesive excellence, was not achieved in World War II. The scientific methods of psychological and intelligence testing to post the best-qualified into more technical fields denied the regiment the ability to perform its expected role. This surprised the leaders of the British Army as they came to understand the effect of such classification on the quality of their infantry.

David French provided perhaps the best modern summary of the concerns and outcomes of the selection process for the infantry of the British Army. He noted the army was rarely the choice of volunteers, who preferred the Royal Navy and the Royal Air Force in far greater numbers than they could take. "But the most serious problem was the way in which manpower was allocated to the infantry. . . . The policy of posting the worst men to the infantry was contrary to its real needs. Infantrymen had to be fitter, to possess more initiative, endurance, and leadership skills than other arms because their job was more arduous, dangerous, and continuous."[38]

The US meta-narrative prior to World War II was an assumed native male excellence in battle. The performance of the US citizen in the Revolutionary War, the Civil War, and World War I cemented this assumption into both subjective and objective reality[39] for the expected performance of a future mass infantry army. Works such as George Marshall's *Infantry in Battle,* and movie depictions as in Gary Cooper's portrayal of Sergeant York and his exploits, sustained the narrative assumption through stories: "the more culturally specific the masterplot, the greater its practical force in everyday life."[40] The problems of the quality of the US infantry were a surprise and based on the same difficulty as for the British Army—the classification system through scientific methods of psychological and intelligence testing to post the best-qualified into more technical fields, the air forces, and the navy.

Both during and after the war, the failure of the narrative was met with a mix of recognition of the problem of infantry quality and one of vehement denial. During the war, stating his public expectation that US infantrymen would stand in battle on an equal basis with their German opponent, Lieutenant General Lesley McNair, the commanding general of the US Army Ground Forces, from 1942 until his death in France in 1944, provided statistical evidence to the leadership of the army on the quality shortcomings of the infantry. He noted that the infantrymen, by November 1943, were shorter in height, lighter in weight, and possessed the lowest average education and intelligence test scores of any combat specialty.[41] As each division embarked for duty overseas, McNair provided a training report card to the theater commanders concerning the state of manning and the level of combined arms training of each combat division. These reports highlighted the training deficiencies of the divisions and the fact that too often their infantry were in the units for too short a time for adequate combat preparation. He found that the process of stripping the infantry from divisions in training to man the next division embarking for overseas diminished the combat capabilities of both the stripped unit and the receiving unit.[42]

McNair's wartime assessment was echoed later by Russell Weigley, who noted that the US Army "habitually filled the ranks of its combat infantry with its least promising recruits, the uneducated, the unskilled, the unenthusiastic. Those left over after the Army Air

Forces, the marines, the navy, the paratroopers, and the technical branches had skimmed off the best of the nation's military manpower were then expected to bear the main burden of sustained battle."[43]

Michael Doubler, in *Closing with the Enemy: How GIs fought the War in Europe, 1944–45;* Peter Mansoor, in *The GI Offensive in Europe: The Triumph of American Infantry Divisions, 1941–45;* and Robert Rush, in *Hell in Hürtgen Forest,* all provided a counterpoint concerning the quality of US soldiers and their performance in battle. In Doubler's judgment, it was an identifiable American entrepreneurial spirit that ensured the success of the US infantry, as opposed to the less adaptable and more cautious and authoritarian German and Soviet Armies.[44] He argued that the superiority of US soldiers in battle over the German enemy was not due to any perceived material or manpower numerical largesse, but to this intangible, superior American characteristic.[45]

Mansoor argued for organizational adaptability and sustainability of the divisions as the key to success, not materiel, noting that apologists for the defeated German Army adhered to the belief that victory came through the massive industrial advantages of the Allies over the "superior but hopelessly out-numbered forces of the *Wehrmacht.*" Noting that the US Army failed to provide a "fair share of quality recruits and replacements" for its infantry divisions, and paid little attention to cohesiveness and small-unit training below the battalion level, this did not amount to lesser capability against their German counterparts.[46]

Rush, in contrast, wrote to address the perception of an increased German individual and small-unit combat capability. He sought to compare and "describe the American infantryman and German Landser," and "to deal even handedly with a subject that is obviously controversial, namely the relative performance of the U.S. Army and its German adversary in Western Europe in fall 1944."[47] He focused on the fortunes of a single regiment, the Twenty-second Infantry Regiment of the Fourth Infantry Division, a veteran division that landed in the first wave on Utah beach at Normandy on D-Day. The division and regiment engaged in battle in the Hürtgen forest during eighteen days from November 16 to December 3, 1944.[48] Having complete access to the daily morning reports of the Twenty-second Infantry Regiment, he "could not achieve the same level of resolution on the German organizations."[49]

As a narrative of the strengths and weaknesses of US Army infantry battle performance, Rush described the shortcomings of the battlefield personnel replacement process. He noted: "There is no question that the German Army stopped the Americans at the West Wall. The question is how."[50] The fact that its commanders judged the Twenty-second Regiment—after eighteen days of combat with continuous replacements, against a weakened and demoralized German infantry—incapable of further combat and withdrawn from the continuing battle, reinforced the underlying hypothesis that a perception of superior German soldier and small-unit combat skill and perseverance existed.[51]

Perhaps the most singular of works assuming the fact of the perception of German individual and small-unit combat superiority was Trevor Dupuy's direct comparison of US, British, and German combat performance during World War II in *Numbers, Predictions, and War: Using History to Evaluate Combat Factors and Predict the Outcome of Battles*. Dupuy, writing during the Cold War, developed a mathematical model from his analysis of historical combat narratives of the Allied and German armies in battle.[52] He concluded that this analysis of historical combat demonstrated the tactical superiority, through casualty production ratios, of the German Army in defense against any Allied formation. This conclusion was not supported by the reports of either Allied or German casualty summaries for the individual theaters of war.[53] A key assumption of the Dupuy model failed, but his work was powerful evidence of the existence of the perception among their opponents that the German soldier and small unit held a demonstrable combat superiority.

Following World War I, the leaders of the reduced German Army sought to change the meta-narrative of the historical mass German infantry armies. General "Hans" von Seeckt looked to the period before the Great War, and sought to build a new army of leaders based on high-quality selection of enlisted and officers. Defeat presented a paradigmatic change to the societal narrative of German battle excellence.[54] Looking past the defeat in 1918, the German Army developed a new method for a war of movement. It relied upon the traditional expectations for high quality and the intentional development of high-quality and cohesive behavior in the small-unit leaders and the individual soldiers. These restored and enhanced stories drove a master plot connecting the past to the army of the future and were the basis for the perception of excellence.

German narratives did not, as a normative study, address the individual soldier, his small unit, or the development and sustainment of cohesive behavior in battle. As noted by Wolfram Wette, the principal focus has been the Nazi elites—political appointees, generals, admirals, and other senior officials. Greater than 99 percent of the soldiers were not in this elite group. "The history of German enlisted men in World War II has remained largely uncharted territory. For a long time, ordinary servicemen, popularly referred to as *Landser,* were simply overlooked. . . . Soldiers who do not belong to this elite function in this system, and in the thinking of those who run it, act merely as agents executing commands—essentially, that is, as parts of a machine."[55]

New evidence presented by authors Christoph Rass and Felix Römer provided a more nuanced picture of the soldier than previous research. Rass wrote two works presenting statistically relevant demographic data, detailing information about individual German soldiers in *Menschenmaterial: Deutsche Soldaten an der Ostfront—Innenansichten einer Infanteriedivision, 1939–1945,* a social and statistical history of the 253rd Infantry Division (*Menschenmaterial*) between its formation in 1939 and surrender to the Red Army near Prague in 1945. René Rohrkamp, in *Deutsche Soldaten, 1939–1945: Handbuch einer biographischen Datenbank zu Mannschaften und Unteroffizieren von Heer, Luftwaffe und Waffen SS,* provided for the first time a unique and statistically relevant set of demographic data for more than 18,000 individual soldiers of all services from the *Wehrkreis VI,* Rheinland-Westphalia region.[56]

Rass provided insight into the social support structure of the German infantry division and its subordinate small units. German policies of recruitment from a single region, maintaining the common origins and dialects of the German soldier, enhanced the internalization and legitimation of the social bond built upon entering a combat unit.[57] Knowing that through selection, assignment, battle, and wounds the German soldier expected to return to his home unit was a powerful tool for the creation and sustainment of cohesive behavior in battle.[58] The soldiers of the 253rd Infantry Division and the more than 18,000 men of the database provided a picture of the soldiers of the Rhineland-Westphalia region. Coming from a stable, lower-working-class population, with an upward-mobility potential into the lower middle classes, these soldiers represented an unremarkable homogeneity of German society with Christian religious confession dominated by Catholic and Protestant groups. The soldiers of these divisions represented a uniquely cohesive group.[59]

Felix Römer, in his works "Milieus in the Military: Soldierly Ethos, Nationalism, and Conformism Among Workers in the Wehrmacht" and *Kameraden: Die Wehrmacht von Innen,* provided an unambiguous look at the contemporary attitudes and expressions of German soldiers of World War II. Römer used US Army interrogation questionnaires and transcripts of eavesdropped conversations of more than 3,000 German prisoners of war, during 1942 to 1945, at Fort Hunt, outside Washington, DC. The statistical relevance of the soldier commentary lies in the volume of transcript material, and its coverage of multiple cohorts of prisoners captured in more than three years of recording, "so that the length of their service in the Wehrmacht varied between several years and a few months."[60] He noted that "for the most part their concept of patriotism corresponded to rather traditional nationalistic views than to National Socialist theories."[61]

Römer's analysis demonstrated that this military socialization was a specific, internalized factor in the soldier's view of himself and his service:

> The Morale Questionnaires demonstrated that the shared pride in the abilities of the Wehrmacht and its martial virtues was deeply rooted throughout German society. It combined to form a collective military morale that vaulted traditional dividing lines in the German armed forces and contributed elementarily to the motivation of the troops.[62]
>
> . . . Military morals with their key categories of fulfillment of duties, bravery and male toughness constituted for them an ideal that they wanted to conform to . . . the loyalty of the troops fed not only from the solidarity of the primary groups but also related to the Wehrmacht as an institution, which conveyed and symbolized those military morals on which the soldiers depended if they wanted to prove themselves in front of their comrades.[63]

New works such as *The German War* reinforced this picture, describing the strongest foundation for the patriotism and performance of the German soldier in his connection through family experience in the prior war, not with the Nazi regime.[64] Prior to the evidence presented by Rass and Römer, the primary sources for soldier viewpoints were the subjective memoirs available following the war, soldier letters, as well as official reporting in a *Kriegstagebuch,* or unit war diary. The problem of reliance on war letters and memoirs has been well documented. One can find evidence for any point of view from the war letters remaining from World War II. In his edited work *Der Krieg des kleinen Mannes: Eine Militärgeschichte von Unten,* Wette wrote: "The history of the soldiers of the Second World War is, in the end, for the historical science, up to now a vast unknown territory. The simple *Landser* exists generally only anonymously as an element of a casualty list or a table of organization . . . seen in a double role as a perpetrator and victim. We render the concrete truth about the war a service when we return 'the unknown soldier' his face and his name."[65] He noted that letters remained the primary source for placing a face on the soldier, but that the use of these same letters was fraught with danger for the historian. "The sole tolerable source, which the soldier left to posterity, was usually the letter from the front. Moreover, the communication medium of the letter from the front has peril for the historian."[66] The soldier faced the problem of the military censors. "Given direct and indirect influences on what the soldiers could say about their feelings and hardships caution should be kept in mind when using them as evidence."[67] As less than 1 percent of the letters written during the war by German soldiers and their families still exist in archival collections, the samples available for examination were scarce and did not achieve any form of structural statistical relevance. The new information made possible by the Rass and Römer research opened new avenues for the evaluation and comparison of the German infantrymen and their Anglo-American counterparts.

About the Book

The book follows a chronological methodology with three national chapters (United Kingdom, United States, Germany) in each part. Part 1 follows each national army through the interwar years up to the reinstatement of conscription. The chapters in this section analyze the history of the obligation to serve in wartime, the selection of the individual, and the institutional intent of the armies to create and sustain cohesive infantry small units, prior to the opening of the war. Part 2 focuses on the intent and effects of conscription and standards for induction and classification that informed the selection of soldiers for the infantry in the opening period of the war. Part 3 follows the war to its conclusion and the outcomes of both selection and classification, and the institutional intent to build and sustain cohesive behavior in the infantry small unit.

Infantry soldiers, when placed into the environment of the battlefield, regardless of their selection process, learned to fight and survive, and to rely on their comrades—to find cohesion in the shared hardship and danger of the front line. They were the anvil against which the nations threw their massed mechanization, artillery, and aircraft against. Theirs was the mission to close with and destroy their enemies. The citizen-soldier in all of his forms and tasks was the decisive arbiter of national power during World War II. Inside the small infantry unit, in the midst of so much that was dreadful, men brought together by the will of their nations found faith in each other, fought for one other, and lived and died for their friends. This study recognizes the soldiers, airmen, sailors, and marines of World War II who answered the summons, who placed their lives into the fire of combat, who lived and died, won and lost. Regardless of the ultimate outcome, the men of these nations fought for their comrades, found solace in the shared bonds of their fate, and endured all as members of a fraternity of the few—the infantry combat soldier.

Notes

1. Ellis, *Brute Force,* 532, emphasis in original.
2. Ibid.
3. Beevor, *Ardennes 1944,* 99.
4. Wegner, "Erschriebene Siege," 287–302.
5. Detwiler, *World War II German Military Studies,* 1–9.
6. Ungerson, *Personnel Selection,* 7; Palmer, *Study no. 5,* 3–4; Kreidberg and Henry, *History of Military Mobilization in the United States Army,* 639–643; Lerwill, *The Personnel Replacement System in the United States Army,* 312–319.
7. TNA PRO WO 216/66, Note from Inspector-General, Training, to C.I.G.S., August 4, 1942; US Army Ground Forces, *Study no. 11,* 42.
8. The US Army Ground Forces formed in 1942 to train and prepare all divisions for combat operations.
9. Palmer, Wiley, and Keast, *The Procurement and Training of Ground Combat Troops,* 65.
10. Ungerson, *Personnel Selection,* 47.
11. Ibid., 19; Palmer, *Study no. 5,* 2.
12. Houle et al., *The Armed Services and Adult Education,* 169–190; Hershey, *Special Monograph no. 16,*159–163; TNA WO 32/9960 E.C.A.C./P(41)111, "Paper by A.G. on Use of Manpower in the Army," pt. 1, "January 1939–December 1941," December 5, 1942, appendixes A, C; "Paper no. 3 by A.G. on Use of Manpower in the Army, June 1942–May 1943," June 1943, appendix A; Ungerson, *Personnel Selection,* 63; Pigott, *Manpower Problems,* 15, 54.

13. BAMA ZA 1/1779, Personnel and Administration (Project 2b) P-005, Teil V, Simoneit, *Die Anwendung psychologischer Prüfungen in der deutschen Wehrmacht. (Deutsche Wehrmachtpsychologie von 1927–1942),* 30–32.

14. Absolon, *Die Wehrmacht im Dritten Reich: Band V,* 116–117.

15. BAMA ZA 1/1778, Personnel and Administration (Project 2b), P-006, Teil I, *Das Rekrutierungssystem der deutschen Wehrmacht in Frieden und im Kriege;* Anlage I, *Durchführung einer Musterung und Aushebung,* 73–75, 97.

16. Ibid., 98.

17. Stouffer, *The American Soldier,* vol. 2, 59.

18. Willer, "Groups Reward Individual Sacrifice," 23, 33–43; Siebold, "The Evolution of the Measurement of Cohesion," 5, 23–26; Siebold and Lindsay, "The Relation Between Demographic Descriptors," 109, 118–128; Milgram, Orenstein, and Zafrir, "Stressors, Personal Resources, and Social Supports," 185, 190–199; Moreno, "Note on Cohesion in Social Groups," 176; Jones et al., *War Psychiatry,* 15–16.

19. Von Clausewitz, *On War,* 185.

20. Marshall, *Men Against Fire,* 42.

21. Gray, *The Warriors,* 40, 46, emphasis in original.

22. Ward, *World War One British Poets,* 12; Archard, *Alun Lewis,* 33–34.

23. Fritz, *Frontsoldaten,* 157.

24. Gray, *The Warriors,* 40.

25. Shils and Janowitz, "Cohesion and Disintegration in the Wehrmacht," 283, n. 3; Moreno, "Note on Cohesion in Social Groups," 176.

26. Shils and Janowitz, "Cohesion and Disintegration in the Wehrmacht," 281; Cooley, *Social Organization,* 23.

27. Shils and Janowitz, "Cohesion and Disintegration in the Wehrmacht," 284.

28. The standard or "line" infantry division was a notionally pure infantry division with two or three infantry regiments and supporting elements, such as artillery, engineers, and logistics. Anglo-American infantry divisions of World War II were larger than their German counterparts. A British infantry division of 1944 totaled over 18,000 men; a US Army infantry division of 1944 numbered over 14,000; German infantry divisions in 1944 averaged 12,000.

29. French, *Raising Churchill's Army,* 189; US Army Ground Forces, *Study no. 12,*, 13; Kroener, Mueller, and Umbreit, *Organisation und Mobilisierung des Deutschen Machtbereichs,* 953.

30. TNA PRO WO 365/203, *W.E. (War Establishment) Analysis by Arms. A.G. Stats (War Office),* S/243/43, *Analysis of Arms of W.E. of Standard Formations,* March 26, 1943; US Army Ground Forces, *Study no. 8,* 71, 72; Kroener, Müller, and Umbreit, *Organisation und Mobilisierung des Deutschen Machtbereichs,* 969; *Kriegstärkenachweisung Schützenkompanie* (KStN) 131e, January 2, 1941, Record Group 242, T-283, Roll 123, Frame 4730251-253; *Schützenkompanie neuer Art (n.A.)* (KStN) 131n, 1.5.1944, Frame 4730259-261, NACP; BAMA RH 2/1331 *Schützenkompanie neuer Art (n.A.)* (KStN) 131n, for January 10, 1943 and January 12, 1943.

31. Frederickson, "From Exceptionalism to Variability," , 601.

32. Wette, *The Wehrmacht,* 256.

33. Müller and Ueberschar, *Hitler's War in the East,* 34.

34. Ibid., 73.

35. The German army, or *Heer,* was one component, the largest, of the arms of service collectively described by the term *Wehrmacht,* which included the army (*Heer*), the navy (*Kriegsmarine*), and the air force (*Luftwaffe*).

36. Diamond, *Guns, Germs, and Steel,* 17.

37. See Abbott, *The Cambridge Introduction to Narrative,* 46–48, for discussion of master plots.

38. French, *Raising Churchill's Army,* 64, 70–71.

39. See Berger and Luckmann, *The Social Construction of Reality,* 128–129, for discussion of socialization and subjective and objective reality.

40. See Abbott, *The Cambridge Introduction to Narrative,* 47, for discussion of master plots.

41. Memorandum for the Chief of Staff, US Army, subject: Utilization of Available Manpower Based upon Physical Capability, tab D, "Characteristics of Infantry," December 21, 1943, WD, AGF (1916–1954), Ref. Files L. J. McNair (1940–1944), RG 337.3, NACP.

42. US Army Ground Forces, *Study no. 12,* 1.

43. Weigley, *Eisenhower's Lieutenants,* 27.

44. Doubler, *Closing with the Enemy,* 5–7, 282.

45. Ibid., 3.

46. Mansoor, *The GI Offensive in Europe,* 12, 267.

47. Rush, *Hell in Hürtgen Forest,* xv.

48. Ibid., xv, 280, 301, 305.

49. Ibid., xvii.

50. Rush, "A Different Perspective," 499.

51. Rush, *Hell in Hürtgen Forest,* 280, 292. The figures provided in his Table 5 represent the table-of-organization strengths of the opposing infantry divisions at full strength. While the Twenty-second Regiment entered combat with 90 percent of its infantry strength and received replacements throughout the eighteen days of battle, German infantry divisions at this time held 50 percent or less strength in infantry, especially when incorporating into infantry units a variety of replacements from noninfantry formations at this late stage of the war.

52. Dupuy, *Numbers, Predictions, and War,* xi.

53. Ibid., 97–104. Dupuy noted here: "These figures have been adjusted to allow for the increased effectiveness of defensive firepower, so that they suggest that on the offensive as well as the defensive two German soldiers could on the average fight three Allies to a standstill." In their work on the war in Italy, Graham and Bidwell, *Tug of War,* 403, noted that the balance of Allied casualties in Italy against defending German forces rather favored the Allies. Overmans, *Deutsche militärische Verluste im zweiten Weltkrieg,* 336, tab. 73e.

54. See Kuhn, *The Structure of Scientific Revolutions,* 111–113, for discussion of paradigmatic change in crisis.

55. Wette, *The Wehrmacht,* 176–178.

56. Rass and Rohrkamp, *Deutsche Soldaten,,* 41.

57. See Berger and Luckmann, *The Social Construction of Reality,* 128–129, for concepts of internalization and legitimation in institutions.

58. Rass, *"Menschenmaterial,"* 50.

59. Rass, "Das Sozialprofil von Kampfverbänden," 686.

60. Römer, "Milieus in the Military," 138.

61. Ibid., 143.

62. Ibid., 136.
63. Ibid., 148–149.
64. Stargardt, *The German War,* 17–18.
65. Wette, *Der Krieg des kleinen Mannes,* 13.
66. Ibid., 20.
67. Kilian, "Kriegsstimmungen," 251–252; Vogel, "Der Kriegsalltag im Spiegel von Feldpostbriefen," 200.

PART 1

Finding Volunteers: The Making of the Interwar Infantry

2

The United Kingdom: Lowering Selection Standards

NEAR THE END OF WORLD WAR II, GENERAL W. J. SLIM, commanding the Fourteenth Army in Burma, noted "the basic fact that success in a campaign depends first on the quality of the infantry. . . . No army which has not got first class Infantry can succeed in war against a well-equipped, stubborn enemy."[1] The manner and method of the selection and assignment of junior officers and enlisted personnel that informed General Slim's comments were the outcome of policies set long before 1945. The core of his concern lay first in the interwar policies for the selection and assignment of enlisted personnel—other ranks, in British Army terminology—that defined service for the British citizen-soldier. The historical obligation for armed service within the United Kingdom informed the connection between volunteers in the interwar period and the relationship and effect of these policies on the laws and procedures that led to the involuntary call to the colors in the United Kingdom in 1939.

Questions about the nature and capability of the British infantry force remained controversial within the British Army throughout the war. It is my contention that decisions made both before and during the war diminished the quality of the infantry and its expected cohesive behavior in battle. Further, both the political and the military leadership of the United Kingdom recognized this fact, with the consequence that the funneling of men with lower physical and intellectual qualifications into the infantry, as determined by intelligence and other tests, became the norm.

Traditions and the Obligation to Serve

English society came late to the notion that being a citizen involved the right of the government to coerce one's service, even unto death, in the armed forces of the nation. The "institution" of conscription took many generations to become reality in the United Kingdom. This did not mean that there was no tradition of impressed or obligated service. As far back as 1066, the Anglo-Saxon *Fyrd* defeated at Hastings consisted of landholding freemen who were called to duty to defend their community. Henry II, with his "Assize of Arms" in the twelfth century, defined wartime service for each class of free men and serfs. Conscripted armies formed an essential component of the English Civil War armies. During the Napoleonic Wars, English militiamen selected via lottery augmented the overseas army.[2]

For the British citizen and subject living prior to the era of modern war, there were several modes of service possible or expected.[3] The first was the basic reliance on militia units as the core, native, armed resident response to emergencies. Second were the permanently armed retainers of the major landholders, available to the landed aristocracy by reason of their ownership of the land on which the soldiers and their families lived. Third were those impressed for service by king or parliament as a result of a campaign or war. Finally, as a result of the need to defend the permanent crown colonies, there emerged for the first time a small peacetime regular army, capable of rapid reinforcement from the soldiers available in the militia, who along with the retainers and impressments made up the additions to the regular army for expeditionary uses in Europe or the colonies, or to repel invasion, alongside a large navy.[4]

The levy of soldiers for expeditionary service fell most heavily on the rural landless peasants, described as laborers or husbandmen, while "the community clearly desired to protect 'useful' citizens (and we may assume their index of social utility was roughly possession of property beyond the merest level of family subsistence) while disposing of its petty criminals, tavern riffraff, nuisances and unemployables."[5] The basis for selection was the simple exclusion of the physically unfit, the aged, and the young, and left to local discretion; "the result was a flood of contemporary complaints about the vicious behavior, low character, poor physical condition and general uselessness of the pressed footsoldier. . . . Soldiers [in the 1570s] in general were chosen from among rogues, ronnagates . . .

drunkards, vagabonds, idle fellows, seditious quarrelers, privy pickers and goose stealers."[6] Leaders such as Sir Edward Cecil, commanding the Cadiz expedition in the 1620s, declared of his own men that they were so little capable with their arms as to present a positive danger to their fellows and so addicted to drink that only if the enemy were near a wine supply would they acquit themselves well in the attack.[7] And yet these soldiers could fight well even while going down to defeat, as they did at the battle of the Isle of Ré near LaRochelle in 1627, standing their ground in the face of death, and performing better than their detractors told.[8]

Religious faith remained a powerful motivation for soldiers. The performance of the volunteers and conscripts of the civil wars and in the expeditions of Cromwell's army and the Protectorate attest to the intensity and ardor engendered by religious extremism, such as the enthusiastic destruction of the Roman Catholic Church structure of Ireland. The regular army of the Protectorate was one of the finest in Europe, but engendered a withering hatred of all things, taxes perhaps above all else, that such a standing force entailed. Cromwell hoped "that the militia may be settled with such wisdom and full security that it may not be in the power of any to make use thereof against the people, or the successive Parliaments, either to destroy their being or their freedom."[9]

An era of revolution, beginning with the rebellion of the British North American colonies of 1776 and the French Revolution of 1789, ushered in the most profound changes in the evolution of the relationship between the nation-state, its armed forces, and the obligation of the citizen for service in arms. Through war, "the nation-state became full-blown both in the sense of a shared community of purposes, privileges, and benefits, and in the sense of a 'peculiar people' exercising its right of self-determination . . . already quite consciously held by the English for many centuries."[10]

War in Europe following the 1789 outbreak of the French Revolution brought about a paradigmatic change. The eighteenth century saw the flowering of an age of limited warfare, characterized by a reluctance to commit armies to decisive battle due to the recognition on all sides of the difficulty of replacing professional infantryman, whether native or foreign, because of the enormous expense of recruitment and training in the tactics of the day. The soldiers of the French Republic surprised the powers of Europe, both with an ardor more reminiscent of the religious wars of the seventeenth century, and in the enormous numbers they were prepared to commit to battle. The revolutionary enthusiasm of the *levée en masse,* introduced in 1793, and the concept of citizen obligation to the nation in wartime service fueled the Napoleonic armies such that by 1813 these armies had conscripted more than 1.3 million men, and in 1813–1814 another million more.[11]

The great shock brought to the dynastic societies of Europe was the overthrow of centuries of agreement that state war was a limited affair. Carl von Clausewitz described this change as the appearance of an absolute form of war, one with the political aim of final victory. Napoleon ushered into European consciousness war that ignored the limits of the eighteenth century and drove his armies to destroy his opponents and occupy their land, rending them powerless and dictating peace terms in their capitol. Every nation in Europe, save Great Britain, accepted that emulation was the way to counter the citizen-soldier and his "enthusiastic" acceptance of service as a condition of citizenship. Most sought ways to open their societies enough to create the enormous armies needed to defeat Napoleon in 1814 and 1815. The Treaty of Paris, which ended the Napoleonic Wars in 1815, sought to return limits to war in the manner of the 1648 Treaty of Westphalia. As Clausewitz noted, however, once unleashed, the notion of absolute war and the enormous armies it required provided a standard with which to judge future wars.[12]

For British society, on the other hand, the process of growing the British regular army to almost a quarter-million men to fight Napoleon made no significant changes to the previous methods of recruitment or to the militia system.[13] Chronic underemployment and better compensation eased the problem of finding soldiers, but in the end the officers remained rooted in the ruling class and the soldiers were those most expendable and least employable. Despite the victory over Napoleon, the British Army was not popular, its discipline considered callous and ferocious, while the words "curse, hang, and flog" summed up the outlook of British society toward the army.[14] George Finer allowed that while these "scum of the earth were stolid infantry," the result of victory in 1815 was "perhaps a little unfortunate . . . it was an obsolete army, and its victory merely confirmed British ruling-class complacency which was to last for another three generations."[15] Wariness continued to be the English attitude toward large standing armies. Fearing the effect on individual and political freedom, conscription was a wartime expedient, as George Flynn noted, and "viewed less as a part of citizenship and more as a burden imposed by the government."[16]

Burgeoning populations in Europe and the United States and an ever-expanding state capability in the industrial revolution of the latter half of the nineteenth century fueled the exponential growth of standing armies in France and Germany, in particular. The population of the United Kingdom expanded from 41 million in 1871 to 45 million in 1911; Germany, within the boundaries of the unified German state of 1871, grew from 41 million in 1871 to 57 million in 1900 and to 65 million in 1910; from 1891 to 1910, France grew from 37 million to 39 million; and the United States grew from 50 million in 1880 to over 75 million in 1900.[17] Improved public health and nutrition increased the robustness and availability of citizens. The German unified state introduced common education requirements. Governments increased their capability to find resources through taxation and the identification and location of citizens through the introduction of the national census. This accounting enhanced the ability to locate the recruits, ensuring their availability and providing for their feeding, housing, equipping, and transporting to the battlefield in ways never possible in the eighteenth century.[18] Railroads, steam-powered warships, and innovations such as the introduction of smokeless powder sparked an upheaval in the management and focus of industry and society and in the corresponding strength and mobilization, if not mobility, of the armies.

Ninety-nine years separated the victory of Viscount Wellington and General Gebhard Blücher at Waterloo in June 1815 and the beginning of World War I in Europe in August 1914. While the British Army returned to a volunteer force after 1815, the great powers of continental Europe maintained standing armies influenced and modeled on the successes of Prussia-Germany against the Austrian Empire in 1866 and the French Empire in 1871.[19] Following the Franco-Prussian War (1870–1871), which unified the German state under the Prussian monarchy, Germany increased its field and reserve army from 1.3 million men to 3.4 million. In 1897 France, in turn, increased its field and reserve army from 1.75 million to 3.5 million.[20] The British Army, however, was not the guarantor of the state as in Germany and France. The Royal Navy was the cornerstone of national defense. The British Army received an increase in neither size nor role, nor was any form of conscription acceptable in peacetime.[21]

In the United Kingdom, it was the primacy of the navy that served as the basis for the grand strategy for the defense of the empire. The purpose of the British Army as laid down in 1888 by then–Secretary of War Edward Stanhope was "the maintenance of public order; Indian defence; garrisoning fortresses and coaling stations; and, finally, home defence."[22] The role of the Royal Navy as reiterated in 1896 by the Colonial Defense Committee was "the maintenance of sea supremacy . . . assumed as the basis of the system of Imperial Defence against attack over the sea. This is the determining factor in shaping the whole defensive policy of the Empire."[23]

The Second Boer War (1899–1902) revealed vast inefficiencies within the British Army bureaucracy and the national preparation for war. The army rejected 40 to 60 percent of applicants for military service as unfit. With the cost of war at "£200 million and 22,000 lives to subdue an enemy whose entire force never equaled 90,000 men,"[24] criticisms of tactical leadership and fiscal management of the army were even harsher. Fears arose that the industrial revolution, and the prosperity it generated, had softened the nation. To counteract an apparent moral and physical deterioration, some leaders in Britain called for conscription to rejuvenate the Victorian ideal of vigor, seeing this spirit as a means to ensure the defense of the nation. Calls for coerced national service, however, met strong opposition. Many sectors of the civil population preferred volunteerism for the rebirth of national spirit, giving rise to organizations directed at youth such as the Boy Scouts. Among the educated and upper classes in the United States and the United Kingdom, however, compulsion in national service appeared to offer many benefits for society. While opposed by the laboring class, these groups saw in conscription a method to elevate society through social management in education, sanitation, and morals.[25]

Field Marshal Lord Frederick Sleigh Roberts, one of the most celebrated British generals from the Boer War, was skeptical of the call for compulsory service. He believed that volunteers were best when inspired to join, disparaging conscription as unpalatable to society and asserting "that the man who voluntarily serves his country is more to be relied upon as a good fighting soldier than [if] he is compelled to bear arms."[26] Lord Roberts, however, changed his mind; after revisiting South Africa, he advocated compulsory service as the most efficient means to conduct a continental war, and he became president of the National Service League in 1905.[27]

Voluntary service remained the norm, however, as reforms begun under Lord Richard Burdon Haldane in 1905—building on the prior Cardwell-Childers reforms (1868–1874) and with the reorganized General Staff—split the regular army between service overseas and home defense. Haldane recast the old militia into a special reserve, requiring six-year terms with all but six months on inactive status and with members liable for service overseas. He placed the existing organizations of the Yeomanry (auxiliary cavalry) and Volunteers (infantry) into the Territorial Army.[28]

The Territorial Army was the product of the consolidation in 1908 of several part-time military organizations that appeared over a long period of British history. The militia (up to 138 infantry battalions) consisted of volunteers who served full time for a few months' initial training period and then, during an enlistment of five years, trained each year for a few weeks. Like the regulars, they were not posted away from their regiments without their consent, but unlike the regulars they could not be compelled to serve overseas. While no cavalry arm existed in the militia, a part-time reserve mounted force existed in the form of forty-six cavalry regiments, known as the Yeomanry. Volunteers (up to 86 battalions) were men who did not wish to leave their civilian occupation, or were unable to afford leaving it, and still desired to serve. Lord Haldane connected all of these reformed units to a regional regimental headquarters for continuity of tradition and local recruitment. With sufficient volunteers trained in peacetime, he deemed compulsory service of the German and French model unnecessary.[29]

The Regiment

The term *regiment* resonates through the centuries in the military traditions of the British Isles. No other single word can convey the concepts of unity, esprit de corps, comradeship, fellowship, and shared hardship. These characteristics instilled a common belief in an unbreakable bond and experience for the British soldier—the development of a common small-unit identity.[30] John Keegan noted that the regiment was "the most significant of British military institutions, the principal vehicle of the nation's military culture."[31] The Cardwell-Childers reforms defined the modern concept of the regiment. While the regimental tradition far predated Cardwell-Childers, with some regiments hailing to the time of the Restoration (1660) and the

Glorious Revolution (1668), these measures, when coupled with those made after the Boer War through Haldane, guided the basic structure of the infantry regiment's volunteer selection from 1881 through World War II. There was little that was systematic, however, in the regimental system. Brian Bond criticized the reforms, noting that "despite all these assets the regimental system lay at the root of the malaise of the inter-war Army in the effects it had of narrowing horizons and obstructing reform."[32]

Besides eliminating flogging and rare punishments like branding, key Cardwell-Childers reforms included both reduction in the length of service and localization. Length of service decreased from twenty-one to twelve years, divided as five to seven years active and the remainder in the reserves. Prior to these reforms, while regiments bore the name of a county affiliation, all regiments recruited on a general service basis and sought recruits from across the country. The reforms intended that the regular army infantry recruit from a single county regional base, incorporating this affiliation through the reserves and the Territorial Army, including Scots, Welsh, and Irish county connections.[33] The King's Royal Rifle Corps, the Rifle Brigade, and the Foot Guards, along with the Royal Artillery, Royal Engineers, Cavalry, and all other departmental corps (medical, military police, army pay, and ordnance corps), recruited on a national or general service basis with some later adjustments to localization for the Royal Artillery.[34]

Local regimental affiliation tied the army to the people of the United Kingdom. Through this connection the army sought to create and sustain esprit de corps within the unit and a bond with the citizens of its region. The regimental system was the cornerstone of the army's obligation to the soldier, requiring the soldier's consent before any assignment out of the regiment. Prior to conscription, the soldier found in his regiment others like himself, from the same county or region or identity. It was a home to which the army committed itself for purposes of morale, recruitment, and cohesive combat performance. The army considered regional recruitment to be a major contributor to unit and individual morale.[35] It cannot be stated, however, that localized recruiting met the needs of the manpower requirements of the regiments. Few found the ability to recruit successfully within their county or regional boundaries. Localized became regionalized, and with World War I and its extraordinary casualties, nationalized recruiting became the norm going into World War II.[36]

The regiment represented a unique and idealized narrative, a form of metanarrative or master plot, that inspired the belief that wherever or in whatever form he served, the basis for the infantryman's cohesion and esprit de corps and presence in battle was that produced by his association and assimilation in the regiment.[37] The narrative proof of this belief was the vision for the conscripted millions of infantrymen of World War I.

Selection for National Service During World War I

At the outbreak of World War I, Britain relied on volunteerism to fill the ranks of the regular army and the British Expeditionary Force in France. On August 1, 1914, Lord Kitchener, the new secretary of state for war, called for the creation of the New Armies, with eight divisions intended for service in France, and expanding to seventy divisions. The organization of these new armies bypassed the Territorial Army and its established methods for recruiting and supplying soldiers to formations in France. Not neglected, the Territorial Force maintained a role in both the fighting overseas and home defense. Members of the Territorial Force were not sent overseas unless they volunteered, until the Military Service Act of 1916 removed that right, and 318 territorial battalions saw service as against 404 battalions raised for the new armies.[38]

Volunteers flocked to the colors in August–November 1914. Physical standards rose and fell with the boom and bust of volunteer recruitment. The height standard rose from 5 feet 4 inches to 5 feet 6 inches in September 1914, reduced to 5 feet 4 inches again in October, and lowered to 5 feet 3 inches in November, with an exemption for the so-called Bantam battalions accepting men shorter than the standard height, and indeed less than 5 feet.[39] By December 1914, the army received almost 1.2 million volunteers and issued a requirement for an additional 100,000 per month in 1915. Recognizing an inability to sustain this level of volunteers, the army raised the upper age limit for volunteers to forty in May 1915.[40]

Voluntary enlistment came with two serious flaws. First, after January 1915, the lack of volunteers threatened the ability of the army to remain in the field. Second, unregulated volunteerism misdirected many of the men needed to maintain the flow of materiel and munitions to the soldiers in the army.[41] Throughout 1915 the government tried with limited success to expand volunteering, with very limited success. The final attempt at volunteerism, the Derby Plan, failed during October 1915.[42] Conscription became law on January 27, 1916.[43] The new law allowed for greater control of manpower in the war economy and created mechanisms to monitor the effect of unlimited draft calls on industrial production and to standardize physical requirements. Not until December 19, 1916, however, did the new director of national service for the first time attempt to gain effective control of both military and civilian requirements for manpower.[44] Personnel shortages were so great by the time of the US declaration of war on April 2, 1917, that the "War Cabinet believed that the Allies should remain on the defensive in France through 1918 to enable the Americans to build up their strength."[45]

These manpower problems and shortages resonated with the infantry crisis suffered during 1917 and 1918. While the army grew from over 3.3 million men in 1916 to over 3.9 million in December 1917, its peak strength, the reservoir of Category A1 men—those both medically qualified and fully trained—fell from 400,000 in September 1916 to 130,000 in December 1917.[46] In addition, the proportion of the services devoted to combatant arms rose from 18 percent of overall strength in the British Expeditionary Force in September 1915 to over 32 percent by March 1918. Parliament lowered the age of conscription in January 1917 to eighteen years and seven months, but allowed no combat soldier until age nineteen into the theater of war.[47]

The effect was a continuing decline in the density of infantry formations holding the line. The infantry strength stood, on average, 15 percent below establishment by January 1918, with an approximate strength of 1.75 million, "at the very moment when Central Power offensives in both France and Italy were increasingly likely."[48] Further, in light of indications late in 1917 that allocations provided a mere 100,000 men against a War Office estimate of a need for 600,000 Category A men during 1918, General Alexander Haig sought to disband some divisions to keep others up to strength. The War Cabinet preferred the disbandment of battalions from existing divisions and consolidation. This was the path chosen to shore up the strength of existing infantry divisions, reducing each brigade by one battalion, and thus disbanding 147 battalions. In the end, however, conscription maintained the strength of the British Army in France at just over 2 million men during 1918, despite losing over 280,000 killed, wounded, and missing from August until the armistice, not including the losses suffered in the abortive German offensives of March and April 1918.[49]

Tactical experience placed the highest value on the man most physically fit. It also placed the greatest possibility for this fit man to be killed, placing the highest-quality man in the most lethal position in France—the infantry. The largest number of men classified for overseas service went into the infantry divisions. Of 2.5 million men examined for service in the final year of the war, however, selection criteria found over 1 million unfit for combat service.[50] This highlighted the British Army's intent to place the most qualified men into units with battle responsibility. The combination of conscription and volunteerism proved just sufficient. However, the number of men who volunteered in 1914 and 1915 outnumbered the totals recruited through conscription in 1916 and 1917, despite the damage to the economy, "while the number of recruits attested in September 1914 alone (426, 901) was only 30,000 less than the total number of conscripts obtained in [all of] 1918."[51]

When conscription ended on April 30, 1920, the realization came that it had started too late in 1916 to bring coherence into a national scheme of labor and manpower management that oversaw the needs of both the army and the economy.[52] The United Kingdom and its Dominions lost over 1 million dead and 1.5 million wounded, or nearly 10 percent of its male population under age forty-five. Those born in the years 1892–1895 were the most liable for service, being of the ages of nineteen to twenty-two during the war, and 35–37 percent of their number were killed in the war.[53] These lessons enlightened the conscription laws of World War II.

Volunteer Standards, 1920–1939

"The first fifteen years in Britain after the end of the Great War were marked by an intense pacifism and an overriding preoccupation with solving the problems of economic dislocation and depression."[54] In 1919, the War Cabinet approved reductions in spending with the assumption that a great war within the next ten years was unlikely, and on this basis there was no requirement for an expeditionary force. With this intention, the ten-year rule became the basis for British military planning until 1932, and a new five-year rule went into effect for planning purposes. Against this backdrop of fiscal austerity, the British Army's capacity to prepare for war remained diminished until the likelihood of war raised its head again at Munich in 1938.[55]

Criteria. The initial criteria established for the voluntary enlistment of an infantry soldier beginning in 1920 reflected the experience of close combat and the high physical and mental requirements of the Great War. The inability to find sufficient volunteers to meet those standards following the war echoed the pacifism of the era and British society's prevailing view of the infantry and the army in general.[56] The army in 1929 attributed the negative trend in enlistments and the falling numbers of infantry recruitment "to the fact that no trade is learnt in the infantry. Possibly the heavy casualties incurred by the infantry in the war prejudices parents against their sons enlisting in the infantry, but it is not possible to prove to what extent this prejudice exists."[57] Exacerbating the problem between 1934 and 1938, the quantity of twenty-year-old males in the general population was at its lowest point as a result of the casualty rate during the Great War. The postwar report on manpower lamented the severe reduction of the available men coming into military recruitment age. Looking back from 1945, it became apparent that even in the worst years of the Great Depression and its prevalent unemployment, service in general, and infantry service in particular, was neither attractive nor in line with a prevailing belief in the "Peace Ballot," i.e., that no war was forthcoming.[58]

Solutions discussed by the chief of the Imperial General Staff, Field Marshal George Francis Milne, in 1929 included the diversion of any qualified recruit into the infantry, because "a large percentage, if they cannot join the arm of their choice, will not enlist at all."[59] The difficulties in maintaining the strength of the infantry led to the decision to reduce standards in order to make the numbers. With just 36 percent of potential enlistees meeting the standard, compared to 57 percent in 1912, the risk in a temporary reduction in the height, weight, vision, and dental standards was acceptable "so long as the army cannot be maintained at the present level."[60]

In November 1935, the adjutant general of the army, General Sir Harry Knox, stressed to the chief of the Imperial General Staff, Sir Archibald Montgomery-Massingberd, the importance of support from the political leadership, suggesting to both the parliamentary undersecretary of state for war, Donald Howard, and the permanent undersecretary of state for war, Sir Herbert Creedy, that the secretary of state for war, Edward Frederick Lindley Wood, take the matter to the cabinet.[61] Knox's report stated that if current trends continued, by 1940 the deficit for infantry alone might total over 25,000 regular soldiers and that drafts would "be utterly impossible."[62]

The quartermaster-general, Sir Reginald May, during November 1936, reported the likelihood of significant improvements in recruiting if accommodations and allowances for items such as food while on leave were made to equal those of the Royal Navy and Royal Air Force. The differences perceived between the British Army, its sister services, and its counterparts in the empire were often small, but meaningful in his view. This was notable in the fact that, while the armies of the Dominions and the workhouses under the London County Council all supplied butter to their charges, the soldiers of the imperial forces were supplied with margarine. He stated his opinion that "we could hardly keep our own servants if we fed them on margarine."[63]

In December 1936, the new secretary of state for war, Viscount Duff Cooper, recommended to the cabinet improvements such as vocational training for soldiers; a free kit for every recruit, including a helmet for soldiers going to India; four meals a day for the army, thus bringing them in line with the navy; and consideration for changing the civil service rules to allow an ex-soldier's years of service to count toward a pension. Reactions from the cabinet were either negative or for referral to additional study, and funding was denied for any of the proposals the army found important for recruitment:

> The Chancellor of the Exchequer [Neville Chamberlain], while appreciating the seriousness of the situation, and sympathizing with the Secretary of State for War, said as yet there was no agreement between the War Office and the Treasury. . . . The Treasury had had to ascertain the reactions which the War Office proposals would have on the Navy and Air Force, which, in some of the items, added enormously to the expenditure involved. . . . The Secretary of State for India [Lawrence Dundas, 2nd Marquess of Zetland] explained fully to the Cabinet the political and financial reasons why the proposals were impossible of acceptance by India.[64]

Physical and educational/intelligence standards. The primary standards for recruits changed in response to the success or failure of recruiting each year. The Army and Air Force Act, reenacted by parliament each year, provided the basis for service and the strength of the army. The regular army, liable for service overseas and with an authorized strength of 210,000, had just over 197,000 men by 1936. The Territorial Army in 1936, not liable for overseas service and authorized for approximately 182,000 men, maintained an on-hand strength of 129,700.[65]

The height standard for the infantry was particularly volatile, lowered in 1928 from 5 feet 3 inches to 5 feet 2 inches, then raised to 5 feet 3 inches again in 1930–1931, raised again in 1932–1933 to 5 feet 4 inches due to a good recruiting year, and settled at 5 feet 2 inches in 1936 as recruiting appeared to be bottoming out.[66] Vision and dental standards changed as well, dependent on recruiting success in a particular year. Standards discouraged dentures, but by 1936 this changed to reject just those applicants with no, or few, teeth.[67]

The Royal Army Educational Corps tested incoming recruits to assess their basic level of intelligence, and results ranged from A (the highest) through E (the lowest or illiterate). These five levels provided a basic intelligence-quality scale for physically qualified recruits received at the regimental depot for training. Recruiting shortfalls, including the loss of 600 recruits per year on intelligence grounds, caused the adjutant general to recommend the elimination of intelligence tests for the line infantry. He proposed to accept those deemed illiterate, or Class E, and grant them special educational assistance in the regimental depot in hopes of retaining up to 400 of these men for some useful employment in the infantry. The chief of the Imperial General Staff, 1936–1937, Field Marshal Sir Cyril Deverell, acknowledged in his handwritten reply:

> I think it would be a great mistake to cancel the education test for Infantry of the Line owing to its effect on morale of the Infantry as a whole. The test is required in any case to grade the men. I have no objection to any man who is not an imbecile being enlisted, i.e., I do not consider rejection on educational grounds alone should be a *sine qua non,* but I should not proclaim it very loudly. I agree that such men may go to the depot in the ordinary way.[68]

The permanent undersecretary of state for war, Sir Herbert Creedy, replied with the suggestion to accept illiterate recruits without lowering the educational standards by following the same protocols for admitting those recruits below physical standards who showed promise of being able to respond to special training. If both the medical officer and recruiting officer agreed that the recruit possessed normal intelligence despite his illiteracy, then an appropriate notation allowed his acceptance.[69] Thus, at the same time that recommendations for increasing accommodations for the soldiers to increase retention and recruitment were denied, proposals to reduce standards for recruits were the norm.

The commanding general, Southern Command, Salisbury, Lieutenant General Sir Archibald Wavell, on December 14, 1938, noted the effect of the reduction in intelligence and physical standards.[70] Responsible for creating and training the new motor battalions and a mobile division, Wavell noted his concerns with the quality of the infantry needed for the anticipated high-mobility operations expected of these units. The mobile and motor divisions were the outcome of years of interwar experimentation beginning in 1922 and culminating in the creation of these organizations in 1938. The army organized the mobile division in 1938 and envisioned it as a key part of British Army reorganization that included the 1938 model infantry division and the motor division. Built around a tank brigade and a mixture of light and cruiser tanks, the army intended the mobile division for deep, offensive maneuvers into enemy formations. The motor divisions, in turn, used their mobility to consolidate gains made by the mobile division. The mobile division became an armored division when war came. The bifurcation of "infantry" tanks in army tank battalions supporting infantry formations, as well as tank regiments (battalion-size formations) with a different armored-force mission with light and cruiser tanks, underlay British Army mobile doctrine at the opening of the war.[71]

Wavell believed that the quality of the infantry assigned to these units decreased expectations for their functioning. He reported this to the permanent undersecretary of state for war, Sir Herbert Creedy, noting "that the low standard of education of the recruits trained for and posted to these Motor Battalions is making the preparation of these Battalions for their duties in the Mobile Division a matter of very great difficulty. At the present time the standard of education required from recruits on enlistment for the Rifle Depot . . . in effect means that any recruit who is not actually mentally deficient can be enlisted."[72] He provided statistical evidence of educational deficiencies for three squads enlisted for the depots of the King's Royal Rifle Corps and the Rifle Brigade, suggesting that these men should have "a high educational standard similar to the recruits enlisted in Cavalry Light Tank Regiments and Royal Tank Corps units, rather than the lower standard which it has been decided to accept for infantry."[73]

The British infantry found itself in hard times by 1939. Parliament mitigated the inability to reach the army's recruiting numbers, irrespective of quality, with a 1937 decision to begin a program of rearmament, which accelerated following the Munich crisis of 1938.

The acceptance of a lower standard for the line infantry in relation to the other needs of the army was a key characteristic of this period leading up to war in the critical year of 1938.

The Making of the Interwar Infantry

If the nature of the infantry battle was to close with and destroy another human being, the task of the institution of the army was to create the conditions for an infantryman to place himself in harm's way, even at the risk of his life. After selection for the infantry, an individual found himself confronted with an organization whose primary function was to provide him the skills needed to destroy other humans, including the disciplinary—the coercive—motivation to ensure that he performed that function in the face of the most extreme form of face-to-face infantry combat. The infantryman's selection and characteristics determined his ability to perform his role within his primary group. Thus, from the perspective of the individual infantryman, the expression of the army's intent, i.e., the basis for cohesion, came from the intersection of two factors. The first factor was his socialization—the internalization and legitimation of the ideals and standards of the army. It was the intent of the army that soldiers embraced these ideals, that they desired to meet those institutional goals as an army-wide societal and group benefit. Second, the individual infantryman faced the worldview of the army from the perspective of his position in his small unit and his socialization of the norms and expectations of his peers in the daily routine of the army. This is the perspective that kept the infantryman in the line and willing to risk his life for his friends. His belief that his performance mattered to them—and that they perceived his concern for them in the combats to come—was such that he would not let them down through a failure to meet their standards.

External and Institutional Cohesion Measures

From the point of recruit selection, the army turned to its institutional requirements, its intent being to provide the new infantrymen with the necessary skills and motivation. These included the infantryman's training within his small-unit organization, the provision of leaders to these small units, and the coercive effects of the disciplinary system. This intent held the assumption that in these spaces, both social and coercive, cohesive behavior was both the expectation and the result. These were the principal external conditions experienced by the soldier in the interwar years that defined his relationship with the institution of the army and with the members of his primary group, absent the combat motivations to come. The regiment was the basis for the social environment and unit through which the expectations for the behavior of the infantry were learned, lived, and expressed. As the role of the regiment diminished, the assumption of these impacts on the infantrymen, the assumptions about the effect of the regiment, came into question. The acceptance of the downward selection criteria for the infantry in intelligence and physicality acted against the metanarrative assumption that no matter the criteria, the regiment provided the ability to overcome any individual deficiencies for the infantrymen. The role of these factors on the cohesive behavior of soldiers in their small units during the battle conditions of wartime expanded to include the method of rewarding combat behavior consistent with the goals of the institution, as well as how the soldier experienced care upon being wounded in battle and how a replacement arrived in his small unit. These were the most personal impacts and gave insight into how the institution appeared to fulfill its role in giving the infantryman what he believed most important—his chance of survival.

Training

After World War II, the British Army conducted a review of the manner and method for the training preparation of the army throughout the war. The report noted the broad purpose of this task as a process through which instruction imparts military knowledge. It included the vast range of organization, teaching and teachers, weapons and doctrine, duration of training, and accommodation and creation of ranges and areas for individual and collective maneuver—"a machine which provides the Army with men and women of a required standard of training, when it wants them and where it wants them."[74] The British Army recognized the enormity of the task: the turning of raw recruits into fighting soldiers, the creation of officers to lead them, and the entire apparatus of the support and technical echelons to keep the fighting arms in battle. During the interwar period, the army maintained a fundamental doctrinal orientation toward this, which won the Great War—the methodical battle on the French model.[75] And while the British Army never embraced the absolute rigidity inherent in this model, its doctrine "ensured that troops could not maneuver more rapidly than artillery could lay on fire support . . . by placing authority in the hands of senior commanders remote from the scene of the action, and by severely constraining the initiative of their subordinates."[76] In this model, the infantry was the primary arm supported by close firing and massive supporting artillery by preceding the advance with an orchestrated creeping barrage. With the notable experimentation in the mobile and motor divisions, the infantry remained the focus of success in battle. Doctrine noted in 1935 that "practically all success in war, which is won by the proper co-operation of all arms, must in the end be confirmed by infantry, which, by closing with the enemy, compels his withdrawal or surrender. . . . It is the most adaptable and the most generally useful of all arms."[77] Thus confirmed in the Great War, the interwar army trained for the coming war in the essential manner of the victory in the previous.

Pre-service training. There was recognition that the training of individuals might advance "if [they] had already undergone some form of military instruction before arriving in the Army." There were four primary ways to achieve this indoctrination prior to the introduction of general conscription in 1939. The first was the Army Cadet Force, containing both closed units in some secondary schools and open units, not school-based, but "mainly of working lads who voluntarily gave up their spare time to cadet activities."[78] Second, the Officer Training Corps, renamed the Junior Training Corps after the opening of the war, was in the public schools; the ages for both it and the Army Cadet Force were between fourteen and eighteen. Third, army technical schools were run by the army for boys between the ages of fourteen and fifteen and a half who intended to have a military career, and were instructed in military trades until call-up. Fourth, volunteers joining the army before call-up age were enlisted into Young Soldier Battalions of the arm for which they volunteered. While these methods were for those under age eighteen, a final eligibility came in the form of the Senior Training Corps in universities, which was compulsory for those with deferred service.[79]

Significant numbers of young men and boys trained in these organizations throughout the war, with the overall object "to inculcate in lads destined for the Services a grounding of character, resourcefulness and all the other qualities which would be useful afterwards as junior leaders."[80] The Army Cadet Force ended the war with over

150,000 cadets, with a peak enrollment of 175,000 in over 500 units in 1943–1944. The Junior Training Corps ended the war with 31,700 cadets in 177 units, while the Senior Training Corps averaged over 10,000 cadets per year. The army technical schools graduated an average of 614 tradesmen-qualified cadets per year. The army believed the greatest advantage lay in the Young Soldier Battalions, in the preparation through section-level training in the infantry, due to their status as volunteers before call-up.[81]

Small-unit organization and training. The War Establishment gave the organization of the rifle section in 1938 as eight men, consisting of a noncommissioned officer (NCO) and seven men including a single euphemistically named light machine gun, the magazine-fed Bren gun or automatic rifle.[82] The rifle section was a subcomponent of the platoon, of which there were three in each of four companies in the infantry battalion. The principal document that described the employment and training for the rifle section was titled *Infantry Section Leading* and established in 1938 the doctrinal standards against which the wartime army began the war in organization and training preparation. In terms of the social structure in which a soldier existed, the section in the British Army formed the "fire unit of the infantry and is the largest body that can be personally controlled by its leader throughout the battle"; it was therefore the "basis of infantry organization."[83] This was the foundation of the social construction of the primary group for the infantryman of the British Army.

"The most important function of basic training was to transform a civilian into a soldier by inculcating into him certain habits of behavior and mental qualities that the military authorities deemed essential," discipline above others.[84] This purpose achieved, the new soldier accepted the truth of his life in the army and his "part of a disciplined hierarchy stretching from the monarch at the top to the newest-joined private soldier at the bottom."[85]

The training of the section leaders was the responsibility of the company leaders, often under the auspices of the battalion commander. Training began at the individual level and progressed to the section and platoon and then company. Doctrine considered the platoon "the basic unit of infantry tactics."[86] It stressed the importance of all members of the section being capable with either the rifle or the light machine gun, the Bren gun. Doctrine further laid out the various forms of combat action in terms of attack and defense, as well as the types of supporting weapons the section might expect in the form of artillery, infantry tanks, and aircraft.[87]

With conscription ended in 1920, the army relied on volunteers in the interwar period. Training the volunteers at this time took two forms, both conducted at the regimental depot. First was that for recruits. Second was that encountered when the recruit joined his own unit in his section and platoon. Thus each regiment accomplished the entire scope of the preparation of the civilian for battle, first in the training of the recruit and his inculcation into its institutional norms in the training depot, and second in his movement to his actual combat unit and section, his primary group within the regiment. The object of training an infantryman was to produce "a formidable fighting man like an expert hunter—always alert and seeking an opportunity of striking at his quarry . . . confident and expert in the use of his weapons . . . determined, inquisitive and self-dependent, but . . . always remember[ing] that he is acting as one of a team. . . . *Above all he must be highly disciplined,* for by discipline alone can morale be maintained; it is the bedrock of all training. It is the ingrained habit of cheerful and unhesitating obedience that controls and directs the fighting spirit, and is the backbone of a unit in a moment of crisis."[88]

In this manner prior to the opening of World War II, British Army expectations for cohesive behavior of the individual within his primary group were stated as both an ideal and an expectation. It remained the single continuous expression of the group as the recruit moved within his regiment in peacetime from initial training designed to transform a civilian into a soldier, to his assignment in his section of one of the battalions of the regiment. At that point he became the responsibility of his junior leaders, both the NCO who led his section and the officer platoon and company commanders responsible for his tactical training in preparation for battle.

Junior Leaders

For the private soldier, life in the regiment revolved around his daily interaction with the authority and leadership provided by junior NCOs (corporals and lance corporals), senior NCOs (sergeants), and warrant officers. "They were an everyday presence in their lives in a way that officers rarely were. . . . The NCOs had a far more intimate knowledge of their men than the company officers, and formed the vital link that mediated relations between officers and other ranks."[89]

Noncommissioned officers. Discipline began with the warrant officers and NCOs and resided most formidably in the company sergeant major. These men ensured that the other ranks "carried out the orders that were passed down through them from their officers, and they did so without hesitation or question . . . [such] that their company commander's orders were performed to the letter. . . . It was their duty to report any improper behavior," representing the first link in the chain of both formal and informal discipline for the private soldier.[90] Doctrine noted the essence of leading for the infantry NCO section leaders:

> A leader must first of all have the *confidence of his men,* and to gain this he must have confidence in himself. He must be able to make up his mind, and having done so, stick to his decision. He should keep calm. To show doubt and indecision is a sure way of shaking the confidence of his men. A stout-hearted man will always go on trying; and by doing so will instill his own fighting spirit into his followers.
>
> *Loyalty* is an essential of leadership; unless a leader is himself loyal to his superiors, he cannot expect loyal support from his subordinates.
>
> Finally, he must understand *discipline*; he must command the men of his section firmly, but with common sense and fairness. He must give his orders clearly and, having given an order, must insist on its being efficiently carried out.[91]

The selection and promotion of NCOs from the ranks was the responsibility of the commanding officers in the regiment. Seniority did not guarantee promotion. Prospective infantry NCOs were required to be literate, with corporals possessing a third-class army education certificate, sergeants a second-class certificate, and more senior NCOs and warrant officers a first-class certificate. Each NCO rank required passing formal qualifying examinations. New NCOs needed to earn the respect of the other ranks as well as the officers, and to navigate a new, more cautious relationship with individuals who were their peers until promotion. Success meant finding the ability to enforce orders without becoming a martinet, of finding a way to move a unit with a sense of "discipline, humour and common sense."[92] Regardless of personality, however, the NCO counted on the support of his senior officer in the application of discipline. In this sense, one of the primary roles of the NCO was the legitimation of the regulation of the lives of soldiers through the formal discipline system.

A senior officer was more likely to take the side of a senior NCO over a junior officer in matters of unit discipline and administration, as "senior NCOs had the advantage that they not only knew their men more intimately than any newly joined officer, but they also had a better grasp of the daily practicalities of running the sub-unit."[93] The best of the NCOs were able to traverse the varying demands of maintaining the respect of their subordinates and their officers. The focus on supervision and the implementation of the officer's order, rather than leadership, socialized as an officer trait, would resound throughout the coming war.[94] "Battle throws up only a limited number of natural leaders . . . [and] the supply of suitable personnel did not meet the demand. A high proportion of . . . serjeants had been fusiliers or lance-corporals only a few months previously. Platoon serjeants were often required to command platoons in battle and were rarely equal to the task: those that were, were given direct commissions."[95]

Subalterns. After 1920, commissioned officers came into the line infantry as graduates of the Royal Military College at Sandhurst. The social structure of the infantry regiment and the socialization of the officer corps was one of the most singular characteristics of the army in the interwar years, and with the exception of the Great War, of the regimental system as a whole across time. Officer graduates were required to express a preference for a line regiment. Until after World War II, regiments owned significant influence on which young officers were recruited and accepted from Sandhurst. Family connections were important measures. While not as rigid or formal as the German Army system, it was "equally effective in maintaining social harmony in the officers' mess."[96] In essence, to be given an invitation to a particular line regiment required the approval of the colonel or colonel-commandant of the regiment. This meant that graduating at the top of one's class at Sandhurst had less to do with one's assignment than did the connections and approval of the regiment. The disadvantage of course was that those with such a "guarantee" of acceptance had little incentive to excel in the military education and preparation that was the intent of the Sandhurst curriculum.[97]

The preparation of the junior officer for leadership was a subject of intense scrutiny after the Boer War and again after 1918. What did the young officer need to know? Modern war appeared to demand a broader and more diverse education than the practical regimen that appeared at Sandhurst after the Boer War. The object in 1929 was that the colleges "would become 'the university which gives the best general education and which best develops a boy's character and personality.'"[98] As in the German Army, the concept of "character" was a key condition expected of the junior officer. Its definition remained uncertain, but stressed the necessity for leadership skills above technical ones. While officers passed examinations for promotion, there was no generalized list for promotion, and a long wait for advancement was one of the key characteristics of the regimental system. Promotion in the line infantry regiments was based on seniority within that regiment's list of officers. Depending on the regiment, with wide variation, "by 1932 the average age at which a subaltern in a line infantry regiment became a Captain was 32; captains did not become majors until they were 39; majors did not become lieutenant-colonels until they were 47."[99] Even into 1943 the idea of a central promotion list by merit across all line regiments was considered too disruptive, echoing the thoughts of the chief of the General Staff in 1908 that "the cost of doing so would be the destruction of one of the foundations of the regimental system," weakening the cordial relations and esprit de corps essential among the officers of the regiment for their fighting efficiency.[100]

The fundamental preparation of the officer for battle, as with the infantry sections, lay within his regiment, dependent on the zeal, capability, and time of the regiment. "Unit commanders were responsible for training their own officers, but few had themselves been trained how to do so. Whether or not a regimental officer studied his profession seriously and continuously depended largely on his [commanding officer]."[101]

The function of the officer was to lead. "To the soldier, the part of father and mother is (or should be) played by the regimental officer, and his regiment takes the place of his home. The quality of the men themselves, the efficiency and tone of their regiment, and the care bestowed upon their well-being by their officers, are the most powerful of the positive factors affecting the morale of a unit."[102] Supervision was the role of the NCO. Oversight of the private soldier in the British Army was close and continuous. In the 1930s the infantry battalion maintained a ratio of 1 officer to every 27 to 30 other ranks, and 1 NCO to every 7 or 8 soldiers. By 1944 this level of supervision increased to 1 officer to every 22 other ranks and one NCO to every 6.6 soldiers.[103] The other side of the coin of leadership and supervision was the implementation of the disciplinary system imposed on the soldiers and the use of formal discipline to maintain order.

Discipline and Sanctions

During World War I, the British Army tried 31,367 of 115,005 cases of desertion by courts-martial, resulting in 266 executions. Of over 10,000 trials for sleeping on post, two soldiers were sentenced to execution. In addition to these executions, the following death sentences constituted the total of 304 executions in the British Army between 1914 and 1918: 18 for cowardice, 7 for quitting post, 6 for striking or violence, and 5 for disobedience. The experience of World War I was a profound influence on the disciplinary system of the British Army in World War II. The perception of the destruction of so many young Britons, for so little tactical or strategic gain, diminished the luster of the army's reputation and its scope of action in dispensing military justice. The Army and Air Force Act provided for the death penalty in time of war for a number of offenses, including mutiny, treachery, desertion, and sleeping on guard, as well as impeding a provost marshal or violence against a person bringing in supplies. After limiting the number and type of military offenses with a possible death sentence penalty in 1925, and again in 1928, the United Kingdom abolished the death penalty in 1930 for military offenses.[104]

Commanding officer punishments. The British soldier faced a range of possible punishments, and punishers. Leaders within his chain of command held wide latitude for judging what constituted an infraction and what penalty to award an individual transgressor. These varied with the severity of the infraction. Courts-martial punitive awards ran the gamut from long prison terms to reprimand. Commanding officers possessed informal and administrative authority to award deductions in pay, extra duty, confinement to barracks, reprimand, or admonition for infractions of rules and orders, without the need to refer to formal court-martial proceedings. For the soldier, his most prevalent contact with the disciplinary system came at the hands of his immediate supervisors, both commissioned officers and noncommissioned officers.

The British system of courts-martial. There were three types of court-martial in the British Army during World War II: the general court-martial, the field general court-martial, and the district court-martial. (A comparison of the various convening authorities and responsibilities for each level and the manner and type of punishments authorized for the British, German, and US Army court-martial systems appears in Appendix 3.) The British legal system, as in the German system, made a strong distinction between military offenses and serious civil offenses such as murder, rape, and treason, and restricted the military courts from jurisdiction in these latter offenses dependent on where they occurred and if the offender in the British Army was on active service.[105] Other types of confinement for the soldier, including the primary military prison at Aldershot, were field military prisons and detention barracks or assignment to labor companies. The length of a sentence depended on its purpose and on whether there was intent to return a "reformed" individual to service in time to contribute to the war effort, whatever the individual's attitude toward that service.[106]

The Effect

The basic external conditions established by the British Army for the soldier upon his classification and selection for the infantry regiment demonstrated an intent, through the regimental system, to enhance the combat capability and the willingness of the soldier to perform his expected role in combat—to close with and destroy an enemy at the risk of his own life. It is the combination of the external institutional intent to provide both the motivation and the skills to perform the infantry battle task with the classification and selection of the individual that provides insight into the two levels soldiers experience in cohesion: the expectations for service in the larger army and the expectations of the men of the small unit. In the subsequent chapters on the British Army, the experience of combat through these lenses enhances an understanding into how the soldier viewed his service and capability in relation to his German Army counterpart and how he perceived the differences in battle.

Notes

1. Liddell Hart Center for Military Archives (LHCMA), King's College, London, March 24, 1945, Letter of Field Marshal Sir Claude Auchinleck to Field Marshal Alan Francis Brooke, First Viscount. Used by permission.

2. Adams and Poirer, *The Conscription Controversy in Great Britain,* ix.

3. "Modern war" is roughly coincident with the end of the eighteenth century and the arrival and defeat of Napoleon, recognizing the destruction of the old societal expectations that limited the effects of state warfare following the Thirty Years' War. Gat, *A History of Military Thought,* 168, 189.

4. Adams and Poirer, *The Conscription Controversy in Great Britain,* 124.

5. Stearns, "Conscription and English Society in the 1620s," 22.
6. Ibid., 7.
7. Ibid., 7–8.
8. Ibid., 23.
9. Toby Barnard, *The English Republic,* 89.
10. Finer, "State- and Nation-Building in Europe," 144: Tilly, *Coercion, Capital, and European States,* 114–116.
11. Finer, "State- and Nation-Building in Europe," 144–152.
12. Miller, *Global Order,* 31; Ikenberry, *After Victory,* 37–39; Gat, *A History of Military Thought,* 202–205; Von Clausewitz, *On War,* 579–581; Kegley and Raymond, "The Long Cycle of Global War," 274; Palmer, "Frederick the Great, Guibert, Bulöw," 91–95.
13. Paret, "Napoleon and the Revolution in War," 135.
14. Chandler, *The Campaigns of Napoleon,* 42.
15. Finer, "State- and Nation-Building in Europe," 153.
16. Flynn, *Conscription and Democracy,* 3.
17. Massie, *Dreadnought,* 134–135.
18. Tilly, *Coercion, Capital, and European States,* 74–86, 96–116.
19. Finer, "State- and Nation-Building in Europe," 155.
20. Ibid., 155.
21. Hayes, *Conscription Conflict,* 55–56.
22. Ibid., 25.
23. Flynn, *Conscription and Democracy,* 24.
24. Adams and Poirer, *The Conscription Controversy in Great Britain,* 7; Pakenham, *The Boer War,* xiv.
25. Flynn, *Conscription and Democracy,* 13–14.
26. Adams and Poirer, *The Conscription Controversy in Great Britain,* 12.
27. Hayes, *Conscription Conflict,* 45, 112–113.
28. French, *Military Identities,* 11, 247.
29. Hayes, *Conscription Conflict,* 98–110, 136–146.
30. Hackett, *The Profession of Arms,* 223–224.
31. Keegan, "Regimental Ideology," 16.
32. Bond, *British Military Policy Between the Two World Wars,* 58–62.
33. French, *Military Identities,* 28.
34. Ibid., 25–30.
35. TNA PRO WO 32/9846, appendix A to E.C.A.C./P (41)47; TNA: PRO WO 33/1488, Report of the Committee on the Cardwell System, August 1937, 14–15, 38.
36. French, *Military Identities,* 57.
37. Abbott, *The Cambridge Introduction to Narrative,* 46–49.
38. Simkins, *Kitchener's Army,* 44–46, 88.
39. Ibid., 104.
40. Messenger, *Call to Arms,* 130.
41. Flynn, *Conscription and Democracy,* 31.
42. Messenger, *Call to Arms,* 131; Hayes, *Conscription Conflict,* 194–195; Simkins, *Kitchener's Army,* 156.
43. Hayes, *Conscription Conflict,* 208; Simkins, *Kitchener's Army,* 157.
44. Hayes, *Conscription Conflict,* 299.
45. Messenger, *Call to Arms,* 274; Flynn, *Conscription and Democracy,* 34.
46. Ibid., 152.
47. Ibid., 153.

48. Ibid.
49. Ibid., 286–287.
50. Winter, "Britain's 'Lost Generation' of the First World War," 455.
51. Simkins, *Kitchener's Army,* 325.
52. Messenger, *Call to Arms,* 14.
53. Winter, "Britain's "Lost Generation" of the First World War," 450; Keegan, *The First World War,* 423,
54. Dennis, *Decision by Default,* 3.
55. Ibid., 10–11.
56. TNA PRO WO 32/9846, appendix A to E.C.A.C./P (41), 47; TNA: PRO WO 33/1488, Report of the Committee on the Cardwell System, August 1937, appendix A, 35–36.
57. TNA PRO WO 279/65, Report of the Staff Conference held at the Staff College, Camberley, January 14–17, 1929, 42.
58. Pigott, *Manpower Problems,* 2; TNA: PRO WO 33/1376, Report on the Staff Conference held at the Staff College, Camberley, January 7–10, 1935, 15–16.
59. TNA PRO WO 279/65, Report of the Staff Conference held at the Staff College, Camberley, January 14–17, 1929, 43.
60. Ibid., 43–44.
61. TNA PRO WO 32/2984, *Report of Regular Army Recruiting Situation: Methods for Improvement,* October 4, 1935, Quartermaster-General comments of November 8, 1935. The hierarchy for these positions ran from the secretary of state for war, an appointed member of parliament, then to the permanent undersecretary of state for war, a career civil servant normally serving regardless of the party in power, and then to the parliamentary undersecretary of state for war, an appointed position and third in authority and responsibility.
62. TNA PRO WO 32/2984, *Report of Regular Army Recruiting Situation: Methods for Improvement,* September 11, 1936. The term "drafts" described the organization of men into units to be transported for overseas assignment.
63. TNA PRO WO 32/2984, *Report of Regular Army Recruiting Situation: Methods for Improvement,* October 4, 1935, Quartermaster-General comments of November 8, 1935.
64. TNA PRO WO 32/2984, *Report of Regular Army Recruiting Situation: Methods for Improvement,* October 4, 1935; Quartermaster-General comments of November 8, 1935; *Report of Regular Army Recruiting Situation: Methods for Improvement,* December 3, 1936.
65. Pigott, *Manpower Problems,* 3.
66. TNA PRO WO 32/4643 A.G. 10, Notes on the Post War [World War I] History of the Physical and Medical Standard of Recruits.
67. TNA PRO WO 32/4643 A.G. 10, Notes on the Post War [World War I] History of the Physical and Medical Standard of Recruits; TNA PRO WO 32/4643, War Office letter 27/Gen./9582 (A.G. 10), November 17, 1936, 41A, appendix C, "Recruits Physical Standards," to C.I.G.S. Letter of November 17, 1936.
68. TNA PRO WO 32/4354, C.I.G.S. Response to A.G. Note to P.U.S. and C.I.G.S. on 43/Educn/895 Report on the Experiment with "I" [Illiterate] Recruits, November 10, 1937.
69. TNA PRO WO 32/4354, P.U.S Response to A.G. Note on 43/Educn/895 Report on the Experiment with "I" [Illiterate] Recruits, October 22, 1937, Letter 43/Educn/895 (A.G. 10) (Directorate of Recruiting and Mobilization) of November 4, 1937.

70. TNA PRO WO 32/4354, Letter from The General Officer Commanding-in-Chief, Southern Command, Salisbury to the Under-Secretary of State for War, December 14, 1938.
71. French, *Raising Churchill's Army,* 33–37.
72. TNA PRO WO 32/4354, Letter from the General Officer Commanding-in-Chief, Southern Command, Salisbury to the Under-Secretary of State for War, December 14, 1938.
73. Ibid.
74. Gibb, *Training in the Army,* 1.
75. Bidwell and Graham, *Fire-Power,* 83–85; Ellis, *Brute Force,* 3; Bonura, *Under the Shadow of Napoleon,* 212, 222.
76. French, *Raising Churchill's Army,* 43; Place, *Military Training in the British Army,* 36–37.
77. General Staff, War Office, *Field Service Regulations,* vol. 2, *Operations, General,* 7.
78. Gibb, *Training in the Army,* 299.
79. Ibid.
80. Ibid., 312.
81. Ibid., 300–310.
82. General Staff, War Office, *Infantry Section Leading,* 7.
83. General Staff, War Office, *Infantry Training: Training and War,* 21.
84. French, *Military Identities,* 62.
85. Ibid.
86. General Staff, War Office, *Infantry Training: Training and War,* 22–23.
87. General Staff, War Office, *Infantry Section Leading,* 14–18.
88. General Staff, War Office, *Infantry Training: Training and War,* 10, emphasis in original.
89. French, *Military Identities,* 170.
90. Ibid., 173.
91. General Staff, War Office, *Infantry Section Leading,* 1, emphasis in original.
92. French, *Military Identities,* 175.
93. Ibid., 175.
94. French, *Military Identities,* 179.
95. TNA PRO WO 231/14, Notes on the Sicilian Campaign, Colonel T. N. Grazebrook, January 3, 1944.
96. French, *Military Identities,* 54.
97. Ibid., 54–57.
98. Ibid., 71.
99. French, *Military Identities,* 149–150.
100. Ibid., 151.
101. Ibid., 158.
102. Sparrow, *Morale,* 21–22; TNA PRO WO 163/50, memorandum by A.G., "Papers and Minutes of the 1st to 10th Meetings of the Army Council Held During 1941," January 13, 1941, 2.
103. French, *Military Identities,* 148.
104. TNA PRO WO 163/88, memorandum by A.G. for the E.C.A.C., "The Death Penalty in Relation to Offences Committed on Active Service," May 26, 1942.
105. UK Secretary of State for War, *Report of the Army and Air Force Courts-Martial Committee,* 10.
106. McPherson, *Discipline,* 69–83.

3

The United States: Raising Selection Standards

FOLLOWING WORLD WAR II, THE US ARMY CREATED A restricted series of monographs on a variety of topics that evaluated the performance of the US Army Ground Forces (AGF), commanded through much of the war by Lieutenant General Lesley J. McNair. In *Study no. 5: Procurement of Enlisted Personnel for the AGF—The Problem of Quality,* its author, Robert R. Palmer, noted that personnel problems of quality were most significant in regard to the preparation and performance of ground combat units for battle. While US Army doctrine did not make the clear statements about the nature and requirements for the infantryman as opposed to other soldiers as did British Army doctrine, Palmer described the AGF expectations for combat soldiers from the beginning of the war:

> The mobile tactics and open formations of World War II demanded exceptional physical vigor and great mental alertness in individual combat soldiers, and required considerable powers of leadership in commanders of small units down to the squad. The wits, skill, and stamina of semi-isolated riflemen and small-unit commanders determined not only individual survival on the battlefield but also the outcome of the battle itself. . . . It was therefore desirable to select a high grade of manpower for combat units.[1]

The general failure of the AGF through most of the war to select, classify, and post "a high grade of manpower for combat units" was the story of this study. This did not mean there was any lack of understanding of the central role of the rifleman in the infantry small unit in the success of the US Army in its battle history or doctrine. The field service regulations of 1923 noted that "the coordinating principle which underlies the employment of the combined arms is that the mission of the infantry is the general mission of the entire force. The special missions of other arms are derived from their powers to contribute to the execution of the infantry mission."[2] The regulations noted that the principal requirements of the infantryman in battle was a fighting spirit expressed in aggressiveness and initiative. The nature of the infantry combat environment shaped the requirement that all commands ensure the morale of the infantry, knowing that these same commands would call for the utmost expenditure of the physical and moral strength of the infantryman at critical points in battle. This demanded the preservation of that morale, physical strength, and spirit at all other times.[3] Thus, the US Army defined quality as the necessity for individual physicality and spirit in the infantryman in producing the aggressiveness and initiative attendant to success in infantry combat.

In 1928, US Army publications and notes continued to discuss the importance of the infantryman in battle, noting that "the Infantry is the basic arm. All other arms are auxiliary thereto. . . . In the Infantry the individual rifleman is the basic fighting unit."[4] The tentative field service regulations of 1939 continued the language of the 1923 regulations concerning the primacy of the infantry in that "the infantry is charged with the principal mission in battle."[5] General George C. Marshall, appointed chief of staff of the US Army in 1939, noted in remarks to the National Rifle Association, "I do mean, by implication, that the foundation on which a successful war is carried to conclusion, aside from the character and resolution of the people, eventually rests with the infantry soldier, no matter what the scientific developments and clever gadgets developed for making war."[6]

These two qualities, "exceptional physical vigor and great mental alertness in individual combat soldiers," proved most elusive for the AGF during the war in its quest to ensure their application in the selection of infantrymen. General McNair informed the leadership of the War Department of the consequences for the war effort, and the soldiers themselves, of the effect of a selection and classification process that sent men with lesser physical stamina and intelligence into the most physically and mentally demanding form of combat.

The far greater commitment of ground combat forces by the US Army in World War II showed in the killed-in-action figure of 234,874 in contrast to 50,510 battle deaths in World War I (see Appendix 5). The infantry bore over 660,000 of more than 930,000 total casualties in dead and wounded. Of the elite divisions, six infantry divisions of the US Marine Corps suffered 18,623 battle deaths, and the five airborne divisions of the US Army Airborne Command had approximately 6,543 killed in action.[7] The ability to advance to kill or capture, and to occupy the positions and the lands of the German and Japanese enemies in final victory, lay disproportionately on those chosen for the infantry.

Traditions and the Obligation to Serve

Adhering to the British model, the United States possessed a view from its founding that a standing army was a threat to freedom, viewing a small, professional army backed up by the state militias as more congenial to the security of the citizen. Before the American Revolution, the Americans experienced the threat of a standing army to their liberty "in the prolonged presence of British troopers in Boston, the Quartering Acts, and the British ministry's pretense of taxation for imperial defense."[8] The American concept of the citizen-soldier, as a volunteer for short service in the Continental Army or the militia and then returned home, was paramount in the revolutionary ideology and distrust in the creation of a standing army to defend the revolution. The Continental Congress and the state governments recognized the need for soldiers of long service for the duration of the war after the disasters of 1776, but in practice never found support for the numbers of soldiers deemed necessary to sustain the army. Volunteerism failed. Coercing men into service through a draft was "vastly disagreeable" to the revolutionaries.[9] While official drafting began in 1777, the contracting for men for the duration of the war was unusual. The distaste for coercion led to a form of draft that consisted of a quota assignment to each state, which was then passed down to towns, resulting in a unique form of bargaining with citizens to pay bounties to enlist as soldiers for short terms, or to find substitutes to pay for the obligation to others. "A standing army departed so far from the American's ideal of personal freedom that they were unable, in conscience or in fact, to force a man to serve for as long as needed, even while they could explain why he ought to want to do so. To resolve this conflict between the impulse to personal freedom and the call to public virtue, the revolutionaries relied on money."[10]

This trading of personal freedom for the bounty allowed one man to keep his personal freedom in paying another to serve, "while the other has gained money and given up some personal freedom. Both parties clung to this system during the war—it was the recruits' most reliable source of pay; it was the citizens' best way to avoid conscription."[11] By 1782 the Continental Army was at its peak of effectiveness, its professionalism and discipline noted by its allies. It was not a European-style standing professional army in attitude or expectation. "The soldiers have said, as soon as peace was concluded, they would immediately go home, as they then consider themselves free men."[12] Even in the end, while the revolutionaries failed to pay these men their overdue debts in payment arrears, the men went home anyway as soon as they learned of the peace in 1783—confident that their duty was done and service obligation ended.

The people of the United States fought a war of final victory in the American Civil War of 1861–1865 under the new conditions of the emerging industrial revolution. The Civil War, engaging the entire societies of both North and South, caused the adoption of laws for the coerced service of citizens on both the Confederate and Union sides. The effects of both the Union Enrollment Act of 1863, and the Confederate Conscription Act of 1862, as implemented, were not the same as twentieth-century conscription, however, and more in the tradition of the attempts to sustain the Continental Army during the American Revolution. The Union effort found that "it was not conscription at all, but a clumsy carrot and stick device to stimulate volunteering. The stick was the threat of being drafted and the carrot was a bounty for volunteering. In the end this method worked, for while a mere 74,000 drafted men served and another 74,000 provided substitutes, some 800,000 Union soldiers enlisted or reenlisted during the two years after the passage of the conscription act."[13] The Civil War draft laws caused many problems including inefficiency, error, corruption, and bloodshed, as in the New York City draft riots in 1863.[14]

Volunteers, and the small active army, fought the Spanish-American War of 1898 and the Philippine-American War (1899–1902), but also led to the adoption of major army reforms under Secretary of War Elihu

Root. These changes included the creation of a "chief of staff and a General Staff, a command and staff college, and the Army War College to educate officers in war planning."[15] These reforms highlighted the difficulties faced by the army in meeting its manpower obligations across the globe under the new conditions of an American overseas empire. This analysis in turn led to recognition of the need for adequate preparation and forward planning to meet the manpower mobilization and industrial needs of modern expeditionary wars. All of these measures fell within the prevailing progressive concept of conscripted service as in the United Kingdom, noted in the lobbying of a group of Progressive elites, such as Root, Henry Stimson, and Theodore Roosevelt. Conscription in this new form allowed progressivism to claim such obligation as further promotion of societal efficiency schemes.[16]

Regardless of the benefits of conscription, the adoption of state coercion to service in the Anglo-American tradition was about war, and finding enough soldiers to fill the ranks of the fighting formations. In this tradition, the government sold the unpopularity of such service as a positive burden shared by all citizens. Both the British and US governments waited until after a declaration of war to take this step during World War I. Protected by water barriers, coerced service of their citizens was not an option in peacetime, unlike Germany.[17] The long-standing citizen tradition of parliamentary or republican government demanded concessions to public opinion in the United Kingdom and the United States.

> The minimal terms of the democratic conscription bargain are that government will conscript according to some legislated and relatively equitable formula. . . . In the Anglo-Saxon democracies of Britain, the United States, . . . where conscription followed the extension of democratic participation, there was a further contractual requirement, at least for its initial imposition: Conscription could be imposed only during war and, moreover, only during a war of defense.[18]

When the nation considered the need for conscription again during 1917, the army addressed these problems as the basis for the new law in that "there were to be no bounties for volunteering and no hired substitutes; civilians rather than military officers should operate the draft; local citizens should be involved in selection; the tour of duty should be the duration of any conflict; any deferments granted should be on an individual rather than a class basis."[19] Unlike the Civil War, "the Selective Service System in the First World War provided 3.5 million draftees, or 72 percent of the armed forces."[20] US methods, including the testing for intelligence and psychological screening, were advancements that informed both US and British conscription the next time conscription appeared necessary.[21]

With the closing of the "War to End All Wars," the draft ended. The US Army and US Marines suffered the fewest casualties of the Allies fighting in France, given the short duration of their combat service. Tactically, US soldiers and marines fought in the same manner as their European allies, placing their best men in the most dangerous tactical position—the infantry. The United States incurred 110,000 dead and 206,000 wounded, in contrast to almost 1 million British deaths and more than 1.5 million French deaths.[22]

After the war the US Army, as the British, was unable to maintain either its personnel goals or modernization as a result of a general pacifism before the Great Depression, and a general belief that isolation from European affairs was a sounder choice for the United States. From 1921 to 1941, the overwhelming majority of Americans wanted to stay out of conflicts in Europe and Asia. From the Washington Naval Conference of 1922, through the Kellogg-Briand Pact of 1928 to outlaw war, the United States sought to defuse international conflict through disarmament and the end of legal war. Given the geographic barrier of a formidable ocean separating the United States from possible foes, and augmented by capable aircraft to protect US continental and colonial responsibilities, Congress called for reductions in the manpower of the US ground combat forces.[23] Greater concern with the Japanese Empire, rather than German or Italian fascist advances, notably after the expansion of the Japanese war with China in 1937, led to the focus on naval and air defense strategies. In the United States, the road to World War II resumed unnoticed and unremarked from the end of the Great War.[24]

The National Defense Act of 1920

The US regular army retained some 19,000 officers and 205,000 enlisted men in 1919 from its high of over 4 million men in 1918. The Great War fresh in their minds, General Peyton C. March, chief of staff of the army from 1918 to 1921, and Secretary of War Newton Diehl Baker proposed to the Wilson administration and Congress in January

1919 a plan for the creation of a 500,000-man regular army. This was more than five times the force authorized by the National Defense Act of 1916,[25] having as its basis a wartime active force requiring swift reinforcement by reserves with a minimum of three months' training from a proposed system of universal military service.[26]

Congress balked at this proposal, rejecting both the size of the proposed active army as well as any form of universal military training, and repudiating the idea as something more Prussian than American in tradition. Lawmakers preferred the model of a citizen army that characterized American warfare from the time of the revolution.[27]

The result of this debate was the National Defense Act of 1920. Its key features were the authorization for a regular army of up to 280,000 men, a restored National Guard of up to 436,000, as well as an organized reserve, modeled on the national army of 1917. The reserve force was planned to consist of those men leaving the regulars or guard formations at the end of their terms of enlistment.[28] Passing the law did not, however, lead to adequate funding for any of its provisions. The army's strength in 1920 following the passage of the act was 194,378 including the Philippine Scouts.[29]

Further, the act divided the nation into nine geographic regions, or corps areas, corresponding to the location of one regular division, two guard divisions, and three unfunded and unmanned reserve divisions in each, with the National Guard attaining a kind of dual status as "both the militia of the states, under the militia clauses of the Constitution, and a permanent reserve component of the United States Army, under the army clause of the Constitution."[30] Stabilized at a strength around 180,000 between 1922 and 1939, the guard became the largest component element of the US Army. The guard relieved the regular army of duty in curbing domestic disturbances, and it was available for immediate induction into the active army, while accounting for only one-tenth of the War Department budget.[31]

Regiments

With the peacetime army organized into divisions, regionally based within their corps areas, the infantry regiments recruited personnel from within these areas, but in fact all sought volunteers nationally. It cannot be said that the regiment in the US Army held the same central status for the infantry as that of the British Army in the years between the world wars. For the British infantryman of this period, the regiment was a place to which he volunteered and that prohibited his removal without his consent. The regional and historical associations of many British infantry regiments were believed a powerful force for the development of battle cohesion.[32] The National Guard divisions were the closest in terms of regional organizations on the British model, encompassing large geographic areas and numbers of states, but these tended toward a shadow form of the intent and practice of the British regiment.

Nevertheless, the army maintained the traditions of its infantry regiments. It was a hard loss when the army inactivated over 60 percent of the active regiments by 1922. Certain regionally and ethnically based regiments survived the economies of the 1920s—the Forty-Fifty and Fifty-Seventh (Philippine Scouts) and the Sixty-Fifth (Puerto Rico) Infantries. Of the thirty-eight regiments that survived this period intact with their TROOPS reduced, fourteen regiments maintained two of their three authorized battalions, and both rifle and machine gun companies held two instead of three platoons. Congress further mandated the disbandment of five additional infantry battalions in 1930 to provide 2,656 enlisted men to the expansion of the US Army Air Corps.[33] Despite the creation of a chief of infantry, authorized by the National Defense Act of 1920, strength of the infantry in divisions fell from 110,000 and 50 percent of the regular army in 1920 to just 40,331 and 25 percent of the army in 1932.[34] Total regular army enlisted strength by 1932[35] fell to 113,441.[36]

Recruiting Volunteers

Following the Great War, the recruiting fortunes for the army followed the condition of the economy. The army struggled throughout the interwar years to find enough soldiers to fill its authorized or appropriated levels. National unemployment ranged from 25 percent in 1933 to over 19 percent in 1938.[37] With the emphasis on reducing federal expenditures in the 1920s as part of a general trend in government frugality, army enlisted strength declined until greater authorizations appeared in 1936. In the decade of the 1920s, with low unemployment under good economic conditions resulting in fewer enlistments, the army lowered recruit standards, and corps area commanders were authorized to waive criteria for recruits at their own discretion. When high civilian unemployment was the norm from 1932, standards became higher and more stringent, and commanders saw their discretion to waive standards rescinded. Thus the US Army, like the British Army of this period, found it expedient to change standards dependent on the success of a given recruiting year.[38]

Physical standards. Determining beforehand the ability of a citizen to perform the duties required in combat was the object of every entry-level physical examination in all countries. "The primary intent of physical standards in the military has always been to select soldiers best suited to the physical demands of military service. This standard has usually meant the selection of soldiers who at least looked as though they could carry loads and fight well. For most of the past century, weight-for-height has been a key physical discriminator of a recruit's fitness for military service, but . . . these standards were used only to exclude underweight candidates. Weight-for-height standards were relevant when a sizable proportion of draftees and volunteers were malnourished, tuberculous, or had parasitic diseases; underweight was a good marker of such individuals who were clearly unsuited to the physical demands of the military."[39]

Changing height standards brought corresponding criteria for proportional equivalents in weight, especially for those soldiers presenting underweight characteristics.

> Some physical standards have changed easily with the need for soldiers, which suggests that what may be portrayed as a soldierly characteristic may not be solidly rooted in combat necessity. . . . Thus, the minimum height for U.S. soldiers was 66 inches early in the nineteenth century and has progressively lowered, with the least stringent requirements (no minimum height standard during part of the Civil War) coinciding with national emergencies when new recruits were in greater demand. . . .When health screening capabilities were less advanced, height standards served a health fitness purpose; for example short stature could reflect disease and poor physical development. Thus, even after careful review of physical standards during World War II, men less than 60 inches in height were "nonacceptable."[40]

The principal outcome of this rationale was to create height-to-weight ratios that provided for a standard upper figure, and an absolute minimum weight as well as chest circumference. These health concerns, coming down from the Civil War through the Great War, focused on height and its indication of health, as noted during World War I: "Common observation indicates that the Southern men have a tendency to lankiness, and this has often been attributed to infection with malaria, hookworm, and other parasites found prevailingly in the South. . . . Pulmonary tuberculosis, if not readily evident, was screened by low weight. It was particularly suspected in tall persons . . . because tuberculous men averaged ½ inch more in height and 12 pounds less in weight compared to the average of healthy World War I registrants."[41]

The changing height and weight criteria for entrance into the interwar army were similar to those in effect during World War I. Changes in 1932 reflected less selective criteria in reducing the minimum weight for shorter recruits, while retaining a higher minimum weight for taller men (169–184 pounds for 77-inch height). In 1940, anticipating the return of conscription, the height requirement increased to 5 feet (with 107–119 pounds for 60-inch height). This reflected the need for more soldiers and an expansion of the pool of recruits for the mass army for the war anticipated in the creation of the new conscription law of 1940. Major changes to the physical standards appeared in army regulations on August 17, 1940, and were repeated in mobilization regulations on August 31, 1940, which superseded the regulations of 1934. These basic regulations received continuous refinement dependent on the fluctuations of draft calls and the manpower needs of the army through the war. The physical examination took place following the classification test and provided two categories, fit for either general service or limited service. Changes over time in the physical characteristics of US soldiers identified an apparent national trend toward younger, larger, and healthier male citizens during the century following the Civil War.

Intelligence testing, 1917–1927. During World War I, the army defined literacy through its application of the Alpha and Beta tests. "The first mental tests designed to be used for mass, group testing were developed by psychologists for the U.S. Army in 1917–1918. The group tests were modeled after intelligence tests designed for individual use in one-on-one assessment. In developing the mental tests, the psychologists subscribed to the position that one could be quite intelligent, but illiterate or not proficient in the English language. Based on this reasoning, two major tests were developed, the Army Alpha for literate groups, and the Army Beta for illiterates, low literates or non-English speaking. . . . Both tests were based on the theoretical position that intelligence was an inherited trait, and the assumption was made that *native intelligence* was being assessed."[42] These tests were separate from the psychological evaluation, and defined literacy as "the ability to read and understand newspapers and write letters home," and defined illiteracy as the equivalent of not having completed the third grade in a US school.[43] From 1918 until 1927, volunteer recruits did not receive intelligence tests such as the Alpha and Beta.[44] An army doctor, Major Edgar King, working in the Department of Psychiatry and Sociology at the Disciplinary Barracks in Fort Leavenworth, Kansas, in 1924 proposed a method for the early identification and elimination of misfits. Major King was "regarded as the foremost authority in the Army on criminal psychiatry."[45]

The focus of Major King's intelligence and mental testing regimen was the identification of men of low intelligence or criminal minds. The army introduced these two new tests army-wide in October 1927. An intelligence test score indicating a mental age of ten years or below disqualified an applicant, although the recruiter had authority to waive this standard, rejecting a mere 4,055 of 88,254 applicants in the following two years because of a low intelligence score. The mental test aimed to discover psychosis and other character defects, and if failed required rejection of the applicant. Corps area commanders possessed the authority to waive these and other physical defects, if in their opinion the recruit exhibited the potential to make a good soldier.[46]

Interview and classification, 1927–1940. Recruiting eased in the first years of the Great Depression, as those out of work sought employment in the army in large numbers. This interest allowed the raising of standards in 1931 and 1932, with regard to both the intelligence test and physical examinations. The army raised the intelligence test standard to that equivalent for completion of the eighth grade and withdrew the authority of corps area commanders to waive test results and physical defects.[47]

The discovery of any civilian trade qualifications of the inductee was of primary importance. "The regulations spelled out the primary importance of the interview to the discovery, especially, of any actual civilian occupation skills, work history, education, sports, and other skills deemed essential to the proper classification of an individual."[48] A single officer, as in the British Army, conducted each interview, following the psychological examination and the intelligence test. Following the determination of civilian skills that might translate into an army-specific personnel requirement, intelligence was the next most important trait, as the regulation noted:

> a. The intelligence rating is an important aid in the rapid assignment of men.
> b. The tests provide an immediate and reasonably dependable classification of the men according to general intelligence. Their specific purposes are to aid:
>   - In the discovery of men whose superior intelligence suggests their consideration for advancement, special training or special duties;
>   - In the prompt selection and assignment of men who are so inferior mentally that they are suited only for selected assignments;
>   - In eliminating from the Army those men of whose low intelligence renders them either a burden or a menace to the service.[49]

The regulation anticipated the standard adoption of the Alpha and Beta tests. (The army adopted instead a new general classification test at the time of the reinstatement of the draft during September 1940.) The 1934 regulation established five ratings for intelligence, A through E:

> A indicated very superior intelligence (high officer type).
> B indicated superior intelligence (commissioned officer and noncommissioned officer type).
> C indicated average intelligence (60 percent of all soldiers).
> D indicated inferior intelligence (15 percent of soldiers—likely to be fair soldiers, "slow" learners).
> E indicated very inferior intelligence (for special labor assignments or discharge).[50]

Thus the manner and method of the classification of soldiers that prevailed during the coming war lay in the planning for mobilization and for an anticipated return to conscription. Classification filtered first those with a civilian trade skill, and second, intelligence (based on test scores). Physical qualifications of the individual came into consideration afterward.

On this basis, and under a continuous process of evaluation, the US Army produced more regulations on mobilization on January 15,

1934, establishing the fundamental methodology for the reception, testing, and classification of inducted men. It created the process whereby the psychological examiner administered the intelligence tests, followed by an interview and subsequent classification. The physical examination occurred after the process of psychological, intelligence, and classification interview, anticipating that most would be qualified physically for service.[51] In addition, a further mobilization regulation was the text used for the specific classification of enlisted men to a specific job category for training.[52] This process established the basic procedures for classification that carried forward into the coming war under the Selective Service Act of 1940.

Conditions Impacting Recruiting

"Purchase discharge" was one unique aspect of the interwar army that affected the retention of quality enlisted men. This policy allowed an enlisted man to buy his discharge from the army after one year's service for $120. An unfortunate result of this policy was the loss of soldiers after receiving technical training. These discharges averaged over 9,000 per year in the 1920s. Discharge by purchase, authorized during peacetime since 1890, remained in effect until suspended in 1940.[53]

The army viewed the purchase discharge rate, though troubling, as a necessary evil of the voluntary service principle. Desertions were another matter. Desertions reached their highest postwar total in 1925, at 7.39 percent. Reenlistments through 1929 were also poor. A 1926 War College study attributed the high level of desertions to high, and well-paid, civilian employment and a national indifference to the problem of desertion. Complaints from field commanders rose as desertions, when combined with purchase discharges, created an adverse effect on the readiness of the forces and appeared to indicate poor quality in the recruits entering the army. The adjutant general addressed the poor quality issue, noting in 1927 that "the demand for labor in civilian activities" made it "necessary to accept every man who could be included under the most liberal interpretation of the regulations."[54]

Fraudulent enlistments also rose. Of particular interest were issues surrounding the enlistment of minors seeking employment, criminals, and men discharged from the army or another service under dishonorable conditions. In response to continuing quality-versus-quantity concerns of the corps area commanding generals, the adjutant general further observed that "men must be obtained. To obtain them a certain price must be paid, and we continue to pay."[55]

The US Army fixed the ration allowance for enlisted men at thirty cents per soldier per day in 1927. The comparable figure for the US Navy and the Marine Corps was fifty-two cents and forty-nine cents respectively. "The fact that we have a very low ration," Secretary of War Dwight F. Davis told the House Military Affairs Committee, "has a bad effect on morale. The situation is unfortunate in having a different ration for the Army from the Navy and Marine Corps . . . [because] the soldier feels he is discriminated against if he sees the men in the other services getting a very much better ration than he has."[56] General Charles P. Summerall, army chief of staff (1926–1930), testified before Congress that raising the ration held the prospect of reducing the desertion rate and increasing reenlistments.[57] Asked if this meant the army underfed its soldiers, the quartermaster general, Major General B. Frank Cheatham, replied that "the question of underfed is a rather difficult one. They would not have starved. The components of the ration have a certain definite number of calories which will keep you in good health, but there is not the variety . . . the ration is not satisfactory; it is not pleasing to the palate."[58] When the Senate balked at the cost of improving the ration, President Calvin Coolidge raised the rate to fifty cents daily by executive order.[59]

The economy acts of the Great Depression were difficult for the morale of the army. The initial act placed officers on one month's furlough without pay in 1931. Enlisted men followed in 1932. In 1933, a 15 percent pay cut for all federal employees included the military.[60] The army was unable to modernize its forces with the minimal exception of the Army Air Corps (AAC), made a combat arm in 1926.[61] Overall, the army sought to preserve manpower, while sacrificing training and modernization.

When the administration of President Franklin Delano Roosevelt proposed cutting 2,000 officers from the army in 1934, the chief of staff, General Douglas MacArthur, responded:

> If you have to cut everything out of the National Defense Act, the last element should be the Officer Corps. If you had to discharge every soldier, if you had to do away with everything else, I would still professionally advise you to keep these 12,000 officers. They are the mainspring of the whole mechanism, each one of them would be worth a thousand men at the beginning of a war. They are the only ones who can take this heterogeneous mass and make of it a homogeneous fighting group.[62]

Mobilization Planning, Classification, and Selective Service Induction

In December 1920, the Army General Staff, War Plans Division, began planning for future mobilization. In 1925 these various mobilization plans emerged from the Joint Army and Navy Board in colored war plans for a number of possible scenarios including combined and individual adversaries. As an example, with "Red" designating the United Kingdom and "Orange" the Empire of Japan, one memorandum noted that "as a tentative Red Orange Plan we have available in an emergency the solution prepared this year by the faculty at the General Staff College. This will prove an excellent basis for working out the War Department Red Orange War Plan as soon as the Mobilization Plan has taken final shape."[63]

The final recommendation settled upon during the period 1923–1925 was for a mobilization requirement of six field armies in the United States, with a target of 2 million men within six months and up to 3.8 million in fourteen months.[64] The chief of staff approved the draft on April 6, 1923, and included it in the War Department General Mobilization Plan of 1924.[65] In order to accomplish this plan, the Joint Army and Navy Board undertook the project in 1926 to create a plan of organization to carry out a selective service act. The basis of this project was a draft bill prepared for the army chief of staff "that embodied the desirable and essential features of the World War Selective Service Act."[66]

While the mobilization regulations remained little changed, the mobilization plans found significant changes between 1934 and 1940 in the Protective Mobilization Plan of 1939. The postwar AGF study on these mobilization plans noted that "the goal of one armored division in an army of 4,000,000 men was unbelievably modest. No cognizance appears taken of the use of armor in the wars in Spain or China in the late 1930s. Expected delay in production and procurement was part of the answer again, but the inescapable fact is that the mobilization planners underestimated armor as woefully as they had airpower. There is no doubt but what the troop basis in the 1938, 1939, and 1940 Protective Mobilization Plan was tactically and logistically unsound. But that the troop basis did provide a point of departure on which changes could be based. The troops of the IPF (Initial Protective Force), if they did not constitute a force inadequate for modern warfare, at least did provide a system for the mobilization of men and equipment actually in existence."[67]

Infantry Quality and the Interwar Army

The quality standards of the interwar infantry force were not different from within the rest of the army's enlisted men. Before 1940, the only measurable improvement to quality was the removal of the discretion of the corps area commanders to waive physical, mental, and intelligence standards after 1932. In 1940, the height restriction was lowered from 5 feet 4 inches to 5 feet. Intelligence and mental standards became more definitive with the introduction of scientific and psychologically based standard testing in the army's new general classification test, fielded concurrently with the introduction of the peacetime draft in September 1940.[68]

Acceptance of minimum quality standards in the enlisted ranks, including the infantry, was the key condition of the period leading up to the reinstatement of conscription in 1940. Notable was the near absence of official policy or commentary about the effect of quality on the potential combat performance of the infantry. General MacArthur's comments on his absolute bottom line, the need for trained officers above any other consideration, was perhaps the most cogent statement of the relative importance of quality—essential in the officer, less so in the enlisted man.

Experience seemed to vindicate the assumption that the interwar enlisted infantry force quality issues were temporary. Both the Civil War and World War I appeared to prove that, in their aggregate millions, American male citizens fought well under almost any conditions. As long as their officers were well trained and motivated to lead them, the American infantryman, despite any classification process, possessed an almost mystical quality of endurance and fortitude in battle.

The Making of the Interwar Infantry

As the British Army, the US Army recognized in its 1923 doctrine the basic nature of the infantry battle task: that "infantry is essentially the arm of close combat," and that "the primary mission of the infantry is to close with the enemy in attack and destroy or capture him on his positions."[69] The morale of the infantry was of special concern in recognition of the greater strain infantry combat placed upon the soldier, and that the mission of the infantry was the general mission of the entire force.[70]

External and Institutional Cohesion Measures

By 1939, the word "cohesion" appeared in the doctrine in regard to the morale of the infantry in battle, noting in words identical to the German doctrine that "man is the fundamental instrument in war."[71] The focus of training the individual lay in understanding the need each soldier has to earn the respect and esteem of his fellows and the most essential task in his training to integrate the individual into a group and to establish for that group a high standard of military conduct and performance of duty."[72]

As in the British Army, discipline in the US Army entailed the ideal of "inculcating in the individual a proper sense of duty, a conscious pride in his unit, and a feeling of mutual obligation to his comrades in the group."[73] Thus the US Army's doctrinal expectations were in line with the creation of primary group norms and standards for the unique demands for both the mental and physical roles of the infantry small unit in combat. Emphasizing further that because of the open order of modern combat, with the loneliness of the modern battlefield for the infantryman, this "requires more than ever a strong cohesion within a unit in order to give it a *sense of unity.* This cohesion is promoted by good leadership, pride in the accomplishments and reputation of the unit, and by mutual confidence and comradeship among its members."[74]

The external institutional measures intended to instill this cohesive behavior were most apparent in the training given the individual, the junior leaders that guided his day-to-day activities, and the formal coercive nature of the military disciplinary system. How an individual responded within his small unit was a combination of his individual view of the army and its standards and expectations, and the expression of those conditions within his primary group, the rifle squad and his comrades therein. Intending to give the individual a sense of how and why to fight, the doctrine was in line with the victory in the Great War—one anticipating the primacy of the infantry and its principal focus on the artillery and other supporting arms required for his success and survival in battle.[75]

Pre-Service Training

Following the Great War, the National Defense Act of 1920 sought to take advantage of the large number of officers and enlisted men with service in the war through continued affiliation with the active army in various training organizations. These included the Officer Reserve Corps, the Enlisted Reserve Corps, the Reserve Officer Training Corps (ROTC), and the Civilian Military Training Camp (CMTC) programs. The renewed emphasis on the National Guard formed the largest reservoir and avenue for the creation of prewar trained individuals.

While very few men ever joined the Enlisted Reserve Corps, the Officer Reserve Corps numbered about 100,000 men between the wars, and saw its ranks transformed as graduates of the ROTC programs on various university campuses entered service with around 6,000 graduates per year. Around 30,000 young volunteers participated in the CMTC summer programs each year, and those completing four years of such training were eligible for reserve officer commissions. While these programs created a large pool of trained individuals of varying ages and skill levels, none were intended to become a form of universal service as the army requested in 1919. As all received their training from the regular army, it both brought the regular army officer and soldier into contact with the civilian community, "and it acquainted large numbers of civilians with the problems and views of the professional soldier."[76]

A final form of civilian contact with military-style organization and disciplined activity occurred in 1933 when the army found itself tasked to undertake the initial formation of President Roosevelt's Civilian Conservation Corps (CCC). Tasked to do so without turning it into a disguised military training program, and with enormous disruption to its already strained training capabilities, "it furnished many thousands of Reserve Officers with valuable training, and it gave nonmilitary but disciplined training to many hundreds of thousands of young men who were to become soldiers and sailors in World War II."[77]

Small-Unit Training and Organization

The army decentralized training and abolished all recruit training depots in 1922 as an economy measure. Centralized basic training was a feature of the Great War army, and the return to the prewar methodology brought with it the same shortfalls experienced prior to the war for units receiving soldiers for basic training. Brigadier John J. Pershing remarked in 1911 that the continuing and complicated requirement for units charged with training both recruits and the normal training of tactical units from squads to battalions created conditions wherein "'driblets' of men arriving at uncertain intervals during the year made recruit training an 'endless chain,' with both trainers and trainees bored by a process that seemed to have neither a beginning nor an end."[78]

From 1922, individual recruits again went directly to their small units, battalions, companies, and platoons. Typically, three or four recruits arrived and began receiving instruction from a noncommissioned officer assigned to them for that purpose. It remained for the initiative of the individual instructor and the unit to determine the length and quality of such basic soldier training.[79] By 1925, the army continued to seek ways to standardize basic training,[80] but the central problem remained.

Budget shortfalls and geographic separation complicated the ability of the army to train basic recruits and to prepare them and their formations for war. "Instead of nine infantry divisions, there were actually three. . . . Regular units [regiments] had to train as battalions or companies. The continued dispersion of skeletonized divisions, brigades, and regiments among a number of posts, many of them relics of the Indian wars, was a serious hindrance to the training of Regulars, although helpful in training the civilian components [National Guard, Reserve Officer Training Corps, and Civilian Military Training Camps]. In the infantry of 1932, the 24 regiments available in the United States for field service were spread among 45 posts, with a battalion or less at 34."[81]

Additional funding provided the chief of the infantry branch the opportunity to improve the training preparation of the infantry as the primary arm of combat after 1935.[82] The army abolished the Tank Corps of World War I and placed responsibility for tanks and their development within the infantry branch. With little money to train and few troops to train with, the infantry nevertheless embarked on a bold program of organizational and weapons programs. The inclusion of tanks in the infantry arm (until 1940) focused on the doctrinal requirements through 1939, which mandated that the role of the tank was to assist the dominant arm, the infantry, to advance; hence the name "infantry tanks."[83]

Experiments with command and control led to the development of the triangular organization of all units from platoons through divisions. The army abandoned the large square divisions of World War I—comprising four regiments organized into two brigades—in favor of divisions with three regiments of three battalions each. These changes intended to make the division, the primary combat organization, lighter and more maneuverable on the modern battlefield came to partial fruition in 1941.[84]

The organization of the infantry squad, the basic field element of the infantry, was twelve men, and added a single Browning automatic rifle to each squad to augment the available firepower. Believing that the core of the infantry in attack and defense was the individual shoulder weapon of the rifleman, the army adopted the semiautomatic M1 Garand to replace the M1903 Springfield bolt action rifle. New and better weapons arrived in the form of a new light machine gun and new 81-millimeter and 60-millimeter mortars.[85] Despite the adoption of these new weapons and organizations, there was not enough money to buy sufficient numbers of them to train before the beginning of the war, nor to conduct large unit maneuvers. Infantry manpower grew by only 17,000 personnel by 1939, to just over 57,000 from its low of 40,331 in 1932.[86]

The Army Air Corps Expansion

Unlike either the British Army or the German Army in the interwar years, the US Army Air Corps was an integral component of the US Army. Maintaining and improving the national airpower capabilities lay within the army's mandate from the National Defense Act of 1920. Experimentation, theorists, and pioneers of airpower kept up America's fascination with airplanes and technology. Among the most prominent was Brigadier General William "Billy" Mitchell and his influential work *Winged Defense* (1925). During the interwar years, the air corps became more and more focused on "the strategic role of bombardment aircraft operating independently of surface forces,"[87] evolving toward an increasing independence in form and function. Unlike the regular army, prior to 1938 the air corps received two special increases in strength. The first, as noted earlier, entailed the elimination of five infantry battalions in 1930 to add 2,656 enlisted men to the air corps, mandated by Congress, without an end strength increase for the army.[88] In 1932, however, Congress approved an increase in strength to accommodate a further mandated addition of 6,240 enlisted men for the air corps.[89]

The US Army Air Corps, while having the same difficulties maintaining its strength due to the loss of technicians to purchase discharges, received relative priority. In response to queries from the War Department concerning the impact of the purchase discharge on air corps readiness, the response of the air corps was ambivalent given the year-over-year loss of from 6 to 8 percent of the air corps force in the years 1934–1937. Noting that while up to 15 percent of enlisted men graduated from the Air Corps Technical School sought purchase discharges during their enlistment after training, the absorption of these men into the private aircraft industry was an overall benefit to the nation. The civilian aircraft industry benefited both at the time, and in the case of a future mobilization of the industry for war, as well as providing excellent advertising for future enlistments in the air corps in the nature of such job prospects. "The advertising obtained from a young man who goes back to his home town spreads the word that a few years in the Army Air Corps have fitted him for an excellent job in civil life. . . . Applicants accepted for such enlistment are required to take an examination and are all high school graduates and many have one or more years of college education. While the disadvantages of the present system are many, it is not believed that a change should be made that will entirely restrict the graduates of the Air Corps Technical School from obtaining discharge by purchase."[90]

On November 15, 1938, President Roosevelt ordered the secretary of war and Chief of Staff General Malin Craig to prepare secret plans to build a 10,000-plane air corps in two years' time.[91] The effect of this order showed in the increased authorized and appropriated enlisted strength of the army from 165,000 personnel in 1938 to 375,000 in 1940.

Out of the Wilderness

The explosion in infantry strength as the war approached did not begin until 1940, and by mid-1941 that strength stood at over 379,000 men, the result of the resumption of conscription in September 1940. Lacking sufficient weapons to train was a cause for significant concern. The modifications in organization and experiments in arms failed to procure the actual weapons needed to train the force.

"As late as 1941 mortar crews went through maneuvers with stove pipes and the crews of light machine guns set up and aimed broomsticks. These harsh facts caused the Chief of Infantry to state on the eve of war that a consciousness of obsolescence in all their arms had seriously damaged the morale of American infantrymen."[92]

The focus of the November 9, 1939, tentative field service regulations was on the primary importance of the infantryman, his rifle, and bayonet.[93] But the opening of the European war in September 1939 blunted infantry hopes for priority in equipment, funding, and manning. The newfound recognition of the threat that enemy aircraft possessed the ability to strike the United States from within the hemisphere demanded primacy for resources to augment the capability of the US Army Air Corps, rather than the infantry or the other ground arms.

Poland's defeat in 1939 brought immediate authorizations, however, for expansion of the regular army and National Guard enlisted strengths to 227,000 and 235,000 respectively. By the end of 1939 there were six infantry divisions and one cavalry division, with total US armed manpower at just over 720,000 personnel (see Table 3.1).

German victories against the armed forces of France, the Low Countries, Denmark, and Norway between April and June 1940 provided the final push for expansion to a force of 1.2 million, including an enlarged and modernized US Army Air Corps. In August 1940 there were ten infantry divisions, two armored divisions, and one cavalry division active. On August 27, 1940, Congress approved the induction of the National Guard into federal service and the calling up of the organized reserves. With this last step, Congress set the stage for the final mobilization action, the reintroduction of conscription during September 1940, and by December 7, 1941, the army had grown into an organization of four numbered armies, thirty-seven divisions (thirty infantry, five armored, two cavalry), and over forty combat air groups.[94]

Table 3.1 US Armed Forces Strength, 1939

Army	Navy	Marines	National Total
508,161 (including National Guard and reserves)	176,787 (including reserves)	35,369 (including reserves)	720,317

Source: Hershey, *Special Monograph no. 16,* 49.

Junior Leaders

Noncommissioned officers. "Traditionally the Army saw itself as consisting of only two elements—officers and men. Unlike the British Army (and for that matter all the other armies of the western world), which saw itself composed of three elements—officers, noncommissioned officers, and other ranks—the United States Army saw the noncommissioned officer as simply an enlisted man, temporarily elevated at his officer's pleasure to a supervisory rank."[95] An artillery officer noted in 1890 that "intellectual ability is not the first prerequisite of a good noncommissioned officer."[96] These ideas set the stage for the role of the NCO into the twentieth century. The perception of the abilities of the NCO in both the interwar years and World War II did not stray far from this sentiment. After the war, and with the experience of the mass army as guide, far greater expectations for NCOs initiated major changes in their training, education, and overall performance expectations. Policies in effect through the interwar and war years, however, limited the ability of the NCO to achieve the measure of respect and responsibility that appeared so necessary after the war.

> An old Army saying, true up until the beginning of World War II, "You leave your chevrons behind you when you move," summed up the bitter irony of a career sergeant's lot as compared with that of the commissioned officer. The officer took his rank with him when he moved from unit to unit during his Army career, whereas the sergeant usually had little choice but to remain throughout his career with the same unit that promoted him, or forfeit his stripes if he transferred.[97]

Promotions to NCO rank were the responsibility of the commanding officer of a unit or location. NCOs in the US Army lost their rank if they left their unit. Their transfer by law and custom was in the grade of private, including overseas. General Douglas MacArthur, while commanding the Panama Canal Department in 1932, noted his displeasure to the chief of staff that this policy applied to the soldier of the line, but not to a sergeant of equal grade in the quartermaster corps, and that "it is a discrimination against the fighting man in favor of the non-combatant. I earnestly recommend that steps be taken to remedy the condition."[98]

The retirement system for enlisted soldiers, those senior NCOs who were the backbone of the army's scattered detachments, was set at thirty years' service. If a soldier failed to reach that year due to disability incurred in the service, such as wartime wounds, and he was discharged prior to the thirty-year mark, his compensation fell to a small monthly disability payment. The effect was for commanding officers to keep these senior NCOs on active duty, although often they were not capable of keeping up mentally or physically with the changing technology and demands of the service as the interwar years moved toward war. Many whose poor physical condition mandated assignment to duties commensurate with their reduced capabilities were kept simply because of the requirement for a thirty-year retirement or nothing.[99]

The NCO was never mentioned in any of the annual reports noted earlier. The NCO's status was at the pleasure of commanding officers and not a responsibility of the army leadership. The US Army NCO nevertheless functioned in the same essential manner as his British counterpart, without the aura of the regimental prestige awarded the British NCO by his officers. Accounts of the NCO in the interwar period tended to a remembrance of the senior NCOs, the first sergeants, whose administrative and leadership abilities were most available and sought after by new officers and relied upon, if competent, by the senior leaders. Their training belonged to their commanding officers. For good or ill, these NCOs were what were available when the first draft notice occurred in 1940. "Although the prewar noncommissioned officers formed a major part of the training cadre that helped to make soldiers of these millions 'with remarkable success, considering the circumstances,' few of them became the noncommissioned combat leaders in the divisions deployed overseas, for most of them were too old, and their numbers too few to match the vast needs of the wartime Army."[100]

Lieutenants. The graduate of the Military Academy at West Point was the lodestone for all officer expectations in training, manner, and leadership for the regular army between the wars. Twelve hundred graduated between the Spanish-American War and the Great War in 1914. Through World War I and the interwar years, these officers carried the tradition of leadership and excellence for emulation in the officer corps. The expansion of World War I was the model for the expectations of the mass army of World War II. The final army was 4 million strong, 13 percent regulars, 10 percent guardsmen, and 77 percent draftees. With the requirement for so many new officers, officer training camps provided the basis for the increase. Forty-eight percent of the officers of the national army came from this source alone of the 200,000 officers serving at the end of the war, from the original 5,791 officers of the peacetime army.[101]

The life of the junior officer after 1918 returned to the experience of those entering the regular army before the war. By 1923 the army was down to 125,000 enlisted men and 12,000 officers.[102] In 1909 the average junior officer expected to spend six and a half years as a second lieutenant, nine years as a first lieutenant, and thirteen years as a captain. In 1933 a junior officer expected to serve ten years as a lieutenant, eleven years as a captain, and spend up to fifty-eight years in service before attaining the rank of colonel. The officers of the interwar period were professionals on long service, if they chose to remain in the army, and Congress modified the promotion system in 1935 to mandate fewer years in service prior to promotion.

The training of these young officers, as in the British Army, was the responsibility of their commanders. Daily routine included inspections and drills in the mornings, and athletics in the afternoons became a major part of the life of an infantry officer in garrison as a needed break from the monotony of routine. Thirty-seven percent of these officers were West Point graduates as well as seventy-one percent of the general officers. The editor of the *Infantry Journal* found that too much of the routine was "eyewash, undue emphasis on competitive athletics, excessive supervision by command and staff, useless paper work, unnecessary reports, requirements that all officers attend all drills, multitudinous administrative checks, daily officer calls . . . time-consuming non-essentials that have nothing to do with training for war."[103] And yet it was a time of preparation for the middle-grade officers, those who fought in the Great War, to come to grips with the changing characteristics of war in mechanization and aviation.

The school system for these officers, centered at Fort Leavenworth, Kansas, and the Army War College in Pennsylvania, provided the basis for the tactical and staff preparation for war, and also set the conditions for the mobilization of the army and nation to fight. For the junior officer, however, these opportunities to grow were far more limited. Much of their time for training was taken up in that done with the civilian components of the National Guard, the ROTC, the CMT, and the very nature of the scattering of the army across posts and stations, in small units, across the breadth of the United States.[104]

As the nation approached its first peacetime draft in 1940, the army numbered 198,335 men, with 14,916 officers in September 1939. "In this peace-time pre-war Army there was, in addition to the official gap between commissioned and enlisted personnel, a breach based upon the general caliber of the men. The official gap, established by law, regulations, custom and tradition, was accepted without question by the vast majority of those men then in the Army. The differential, due to caliber of most of the enlisted as compared to that of the commissioned personnel, *permitted the development of habits of dealing with men in the ranks which were later to cause difficulties. . . .* Drawing of manpower from civilian sources wiped out the peacetime gap based on the difference in caliber between commissioned and enlisted personnel, for the Army was now receiving into its ranks men who were far superior by training, education, and work experience, to men in the commissioned ranks."[105]

The army grew to 8,291,336 in May 1945, with 897,777 officers of all types.[106] The unprecedented and rapid growth of the army resulted in a prewar officer corps and NCO corps overwhelmed by the civilians conscripted into the service.[107] Leading these civilians proved to be the largest challenge that lay ahead for the junior officer of World War II. One of the most difficult of these challenges lay in the administration of military justice to this civilian army.

Discipline and Sanctions

There were three types of court-martial in the US Army disciplinary system provided by the Articles of War as amended by Congress in 1920: general court-martial, special court-martial, and summary court-martial (see Appendix 3).[108] The laws behind the military justice system for both the US Army and US Navy were adopted whole cloth from the British laws in place during 1775 and 1776. The US Army Articles of War were revised in 1806, 1874, 1916, and 1920. "The World War II 'GI' [and his interwar counterpart] was essentially subject to a 160-year-old criminal code that provided no right to trial by peers, that was largely administered by men untrained in the law, and that was closely controlled by a commander whose natural and primary interest was the maintenance of good order and discipline within his command."[109] The author of the *Military Justice for the Field Soldier* (2nd ed., revised in 1944) "unabashedly states the function of courts-martial as an instrument of the commander for the maintenance of discipline [and] he constantly reminds his readers, future court-members, to be aware of the commander's powers and to expect unfavorable reaction from lenient sentences."[110]

Commanding officer punishments. Any officer able to order general or summary courts-martial held the authority to impose minor disciplinary actions on men under their command. These included reductions in rank or rating, confinement not exceeding ten days, solitary confinement on bread and water not to exceed five days, solitary confinement not to exceed seven days, deprivation of liberty ashore, and extra duties. Below the level of this court-martial authority, commanding officers of any detachment, company, or higher command were authorized to inflict administrative punishments including admonition, reprimand, withholding of privileges, extra fatigue, restriction to certain specified limits, and hard labor without confinement, all not exceeding one week, with the exceptions of an award of forfeiture of pay or confinement under guard.

The Effect

The US Army entered World War II with a continuing metaphysical belief in the near-mystical abilities of the American male as a fighting man. This metanarrative was a product of the victories gained by the infantry armies of the Revolutionary War, the Union Army in victory in the Civil War, and the narrative of the Southern soldier in defeat. The Great War cemented this narrative for the US Army in the performance of the American soldier in combat against the German enemy in France. George Marshall sustained this model in his work *Infantry in Battle* (1934). Marshall needed no explanation as introduction or conclusion concerning his examples of American infantry and their extraordinary small-unit genius against the German infantry of World War I. It was a continuing assumption that the manner in which American men fought as infantry of the Great War, and all the wars before, was the norm. It was with this narrative and expectation that the US Army entered World War II.

Unlike the British Army, there was no expectation that the concept and narrative of the regiment was the driving force to instill this capability and cohesive behavior and power in the infantry. This was a narrative expectation for American manhood in wartime since the beginning of the republic. Unrecognized was the effect of the movement to scientific and psychological methodologies for the selection, classification, and assignment of soldiers to the infantry. The low standards set in the interwar years were not seen as indicative of the conscripted citizen in any future war. It was not anticipated that the character of the infantry changed due to the emerging characteristics of modern, mechanized war. No longer were the infantry the dominant element in manpower. No longer were the infantry receiving the flower of American manhood. The best and the brightest were siphoned away to the more technical arms and the air corps, because their attributes were scientifically anticipated. The US Army entered the war with a narrative understanding of its infantry at odds with the effect of the downward selection of these soldiers. The unanticipated effect of this selection created a failure to build or sustain the cohesive behavior in the small unit, the infantry primary group, and the negative impact on the expectations for the infantryman in his small unit.

Notes

1. Palmer, *Study no. 5,* 1.
2. US Chief of Staff, *Field Service Regulations 1923,* 11.
3. Ibid.
4. Background papers to Brigadier General Frank Parker, Assistant Chief of Staff, G-3, lecture delivered at the Army War College, Washington, DC, *The G-3 Division of the War Department General Staff and Its Problems,* September 11, 1928; *Duties and Methods of Operation of the Operations Branch,* G-3, Adjutant General files, 1920–1942, RG 407, NACP; "The Human Element in War," 463.
5. US Chief of Staff, *Field Service Regulations,* 5, 7.
6. Bland and Stevens, *The Papers of George Catlett Marshall,* vol. 1, 689–694.
7. This number is an amalgamation of various sources and uncertain due to the different manner in which casualties were counted among the different airborne divisions.
8. Royster, *A Revolutionary People at War,* 36.
9. Ibid., 64.
10. Ibid., 68.
11. Ibid., 69.
12. Ibid., 343.
13. McPherson, *Battle Cry of Freedom,* 605.
14. Ibid., 609–611; Donald, *Lincoln,* 448–451.
15. Linn, *The Echo of Battle,* 93–94.
16. Flynn, *Conscription and Democracy,* 12.
17. Ibid., 26.
18. Levi, "The Institution of Conscription," 134.
19. Flynn, *Conscription and Democracy,* 12; Hershey, *Special Monograph no. 16,* 9.

20. Flynn, *Conscription and Democracy,* 38.
21. Vernon, *Personnel Selection in the British Forces,* 17.
22. Keegan, *The First World War,* 423; Winter, "Britain's 'Lost Generation' of the First World War," 450.
23. Lejeune, *The Reminiscences of a Marine,* 475–477.
24. Conn, *American Military History,* vol. 2, 414–415; Iriye, *The Globalizing of America,* 73–87; Stoler, *Allies and Adversaries,* 1–6; Overy, *Why the Allies Won,* 29; Griffith, *Men Wanted for the U.S. Army,* 78–79.
25. National Defense Act of 1916, 1.
26. Griffith, *Men Wanted for the U.S. Army,* 10.
27. Ibid., 16.
28. National Defense Act of 1920, 1.
29. *War Department Annual Reports, 1920–1941,* annual reports of the Secretary of War for the fiscal year ended June 30, 1920, Adjutant General files, Statistics Branch reports, Army Adjutant General's Office, 1920–1942, RG 407, NACP, 10, 26.
30. Weigley, *History of the United States Army,* 401–402.
31. Conn, *American Military History,* 411.
32. TNA PRO WO 32/9846, appendix A to E.C.A.C./P(41)47; TNA PRO WO 33/1488, *Report of the Committee on the Cardwell System,* August 1937, 14–15, 38.
33. WD Adjutant General files, Statistics Branch reports, Army Adjutant General's Office, RG 407, NACP, *Annual Report 1930,* 308–309.
34. Mahon and Danysh, *Infantry Part I,* 50–51.
35. US Department of War, *Annual Reports.* Weigley, *History of the United States Army,* disagreed with the figure for 1932; he stated 120,546 for enlisted strength in 1932. Weigley gives his source as the *Historical Statistics of the United States.*
36. US Department of War, *Annual Reports,* 1932, 193; Griffith, *Men Wanted for the U.S. Army,* 135, 233; Weigley, *History of the United States Army,* 568.
37. Rostow, *The World Economy,* tab. III-42, 220.
38. Griffith, *Men Wanted for the U.S. Army,* 102–103, 151.
39. Friedl, "Body Composition and Military Performance," 31.
40. US Congress, 1944.
41. Ibid., 40.
42. Armstrong and Sticht, *Adult Literacy in the United States,* 30.
43. Hershey, *Special Monograph no. 15,* 68.
44. Solas, DeRoum, and Gade, "The Military's Contribution to Our Science and Practice," 171.
45. Wigmore, "Modern Penal Methods in Our Army," 170.
46. Griffith, *Men Wanted for the U.S. Army,* 102–103.
47. Ibid., 151.
48. US Army, Army Regulation 615-25, *Enlisted Men: Initial Classification,* 9–23.
49. US Army, Mobilization Regulation 1-9, *Reception of Selective Service Men,* 10; Kreidberg and Henry, *History of Military Mobilization in the United States Army,* 492.
50. US Army, Mobilization Regulation 1-9, 12, 16.
51. Ibid., 8.
52. US Army, Mobilization Regulation 1-7, *Reception of Selective Service Men,* 17; US Army, Army Regulation 615-25.

53. Griffith, *Men Wanted for the U.S. Army,* 100.
54. Ibid., 101.
55. Ibid., 102.
56. Ibid., 97.
57. US Department of War, *Annual Reports,* June 30, 1941, 116; Griffith, *Men Wanted for the U.S. Army,* 94.
58. Griffith, *Men Wanted for the U.S. Army,* 97.
59. Ibid.
60. Coffman, *The Regulars,* 242; Griffith, *Men Wanted for the U.S. Army,* 128–129.
61. Conn, *American Military History,* 408.
62. Griffith, *Men Wanted for the U.S. Army,* 129.
63. Memorandum for Assistant Chief of Staff, G-1, subject: Six Field Army Program; H. A. Drum, Brigadier General, Assistant Chief of Staff, G-3, "History and Development of the Six Field Army Program as of June 30, 1925," August 8, 1925; War Department, Adjutant General files, Army Adjutant General's Office records 1921–1942, RG 407, NACP, 11; Gole, *The Road to Rainbow,* 22–23.
64. Memorandum for Assistant Chief of Staff, G-1, subject: Six Field Army Program, August 8, 1925, 18.
65. Ibid.
66. Memorandum for the Chief of Staff, War Department, subject: Plan of Organization to Carry Out the Selective Service Act, From: Campbell King, Brigadier General, Assistant Chief of Staff, G-1, February 11, 1926, War Department, Adjutant General files, Army Adjutant General's Office records 1921–1942, RG 407, NACP, 1.
67. Kreidberg and Henry, *History of Military Mobilization in the United States Army,* 492.
68. Capshew, *Psychologists on the March,* 101.
69. US Army Ground Forces, *Field Service Regulations 1923,* 11–12.
70. Ibid., 11.
71. US Chief of Staff, *Tentative Field Service Regulations,* 29.
72. Ibid., 30.
73. Ibid.
74. Ibid., emphasis in original.
75. US Chief of Staff, *Tentative Field Service Regulations,* 5.
76. Conn, *American Military History,* 411–414.
77. Ibid., 414.
78. Coffman, *The Regulars,* 104.
79. Griffith, *Men Wanted for the U.S. Army,* 165.
80. H. A. Drum, Brigadier General, Assistant Chief of Staff, G-3, War Department General Staff, lecture delivered to the Army War College, Washington Barracks, DC, "Organization and Functions of the Operations and Training Division, G-3, War Department General Staff," November 3, 1925, G-3 General Correspondence files, 1920–1942, NACP, 7.
81. Conn, *American Military History,* 410.
82. Mahon and Danysh, *Infantry Part I,* 50.
83. Ibid., 52.
84. US Chief of Staff, *Field Service Regulations: Operations,* 253. The manual included both the triangular division, listed first, followed by the square

organization, seemingly uncertain of the role of the new organization or its implementation.

85. Mahon and Danysh, *Infantry Part I,* 51–56.
86. Ibid., 53.
87. MacIsaac, "Voices from the Central Blue," 633.
88. US Department of War, *Annual Reports,* 1930, 308–309.
89. See US Department of War, *Annual Reports,* 1932, 194, reflecting the additional authorized strength of 6,240 enlisted men for the US Army Air Corps.
90. Memorandum for the Acting Chief of Staff, G-1, from the office of the Chief of the Air Corps, subject: Purchase of Discharges by Enlisted Graduates of Army Schools, September 22, 1938, Correspondence files of the G-3, War Department, 1920–1942, RG 407, NACP.
91. Louis Johnson, Acting Secretary of War, memorandum for General Craig, subject: 10,000 Army Airplanes and Augmentation of the Army, November 15, 1938, War Department, Adjutant General files, Army Adjutant General's Office, RG 407, NACP.
92. Mahon and Danysh, *Infantry Part I,* 54–55.
93. US Chief of Staff, *Tentative Field Service Regulations,* 5–7.
94. Conn, *American Military History,* 436.
95. Fisher, *Guardians of the Republic,* 142.
96. Ibid., 244.
97. Ibid., 230.
98. Ibid., 232.
99. Ibid., 239–241.
100. Ibid., 248.
101. Coffman, *The Regulars,* 205.
102. US Department of War, *Annual Reports,* 1922, 161–163; Stewart, *The United States Army in a Global Era,* vol. 3, 59.
103. Coffman, *The Regulars,* 262.
104. Schifferle, *America's School for War Fort Leavenworth,* 9–10.
105. Doolittle Board, *Report of the Secretary of War's Board,* 9–10, emphasis in the original.
106. Ibid., 15.
107. Wiley and Govan, *Study no. 16,* 25.
108. US Department of War, *The Articles of War,* 4–6.
109. Willis, "The United States Court of Military Appeals," 43.
110. Ibid.

4

Germany: Learning from World War I

THE INTERWAR YEARS FORMED THE CRUCIBLE WITHIN which the leaders of the German Army sought to learn the lessons of the Great War and to plan for the army's rebirth as the traditional defender of the nation. It was during this period that the focus on quality in the standards for the selection, assignment, and training of volunteers and the carryover of this focus with the reintroduction of conscription became paramount. It was the infantry anvil that made up the majority of the soldiers of the army, against which the mechanized and armored formations operated to destroy an enemy. The German Army's aggressive solution to the tactical crisis coming out of the Great War was selection of the best, most physically robust man for the most difficult form of battle—an individual within his small unit capable of overcoming the problems of the empty battlefield and the overwhelming power of modern weapons to destroy the fighting spirit of the soldier. The criterion used for the basis of the comparison of the UK–conscripted soldier and the US–conscripted soldier with his German counterpart was that the selection of the most physically capable man for infantry combat also meant the disproportionate assignment of the most mentally capable to withstand the trial of face-to-face combat.

Traditions and the Obligation to Serve

The geography of the northern German plain formed a key and fundamental historical difference between a German societal, or "national," tradition and that of the Anglo-American polities. States arising in the region of the eastern Baltic Sea, on its southern shores, found no substantial geographic barriers to the movement of armed bands of soldiers (or brigands) or armies. No ocean bordered the cities and states of the northern German plain, leaving them open to the movement and travails an army might inflict on their citizens or subjects, unless an equal or significant army of one's own contested the ground. This basic geographic uncertainty raised the role of the army to that of the physical guarantor of the lives and liberty of the members of the society. Thus a standing army in this region was a prerequisite to survival for states and polities lacking an ocean barrier to invasion and destruction. The legitimate role of the army that grew in the minds and narrative of these peoples of the plain formed from the moment of the inception of the idea of a state. Over time, the value and thus the legitimacy of an obligation to service in such a standing force became foundational to the normal relationship of the members of the polity to the state, and the role of the state to ensure tranquility against the marauding of the armies of others. The lack of an ocean barrier denied the nascent Brandenburg-Prussian state the opportunity of time—to build an army for specific purposes of defense or foreign adventure; requiring an army for immediate use and capability was the outcome of such circumstances.

The foundation of what became the Prussian-German state began during the Wars of Religion, attesting to the intensity of the wars fought following the opening of the Protestant Reformation and the Roman Catholic Counter-Reformation of the period 1520–1540.[1] The Teutonic Order of knights first conquered the area of the southeastern Baltic shore during the twelfth century, and converted its pagan inhabitants to Christianity. By the time of the Protestant Reformation, the Teutonic Order was on the wane following defeats by the forces of the Lithuanian-Polish union after the 1410 Battle of Grunwald. "In 1525, the Order's remaining Prussian lands were secularized by its last Grand Master, Albrecht von Hohenzollern, who converted to Lutheranism and established the Duchy of Prussia as a fief of the Polish Crown. It was Europe's first Protestant state."[2]

In the territories of the German-language principalities of the Holy Roman Empire during the Thirty Years' War (1618–1648), religious enmity generated tremendous destruction of life and property. Soldiers, however, preyed upon either confessional group, Roman Catholic or Protestant, with professional zeal, dependent on the need of the moment for supplies or battle. "The armies themselves, consisting as they did of contractors and professional soldiers who lived on and for war, formed independent, self-conscious entities with their own corporate interests, not always amenable to control by their employers. Moreover, if the contracting system had proved capable of raising large numbers of men and sustaining them, it had only been possible through the most frightful destruction of people and property and the emptying of state coffers. The contracting generals had not liberated themselves from the restraints of living off the land, nor had they led their armies to convincing victories."[3]

After 1648, the continuation of the army of Frederick William, the elector of Brandenburg-Prussia, depended on his estates to levy taxes. The ultimate acquiescence of the estates, bound to the elector through awards of various offices, tax exemptions, and a leading role in military service, confirmed the role of the army as the principal guarantor of the security of the state, ensuring the political and territorial integrity of a Brandenburg-Prussia maneuvering among its more powerful and unpredictable neighbors.[4]

The establishment of a standing army placed the ultimate power to coerce in the hands of the elector.[5] The polity of Brandenburg-Prussia consolidated with the addition of territory from the duchy of Pomerania in 1648 and Ducal (East) Prussia in 1660.[6] By the time of his death in 1686, Frederick William had instituted a system to provide manpower for the army that combined hiring mercenaries and recruiting peasants by regimental regions. In contrast to other armies of this time, "the Great Elector used contractors to find men for his army, but recruitment was done in the Elector's name, and the recruits swore loyalty to him not the *condottieri*."[7] He bequeathed a standing army of 12,000 men to his heirs, who continued the tradition.[8]

Frederick William I, the grandson of the great elector, created the *Kantonreglement,* or Canton Regulation, of 1733. The directive divided the country into demarcated recruiting districts—the cantons—and ordered the enrollment of all male youth within each territory. He thus made official what was heretofore an unofficial practice. Thus he protected both his ability to raise the army in time of war, and to maintain the peasant on the land for the majority of the year through furloughs. The cantons—the state organization handed down to him by his father and grandfather—served as a workable compromise between the need to keep the peasant in the economy and working, and to meet the wartime requirements of military service.[9] "[T]he noblemen who were the officers, or their sons, provided their own serfs and led them into the field for military service; trained them; and controlled them."[10]

"By 1740 [the death of Frederick William I, the Soldier King] . . . the territories had a common administrative structure, a population of some 3,000,000, and an army of 80,000; the ruler, now a king, was absolute, with a bureaucracy that was one of the most numerous per head of population in Europe."[11] While the canton system continued the forcible recruitment of the peasants, it offered a sense of stability. "The peasant and servant at least knew now, from childhood on. . . . They would receive training for eighteen months to two years, and thereafter would be at home with their family and work for ten months of the year. Thus, while their general opinion of military service may have remained negative, at least their fate appeared bearable again."[12]

Enrolling the male Prussian peasantry in military service provided its first sense of connection to the concept of the state. Even while at home working, a peasant was no longer just the serf and a servant of a manorial master. The regulations of the service required that he be protected in his person and property by his regiment, and in this sense was one of the first steps taken toward giving the peasant his initial exposure to civic responsibility—a sense of citizenship—as a direct subject of the king, and ushering in a process leading to emancipation.[13]

Through the wars of Frederick II the Great (1712–1786), this regimental recruiting system for conscripting native soldiers, in combination with the hiring of foreign mercenaries, brought victory and stability to Prussian arms. Frederick knew the characteristics of his soldiers, noting that "those who inhabit cities or who till the soil are exempted from military service upon paying a certain tax for maintaining the soldiers who defend them. For this reason, our armies for the most part are composed of the dregs of society—sluggards, rakes, debauchees, rioters, undutiful sons, and the like, who have as little attachment to their masters or concern about them as do foreigners."[14]

The regulations for the Prussian infantry of 1759 exempted "those who are incapable by reason of any infirmity, as also the sons of all such burghers as are worth ten thousand dollars, are exempted, and no Officer shall enlist them, on pain of being cashiered."[15] The type and age of the soldiers recruited for Frederick the Great's armies demonstrated the close tie between the soldier's characteristics and the tactical intent for his employment. The Prussian army employed a very large number of foreigners in the ranks, and Frederick sought to maintain the core of the agrarian Prussian state on the land. Like his predecessors, he sought to protect and increase the peasant population, "to preserve that species, which is admirable. . . . The peasant class is very important to the state; it forms its foundation and carries its burdens; it has the work and others the glory."[16]

Severe discipline and structure characterized the tactical requirements of the age. "The extent, complexity, and brittleness of the tactical patterns, as well as the difficulty of maintaining an orderly and rapid fire, demanded intensive training of the soldiers, and their most stringent supervision. For the man in the ranks, if not for all his officers, this meant renouncing his will to reason and his total submersion into the unit, the strength and safety of which resided largely in the degree to which its members could act as one. . . . The need for close supervision was heightened further by the very large number of foreigners in the ranks. . . . Fear alone produces the conditioned reactions necessary for carrying out the Prussian drill; and fear and compulsion, linked perhaps with habit and *esprit de corps,* were the factors counted on to keep the men tractable and effective in combat."[17]

The regulations required the selection of "sound and able men," none of whom were acceptable below 5 feet 6 inches in height, and wanted those from 5 feet 8 inches to 5 feet 10 inches in height.[18] Generals of regiments were admonished to ensure that "every company is composed of two-thirds foreigners and but one third Prussians."[19]

> The sixty native Prussians in each company are disbanded for ten months out of the year and the funds thus saved can be used for recruiting, for the war chest [*Die Wartenbergsche Kasse*], and for the captains. These sixty Prussians come from the same region; many, in fact, know or are related to one another. Integrated with foreigners, they make an excellent unit. The cantons spur on competition and bravery, and relatives and friends are not apt to abandon each other in battle. Resident farmers and their future heirs are exempt from military service. Only the younger sons and servants must serve. The subprefects attend the draft and release everybody whose loss would be disastrous for the state.[20]

Frederick believed a sovereign waged war with paid foreign troops. He recognized the weaknesses of mercenaries, and advocated always mixing them with native soldiers, and again highlighted the need for strict discipline to prevent desertion, and the changing composition of the infantry companies, noting in 1768 that his regiments were composed of half native Prussians, and half foreigners enrolled for money. The prevention of desertion to maintain the strength of the army stemmed from that fact that these foreigners held "no particular attachment to our service and wait only for the first opportunity to desert. The question therefore is to prevent desertion. The composition of my troops requires infinite attention on the part of those who command them. It is necessary always to make them observe the most exact discipline."[21]

One of the few surviving regimental records of the era preserved the demographics of the soldiers of the *Hacke* regiment, composed of half foreigners and half native soldiers, and garrisoned at Stettin in Pomerania at the end of the reign of Frederick the Great. More than half of the privates were older than the age of thirty, with the foreigners two years older on average, reflecting contemporary tactical requirements.[22] Steadiness under fire was the prerequisite and spoke to the experience of veterans rather than the eager courage of youths. "The machinelike tactics of the infantry battalion placed a premium on the mechanical performance of close order drill and the manual of arms, but minimized the importance of individual enthusiasm. . . . Ideally, the soldier's individuality was totally submerged in the collectivity of the battalion. In such a context, the seasoned veteran possessed an inherent superiority over the recruit no matter what the latter's spirit might be."[23]

Conscription, 1806–1918

Napoleon shattered the Prussian Army, nineteen days after its mobilization against him, at Jena and Auerstadt in 1806. This same Prussian army, emulated throughout Europe as the eighteenth century came to its end—the tried-and-true exemplar of the ideal of iron discipline—ended its life in traumatic defeat—a Prussian military and morale collapse without precedence.[24]

One factor in the defeats of 1806 noted by the Prussian reformers was the age of the Prussian soldiers and their relative lack of mobility compared with the younger, more enthusiastic armies of Napoleon. Reformers such as Gerhard von Scharnhorst, Ludwig Yorck von Wartenburg, and Leopold von Boyen placed a premium on this mobility, contrasting their armies with the French light infantry and skirmishers, who played havoc with the dense and slow-moving Prussian lines at Jena and Auerstadt. A key tactical lesson was the example of intelligence and physical stamina of these soldiers as an ideal for the new Prussian infantry. They enshrined these concepts in the Boyen Law (1814) for the future of the Prussian Army.[25]

The destruction of the army in 1806 opened a window of opportunity to reform both its organization and the nature of service in emulation. "If a new kind of army were wanted, it could only be created by a new kind of state that would grant the individual a more active role in public affairs. . . . [A] change was needed, since in the long run the enthusiastic and patriotic fighter would always defeat that 'soulless automaton,' that 'lifeless work of art,' the mercenary."[26]

Reformers proposed to King Frederick William III nothing short of a complete regeneration of the concept of citizen participation with the Prussian state in everything from military to economic to educational and social reform.[27] The edict of October 6, 1807, emancipated the serfs, and while the effect in the Prussian territory east of the River Elbe did not achieve any real change in the condition of the peasants, nor more landed rather than landless poor, overall its effect was positive for the nation.[28] The edict confirmed the right to own property by non-nobles as well as the right to choice of occupation. As in so many of these reforms, the actual practice was very conservative and slow in full expression, but "the emancipation of the serfs did more than any other measure to undermine the exclusive, aristocratic character of the old Prussian officer corps and open it to commoners of ability."[29]

This radical, even revolutionary, ideal was difficult for the conservative ruling Junker caste of Prussia to accept. Embracing with reluctance the concept of a mass citizen army, they extended the obligation to all male citizens through universal military obligation, on February 9, 1813. They agreed to this until the end of the War of Liberation. On March 17, 1813, they established the *Landwehr* as a militia force, and the *Landsturm* to back up the army, recognizing a nation in arms for the first time.[30]

But the major societal reforms did not survive the defeat and removal of Napoleon. The swift end of the campaigns of 1813–1815 renewed aristocratic resistance to a nation of citizens-in-arms, leading to the immediate reduction of the reformer's principal precepts until revived with the conscription law of 1850.[31] With the exception of the *Landwehr,* the army returned to its old ways after 1816, neglecting the militia, and restricting non-nobles from the officer corps. The king withdrew a promised constitution by 1819.[32] "The reason why in 1848 so many features of the old structure of society, and therefore so much cause for discontent among the middle and lower classes on the land still persisted, was to be found in the indulgence of the landowners' wishes in the year 1816."[33]

The constitution of 1850, as a response to the risings against the state in 1848–1849, again made universal military service the law in Prussia. Frederick William IV's constitution provided a bicameral legislative body, but it was written and imposed by monarchical fiat, preserving the right of the monarch to disapprove any law written by the legislature, and denying control over any budget measure concerning the army or navy. While acknowledging a liberalizing desire of the Prussian people, it was not a liberal document in practice, but continued protection of persons and property.[34]

Article 60 of the constitution provided a constraint on the size and independence of the army through its requirement that the size of the peacetime army "must be determined by law."[35] The *Landwehr* system, however, was all that remained of the reforms of 1806–1813. Further change, based on the poor mobilization undertaken in support of neutrality in the war Austria lost against France in 1859, sought the restructuring of the army into one based on a standing or active army large enough to maintain a powerful and professional presence in the German Confederation. These were completed in time for the war against Denmark of 1864 and in force by 1868 despite the resistance of the *Landtag*.[36] The reforms of 1860 ended the role of the *Landwehr* as envisioned in the reforms after 1806 and instituted in its place a reliance on the standing active army.

Under the reorganization, active service was three years at age twenty, then with the reserve (*Landsturm*) for four years, and then the *Landwehr* for five more years. Over time the reserve elements became two-tiered based upon age as *Landsturm I* and *Landsturm II*. All elements were under active army supervision, with oversight exercised through the long-standing regional army corps areas.[37] The North German Confederation, formed in 1867, following the Prussian victory at Königgrätz in 1866, instituted compulsory service that same year, and the southern German states, not within the confederation, also created conscription laws between 1866 and 1868.[38] On paper Prussia and the North German Confederation had the ability to call up more than a million men in the event of war. It was on this basis that

Prussia went to war with France in 1870. The kaiser's cabinet set, and the Reichstag funded, the peacetime strength of the active regular army, by law in 1868 (the Iron Budget) at 1 percent of the population until December 31, 1871, then extended it for three years.[39] In 1874, the Septennial Law called for an army totaling 401,659 men, but required renewal every seven years.[40] Again, in 1887 another increase in the standing army's peacetime strength was granted to 468,409 men. Increases in strength occurred in 1892, 1905, 1912, and 1913. Each appeared to coincide with the international crises that led to the opening of World War I. The peacetime active strength of the army on July 31, 1914, stood at 734,000 men.[41]

The army's key continuity throughout this period was the historical recruitment and classification of manpower on a territorial basis in law from 1868, with its antecedents in the *Kantonreglement* of 1733 and the Reform Movement of 1806–1815, and all iterations through 1945.[42] Eight *Armee-Inspektion* (later designated *Gruppenkommando*) were located through World War I in Danzig, Berlin (two), Hannover, Munich, Karlsruhe, Stuttgart, and Saarbrücken, each controlling three or four *Armeekorps,* established within its territory. *Armeekorps* controlled two or three divisions with appropriate artillery, pioneer, and train transport units, while each division consisted of two or three brigades of infantry or cavalry, with two or three regiments each.[43]

Mobilization in World War I included both *Landsturm I* by August 15, 1914, and *Landsturm II* on November 27, 1914. As with British initial efforts at recruiting, these measures fell short of the requirements for replacements as well as damaged the war economy by taking men without regard for the requirements of agriculture or industry. On November 1, 1916, a new organization took control of all war preparation for industry and agriculture in regard to manpower allocation, and from December 5, 1916, all persons aged seventeen to sixty became liable for service. Strength of the army and garrison forces during the war averaged 4.1 million men in the field army and 2.1 million in the garrisons, for a total average of 6.2 million soldiers on duty throughout the war. The enormous casualties sustained drove the number of men who passed through the German armed forces during the Great War to 13.3 million.[44]

Classes called up by year group were those entering age twenty for medical examination and enlistment (*Musterung* and *Einziehung*), while seventeen-year-olds were liable for inscription or signing in with their recruiting district (*Meldung zur Stammrolle*).[45] The Great

War medical classifications were as follows for induction between 1914 and 1918:[46]

k.v. *Kriegsverwendungsfähig:* Fit for active service.
g.v. *Garnisonverwendungsfähig Feld:* Fit for garrison duty in Germany, on the lines of communication, or in the field.
a.v. *Arbeitsverwendungsfähig:* Fit for labor employment.
d.u. *Dauernd-untauglich:* Permanently unfit.

It cannot be said, however, that conscription plans were either effective or efficient. In two waves beginning in September 1916, the German Army released 1.2 million men, and again in July 1917 a further 1.7 million men, back into industry and agriculture. As the war progressed, it became ever more apparent that there were too few workers to both sustain the army and feed the population. Antiwar sentiment built as the workers found themselves unable to feed their families, becoming ever less interested in continued sacrifices for the war effort. "Between January and March 1917 there were a number of strikes in the Ruhr, Berlin, and elsewhere. . . . The unrest reached its height on 16 April with mass strikes in Leipzig and Berlin. . . . In June and July there was another wave of strikes, centered principally on the Ruhr and Upper Silesia."[47] The winter of 1917–1918, *Der Hungerwinter,* demonstrated the extraordinary inability of the German *Reich* to feed itself, leading to the largest of the wartime strikes and the first mutiny in the High Seas Fleet. Following the disastrous strategic failure of the 1918 spring offensives, in an attempt to salvage control of the war production efforts in June 1918, "the Supreme Army Command once again endeavored to turn the tide by renewing its demand for the total mobilization and militarization of the economy . . . compulsory service for all men aged between fifteen and sixty; compulsory service for women; a military order binding workers to their place of work and disciplinary measures to ensure higher output. . . . In this connection it is interesting to note the War Bureau's objection that, whereas the workers were agreeable to military control of industry, the factory owners were not, and 'refused to submit to military authorities.'"[48]

The End of Empire

Prior to the four-year ordeal of the Great War, the Germany of August 1914 viewed itself as a nation confident, calm, and reasoned.

Germany relished great power status, and boasted a world of culture, music, and literature. The Germany of November 1918 was a nation defeated, its throne abdicated, and its community filled with unrest, revolution, and poverty. Hunger, disillusionment, and world reprobation abounded.[49] The field regiments of the German Army, with those members who had not mutinied or deserted, walked across the Rhine bridges in November and December 1918. In disciplined formations—with colors uncased—they marched down the *Unter den Linden* in Berlin on December 11, 1918, for a parade signifying the end of the war.[50] The army leadership, particularly the Great General Staff and the area commands (*Wehrkreise*), continued to function, supporting the establishment of the authority of the German Revolution and the new socialist government that emerged, acquiescing to the terms of the victors.[51]

Versailles Treaty restrictions. The Treaty of Versailles reduced the traditional German Army (*Kaiserheer*) to 100,000 men and 4,000 officers; 500 of the officers were either medical or veterinary officers, with no conscription and no reserves allowed. All soldiers were volunteers enlisting for twelve years, and officers for twenty-four years. There was no permission for heavy artillery, no chemical weapons, no aircraft, and no munitions industry. The German Navy fell to a few corvette-type ships and inshore craft. The treaty abolished the Great General Staff and the War Academy. The victors demanded crippling reparations, as well as significant territorial losses, including the loss of *Reich* territory to Poland, France, and Belgium, among others, and the demilitarization of both banks of the Rhine River. The treaty enjoined Germany to accept the guilt for initiating aggressive war.[52]

The *Reichsheer* of the new government of the German *Reich* replaced the German Army of the Great War, the *Kaiserheer*. A republic, which became known as the Weimar Republic, came into being on November 9, 1918. The new army administration (*Heeresleitung*) and the new navy administration (*Marineleitung*) together formed the new armed forces (*Reichswehr*). The head of the armed forces was the *Reich* president, and all officers and men swore an oath, repeated each year, to defend the new constitution. The treaty further directed that the organization of the army consist of seven infantry divisions of 12,000 men, with three regiments, each with three infantry battalions, and three cavalry divisions of 5,300 men each.[53]

The cabinet of the new republic rejected the terms of the Versailles Treaty twice during May and June 1919, reflecting popular sentiment to achieve a more just settlement. Opposition to the treaty led to the resignation of Chancellor Phillipp Scheidemann and his ministers, although not President Friedrich Ebert, on June 18, 1919. In the military opinion of the army leadership, despite the earlier popular rejections of the treaty terms, if the Allied powers attacked to enforce the treaty provisions, humiliating defeat awaited the nation. The army in 1919 consisted of fewer than 350,000 soldiers, centered in unofficial *Freikorps* formations, and without the ability to prevail in battle if the government chose to reject the treaty. The Allied powers demanded unconditional acceptance of the provisions of the treaty, and on June 24, 1919, minutes before expiration of the Allied deadline, the government of President Ebert signed.[54] Once organized in conformance with the Treaty of Versailles, the army, and indeed the nation, were reconciled neither to the defeat, nor to the concept of the *Reichswehr* as a glorified police force.[55]

General "Hans" von Seeckt and the "new" German Army. General Hans von Seeckt provided the direction and preparation of the German Army during the period of unrest that characterized the turmoil of the end of the Great War and preceded his appointment in 1920 as the leader of the German Army. These experiences hardened the army under his leadership to redress the condition of the armed forces. From the consolidation of the state of Prussia-Brandenburg before 1660, the army believed itself to be the guarantor of the liberty and existence of Prussia and later the German Empire.[56] The goal of the leaders of the defeated army was the restoration of its place of honor in German life as the sole entity with responsibility for the defense of the nation.[57] Seeckt believed the army and state were one: "The army serves the state and the state alone, for it *is* the state.[58]

Seeckt became the chief of the General Staff on July 4, 1919, one week before its dissolution in accordance with the Versailles Treaty.[59] Along with it went the War Academy. The office was henceforth known as the Troop Office (*Truppenamt*), and there the heart of the old General Staff continued to beat. The General Staff sections did not disappear. They found reassignment in contravention of treaty restrictions. Seeckt made it his primary goal to preserve the officers of the General Staff and filled the ranks of the *Reichswehr* officer corps with these men.[60] At its heart was the Operations Section, with about sixty officers retained in its Troop Office; others were moved in whole or part to other places in the government, continuing to function as before. Military History transferred to the Ministry of Interior in the Reich Archives, while the Survey and Map Section also shifted to the Ministry of Interior to become the Reich Survey Office. The Transportation Section crossed to become part of the Reich Transportation Ministry. The methods used by Seeckt to continue the work of the disbanded General Staff were well known to the Allied signatories of the Versailles Treaty, and as violations of the spirit, if not the letter of the treaty.[61]

Seeckt became chief of the Troop Office (*Chef des Truppenamtes*) on November 24, 1919, and then of the German Army Command (*Heeresleitung*) on June 5, 1920, just weeks after the failed army-led Kapp-Lüttwitz coup attempt.[62] "The *Reichsheer* set two goals from the outset, which had to contend one with the other: on the one hand the idea of the *Führerheer,* and on the other the constraints of both internal and external political forces on operational readiness."[63]

On January 1, 1921, the *Reichswehr* reached the Versailles Treaty limit of 100,000.[64] The army settled into the old barracks and routines, following the end of the political crises, with individual battalions charged to maintain the traditions of one of the old regiments. Seeckt ordered the preservation of "the old regiments, with their traditions dating back to Warsaw (I refer here to the famous battle of July 28th–30th, 1656, in which the Poles received such a welcome thrashing) and Fehrbellin [June 28, 1675, victory against Charles XII], and extending to the most recent past. It would be a great and irreparable injustice to the people connected to these regiments of the royal era, if we rip away all connection to the past. For the common people—even if they vote Social Democrat—are conservative. Even today, every provincial newspaper carries stories of the death of an old fighter, and the assembly of his former comrades in the regimental association to pay him the last honor."[65]

Training and selection of General Staff officers and the courses of the old War Academy were the responsibility of the regional commands, or *Wehrkreise.* They worked under the watchful guidance of Seeckt himself and the new T-4 Training Section. These staff officers, as their predecessors, received the same unique treatment as before the war, with its members "called leadership assistants" (*Führergehilfe*) to avoid treaty entanglements.[66] He "initiated a comprehensive program to collect and analyze the experiences of the World War to create a new body of military precepts for the *Reichswehr*."[67] He wrote that "it is absolutely essential to put the experience of the war in a broad light and collect this experience while the impressions won on the battlefield are still fresh and a major proportion of the experienced officers are still in leading positions."[68] Up to 400 officers were at work on these studies between 1919 and 1923. Seeckt issued the new regulation, on combined arms leadership and battle, in two parts, the first in September 1921 and the second in June 1923.[69] The foreword stated: "The regulation adopts the strength, the arms, and the infrastructure of a modern military Great Power as its basis not merely that composed of the 100,000 man German army from the peace treaty."[70]

The result of the studies undertaken by the new *Truppenamt* staff was a reaffirmation of the traditional Prusso-German methods of warfare, which revolved around the concept of an offensive war of movement (*Bewegungskrieg*). These concluded that during the Great War, "poor tactical mobility helped produce a premature culmination point in the West."[71] Seeckt recognized the value of a smaller, mobile army as the principal answer to the stalemate of the Western Front. His belief was that the very bulk of the mass armies of 1914 to 1918, lacking in essential mobility differential, caused the disaster of four years of trench warfare; he noted that "the soldier must ask himself whether these giant armies can ever be maneuvered in accordance with a strategy that seeks a decision, and whether it is possible for any future war between these masses to end otherwise than in indecisive rigidity."[72]

He looked to the war in the east, where he fought with distinction against the Russians, as more indicative of the answers to modern warfare and the future of an army seeking victory. Nor was he enamored of those espousing a view of future warfare as one between masses of machines, cautioning that "we certainly ought not to close our eyes to the development of the motor vehicle and its employment for military purposes. We shall not ignore it, but rather try to lay the theoretical foundations, and, as far as possible, the practical foundations for its use, but we must take care not to neglect existing, tested, serviceable appliances in favor of something that may be possible in the future. . . . The solution to the problem lies therefore in making full use of the products of technical science to extend and modernize what already exists, but not substituting something dead for something alive."[73]

German Army tactical and operational regulations for the employment of arms and men were guides to rational analysis and action rather than hard-and-fast rules. In keeping with a military mindset that placed little value on any form of "textbook" solution, tactical problems sought to encourage creative thought in the resolution of tactical and operational challenges. There were no "school" answers. Student evaluations focused on their application of the army's fundamental emphasis on combining all arms, both in a manner consistent with the enemy and the mission assigned, and in the context of its unique time, place, and circumstance. The value and practice of individual initiative and independent evaluation was paramount. These questions formed the core of the estimate of each situation.

> *Was soll ich?* What am I [ordered] to do? [My mission.]
> *Was kann ich?* What can I do? [The means provided to accomplish my mission.]
> *Was tue ich?* What will I do? [My orders to my units.][74]

Looking to the future. Beginning with a May 6, 1921, commercial agreement with the Soviet Union,[75] Germany moved to ensure its economic survival. In the process, the army sought to safeguard its ability to maintain expertise in both the theoretical and practical foundations of modern, technical war while laboring under the limitations of Versailles. The Treaty of Rapallo of April 16, 1922, solidified the economic and materiel connection between the Soviet Union and Germany. The secret military cooperation clauses of the agreement allowed German industry to build and test aircraft, armored vehicles, and weapons, as well as to train in armor and flying.[76] These arrangements included a "Junkers factory at Fili near Moscow, a poison gas factory at Samara, and shell factories—under Krupp administration at Tula, Leningrad, and Schlüsselberg (Petrokrepost)," and Zlatoust.[77] Training for German pilots took place at Lipetsk in Tambov province and for armor specialists at a Red Army school near Kazan, as well as through exchanges between General Staff officers of both sides to view training and maneuvers.

The entire enterprise was almost derailed with the exposure of the delivery of prohibited artillery munitions to Germany from the Soviet Union at the port of Danzig in the *Manchester Guardian* in December 1926.[78] While the production of artillery munitions was stopped, flying and maneuver training continued. That same year at Kazan, a training program opened in which Germans instructors taught mechanized combined arms warfare to classes of ten German and ten Soviet staff officers. These classes continued each summer until 1933 when Hitler came to power. These courses exposed future Soviet commanders to the latest German thought on armored and mechanized warfare in the concept of a war of movement (*Bewegungskrieg*), as well as allowing German insight into the latest thinking on the Soviet Union's "Deep Operations Theory of Annihilation."[79] Both theories sought to avoid a repetition of the trench stalemate of World War I. Each espoused the value of deep tactical and operational penetration, encirclement through concentric maneuver, and the successive annihilation of enemy forces until victory.[80]

Soldier standards. The volunteer army of the *Reichsheer* changed the entire recruit experience with the intent to develop a keen and willing long-term soldier. A decentralized recruiting system remained in place under the area commands, or *Wehrkreise*. The company, battery, and squadron commanders were the local recruiters. The regional recruiting system, method, and structure continued the organization from 1874 to 1918 despite the post-Versailles restrictions. The small size of the force allowed commanders very selective criteria for enlistment, and by 1928 there were fifteen applicants for every position.[81] The typical recruit was above average physically and mentally. Entry conditions from the order on recruiting for the *Reichsheer* indicated the quality required. These requirements admonished commanding officers, responsible for recruiting volunteers, to restrict acceptance to those with the moral and physical affinity for soldiering with the fundamental abilities characteristic of the infantry. The three physical qualifications for judgment by the physician required a statement that the volunteer was suitable for the army, temporarily not suitable, or unsuitable. The interview preceding acceptance stressed the physical ability of the volunteer and his mental agility in apprehension, adaptability, clear judgment, and concentration. Officer selection took place from the ranks, and the desire of the recruit to serve near home required compliance.[82]

Soldiers took an oath to defend the German constitution and renewed it at least once each year based upon a new order issued on March 2, 1922, replacing the old Articles of War of the *Kaiserheer*. The Articles of War admonished the soldier to defend the constitution of the German Empire or the German states, to live in harmony with his comrades, to not forsake them in battle or danger, and by his own bearing to set an example for the younger, inexperienced soldiers. Soldiers must not mix in politics and it was noted that "during movement warfare the soldier must never forget that the fight is only directed against the armed enemy. The property of peaceful inhabitants, of the wounded, sick and prisoners, as well as the property of the killed is governed by law. To arbitrarily take booty, plunder, to maliciously or willfully damage or destroy the property of strangers in the zone of hostilities, and to oppress the inhabitants is strongly forbidden."[83]

Each infantry division, in addition to its three tactical regiments, possessed a training battalion of three training companies. Recruits arrived at their units in the fall, following that year's maneuvers, further allowing for consistent application of training prior to the subsequent year's fall maneuvers. Initial training for recruits was six months and all received basic infantry training, emphasizing athletics as well as drill, and strenuous marches for conditioning. On observing field training, a US military attaché remarked that "as regards agility and suppleness, the German is very much more highly trained physically than our own men," giving the "impression of extraordinary alertness."[84] The basis for this expectation of alertness lay in the social requirement that all German children attended compulsory school at least through a *Volkschule* level, with German illiteracy in 1900 at 0.05 percent.[85] In the Ninth Regiment (Third Division, *Wehrkreis III,* Berlin),[86] a typical breakdown of soldiers was: 55 percent trained or worked in a business trade; 18 percent trained or worked in a manual trade; and 15 percent came from the farm.[87]

In addition, the conditions under which recruits lived were improved in the *Reichsheer* as "pay was increased, the rigid discipline of a conscript army was tempered, and the living standards of professional soldiers were dramatically improved. Starting in 1920–21, the army barracks were remodeled and renovated into comfortable troop quarters. . . . [E]ach NCO would have his own room, and only four to eight privates would share pleasant, well-furnished rooms resembling college dormitories. . . . Good food would be emphasized, and units would have a full range of recreation and sports facilities, unit libraries, and soldiers' clubs. A whole system of special trade and high-school courses was instituted within the army to teach each long-term soldier a profession to prepare him to succeed in civilian society upon completion of his twelve-year term of enlistment."[88]

***Legacy of the* Reichsheer.** With General Seeckt's forced resignation on October 9, 1926, command of the army passed to General Wilhelm Heye.[89] Seeckt, in his own way, supported and defended the new German constitution from a range of internal and external threats, including the revolutionary upheavals in various parts of Germany between 1919 and 1921. He opposed the Kapp-Luttwitz revolt of March 1920, but refused to support the use of the army against the conspirators. Removing himself to his home for the four days of the revolt, his ambivalence, rather than full support, toward the democratic government became apparent. His intent through his lack of action was to prevent the army from direct engagement against the conspirators, also members of the army.[90] His muted support of the republic during the brief Nazi *Putsch* in Munich, and of Adolf Hitler's arrest in 1923, was again indicative of some ambiguity in his actual intentions and support of German democracy.[91] Seeckt espoused his own political agenda, including a desire for the leadership of the German nation itself. Handed near dictatorial powers in 1923 to put down rebellion, his high-handed and ham-fisted revelations of that ambition lost him the trust of the many politicians who supported his summary dismissal in 1926.[92] Despite his ultimate support of the Weimar government in each case, collaboration did not indicate affection for the republic, nor his personal belief in a democratic method of government. He was, and remained, a soldier of the old German *Kaiserheer,* and the cultural and political narrative of that generation. He viewed with skepticism the manner and mechanisms of representative democracy. In the midst of the struggle over ratification of the terms of the Versailles Treaty, Seeckt's diary entry for May 21, 1919 noted: "Democracy; in these circles that simply applies to chic. In my absence many Jews and several professors have arrived; it doesn't get any better."[93]

In the *Reichsheer,* Seeckt bequeathed to the army of World War II an expression of German historical emphasis on a war of movement. It took a decade to instill in the 100,000-man army the precepts of this manner of warfare and to prepare the ground for the aircraft and mechanization and leaders to bring it to fruition. Its legacy is critical to understanding the later perception of German combat capability. Seeckt saw the army through its most difficult period of disorganization and unrest. He left an army, he believed, prepared for its inevitable expansion in Germany's future.[94]

The Making of the Interwar Infantry

Once selected, the infantryman of the Weimar *Kaiserheer* entered a long-service, professional, all-volunteer army. The emphasis on alertness, physicality, and initiative became a defining characteristic of the "army of leaders." The regulation on combined arms leadership and battle and the later *Truppenführung* each emphasized the critical importance of the individual in his small group—the importance of his bonding and care for his fellows. "Quality" was the watchword in the selection and training of the officers, noncommissioned officers, and men of the army. The traditional German concept for the primacy of a war of movement (*Bewegungskrieg*) became the basis for evaluating the lessons learned in the Great War. The combination of these two factors, common understanding of tactical concepts and quality combat manpower, formed the basis of the perception of the superiority of German infantry and combat arms during and following World War II.[95] The intent of the external and institutional cohesive measures was to prepare the soldier for the test of battle and to ensure his ability to stand in the line of battle and to advance across no-man's land.

External and Institutional Cohesion Measures

The phenomenology of the modern battlefield, with the advent of smokeless powder, and automatic and machine gun weapons, observed during the Boer War (1899–1902), the Russo-Japanese War (1904–1905), and the Balkan Wars (1912–1913), led the German army, as it did in all armies, to search for tactical methods to enable the return of decisive offensive maneuver to warfare. "The increased accuracy, range, and rate of fire of modern weapons threatened to push soldiers in combat to their psychological breaking points more quickly."[96]

Tactical development to 1918. The German concept of *Auftragstaktik* emerged from the competing needs of centralized control and dispersed and decentralized execution to enable offensive movement forward into the face of an enemy's lethal defensive fire. *Auftragstaktik* centered on developing the initiative of subordinate leaders, encouraging them to independent action within the unique conditions of an unfolding tactical situation and not on pre-planned or "textbook" solutions. The elder Helmuth von Moltke, for thirty years the chief of the Prussian General Staff and later the Great General Staff, noted the indispensability of independent action by subordinate leaders from army corps to junior leaders in the face of changing conditions on the battlefield, as noted in his review and critique of the tactics employed during the Franco-Prussian War 1870–1871.[97] The problem of such free-flowing tactics was the loss of control inherent in their execution, and the need for well-trained and seasoned men to carry them out. This played out in the years following the 1871 victory over France in two ways for infantry combat: the intellectual battle between *Normaltaktik,* a term for standardization of tactics to ensure control in the attack, and *Auftragstaktik.*[98]

Normaltaktik, while not dismissing the idea or necessity for open-order formations in some circumstances, stressed the need for control of close, dense formations through drill that ensured the synchronization of the force in action. In essence, it required less initiative and greater use of standard formations such as column and line in compact masses to ensure appropriate control of soldiers in contact with an enemy, especially as a means to move quickly in the spirit of the offensive.[99]

Auftragstaktik, also described as delegated tactics, emphasized the role of individual enterprise to the success of open-order attacks, allowing the conditions on the field of battle to dictate the tactics required for success. These characteristics centered on the unique mission of the unit, the terrain, and the dispositions of the enemy. All of these lay in the purview of the junior leader to exploit for offensive success in the movement across the defensive field of fire of the enemy. The German Army infantry regulations of both 1888 and 1906 favored the concept of open order and delegated tactics. Explanatory works on training and infantry fires focused also on the importance of leadership in open-order and flank attack. Despite these written expectations, significant resistance from conservative officers until the opening of the Great War diminished the ability of the bulk of the army to train to the level required to make open order a legitimated concept in the training of German tactical-level formations.[100]

Auftragstaktik was not the practice of the German armies of World War I under its chief of the General Staff, Helmuth von

Moltke, known as "the younger" when compared to his uncle, Moltke the elder. In the march into France, the need for control overwhelmed the ability or the intent to react to a thinking and powerful enemy.[101]

Intent on a quick victory in France in order to destroy the Russians in their turn, all reserve divisions were activated and employed at the front. These divisions found that open-order and coordination of arms required significant training to master and time to coordinate on the field of battle. The tactical impatience of senior commanders prevented them from taking the time to prepare attacks against even hasty defenses, or allowing their subordinates to do so. Enthralled to the all-important timetable and coordinated maneuver essential to the opening of the Schlieffen-Moltke (the younger) plan that guided the attack into France in 1914,[102] the concept of intelligent initiative was lost.

All pretense of understanding the lethal nature of the modern defense was overthrown by a misguided view that the offensive spirit of the soldiers might be lost to the time taken to bring up machine guns and soften up enemy positions with artillery prior to the infantry attack as Moltke (the younger) himself observed during training in 1906. Maneuver in 1914 devolved into tactics that favored control using mass formations of infantry thrown at French and British defensive positions without artillery or machine gun preparation, resulting in extraordinary casualties.[103] The war of position in the trenches of the Western Front was the testament to this failure.

The answer to the riddle of the trenches appeared in the special selection and training of soldiers skilled at using their initiative to move forward, to cross no-man's land in open, dispersed movement in short bounds, well-coordinated with a brief, powerful artillery attack and machine gun support. The answer lay in the combined arms action and cooperation of all arms.[104] The hard lessons of four years of position war bore fruit for the German infantry in the creation of these specialized assault troops *(Stoßtrupps)* for their central role in the offensives of 1918. The disastrous strategic dispersal of effort during these final attempts to break the stalemate of trench warfare proved devastating and ensured the defeat of Germany. What they proved was that well-trained and supported infantry possessed the ability to breach the trenches in depth. The failure lay not with the spirit of the soldiers, but with the inability of the current technology to exploit their successes given equal foot mobility between the belligerents. These were lessons the army carried with it into defeat in 1918.[105]

Tactical development, 1919–1935. German tactical precepts, following the defeat in 1918, reemphasized the decisive nature of the quality of the individual. They called for intelligent, physically capable, and alert fighters prepared to think and act independently, correctly assessing the myriad situations found in combat. The ideal outcome of selection was that men determined capable of withstanding the physical and mental demands of battle, and with the character to bond with their fellows, received assignment to frontline combat formations.[106] Seeckt guided the doctrinal development of these concepts of the quality characteristics of the soldier in 1929, noting that "the soldier's field of activity is man, who controls science, technology and material. The army is a combination of many men with the same serious purpose. This gives the soldier's profession a quite peculiar bond of unity, a corporate sense which we call comradeship."[107]

The driving concept of a war of movement (*Bewegungskrieg*) was the basis and foundational concept for German Army training and operations. Seeckt wrote in his critique of the fall 1922 army maneuvers "that one must free himself from the shackles of a war of position [*Stellungskrieg*] and embrace more fully the detached operations of a war of movement [*Bewegungskrieg*]."[108] He noted "that the soldier, who seeks a decision in mobility, quickness and spirit of decisiveness, has grave doubts whether armed masses can ever secure a decision, and whether nations in arms can avoid finishing in trenches once more."[109] Seeckt was adamant that "the war of position [of World War I] was an operational perversity. It was a congealment, born out of the lack of power of both warring parties."[110]

Given the development of a guiding concept for a mechanized combined arms war of movement, testing the training and physical components necessary to accomplish this form of warfare was problematic and illegal within the constraints of the Versailles Treaty. To overcome this difficulty, Seeckt sought the establishment of training, experimentation, and technical evaluation bases in the Soviet Union, at Kazan for armor from 1927 to 1933 and Lipetsk for aircrew from 1925 to 1933.[111] The army encouraged civilian corporations and industry to build and support prototype development of aircraft, motor, and tracked vehicles and components for testing in the Soviet

Union. When the announcement came to expand the existing armed forces, and the re-creation of the German Air Force (*Luftwaffe*), for example, its basic doctrinal and training requirements, as well as a cadre of pilots and aircrew, were already in place. Hitler ended this cooperation with the Soviet Union upon coming to power in 1933.[112]

By 1933, a new manual was in order: the new *Heeresdienstvorschrift Truppenführung* (translated as "troop-leading" or "unit command"). The *Truppenführung* represented an essential continuity between the regulations of the *Reichsheer* of the Weimar Republic of 1921–1934, and the *Heer* of the new *Wehrmacht* of 1935–1945.[113] Combined arms coordination with tactical air power and mobile artillery, mechanization, and the overriding importance of the individual, his alertness and mental and physical fitness and initiative, were the keys to avoiding a war of position.[114]

Junior Leaders

The regulations of 1921 noted the doctrinal expectations for leaders, including a strong will and sound character, that tactical precepts were a guide, not binding, and required the leader to have knowledge, proficiency, and an understanding of precise fundamentals as his guide in battle. The foremost leader trait was the acceptance of responsibility as well as the personal care and well-being of his troops in sharing their hardships. "The soldiers know that the leader lives for them and shares the good and bad times with them, thus will they be willing to exert their last effort to achieve success and to suffer disappointment. In misfortune is first revealed the true inner spirit of each soldier from the leader down to the last man."[115]

Noncommissioned officers. Following World War I, *Reichsheer* redefined the nature and role of the NCO (*Unteroffizier*), demanding a higher leadership capability than the tradition of the *Kaiserheer,* whereby the NCOs looked too much to the officer for guidance.[116] Junior officers were those closest to the fighting and took the brunt of the officer casualties. This recognition created the renewed emphasis on the leadership and tactical preparation of the NCO in the interwar army.[117] Preparation of NCOs in the *Reichsheer* was as intensive in their grade as for their officers. To become an NCO required three and a half years of troop experience, followed by several months of regimental duty. Development emphasized leadership two grades up. "The importance of the junior leaders is increased through the dispersed battlefield. The lessons of the battle ground demand independently thinking and acting fighters."[118] While nearly 50 percent of the US Army held NCO rank by 1945, German NCOs never numbered more than 16 percent of the soldiers.[119] In addition, tens of thousands of NCOs received commissions during the war, with eleven prewar former NCOs rising to general officer rank. Through 1943, "over one-half of all NCOs captured by the Western Allies stemmed from the pre-war army."[120] This reflected the emphasis placed on the NCO corps.

> Clear-sighted, deliberate, organization of the noncommissioned officer corps, thorough education and training of its members are crucial for the bearing and performance of the troops. They rank in the first duty of the company commander. The higher commander encourages them in this connection and supervises the uniformity of noncommissioned officer training in his organization.[121]

Junior officers. To become an officer in the *Reichsheer* was a minimum four-year process, subjecting the officer-aspirant (*Fahnenjunker*) to a battery of tests prior to induction to determine if he possessed the education, good character, and excellent physical condition that were the principal criteria for acceptance. Social background was not a restriction for officer selection as in the *Kaiserheer,* although the requirement for an *Abitur,* certificate of graduation from a secondary school, placed most aspiring officers in the middle class and above.[122]

If accepted by the regimental commander, the officer candidate spent fifteen months as a private soldier learning the basics of his profession, ensuring that all officers received experience in the junior ranks of the army. After passing an exam, the soldier received promotion to the rank of officer-aspirant, the equivalent of a junior NCO, followed by another ten months of rotation through the various branch schools, and required passing the exams for sergeant and ensign. He returned to his regiment for a minimum of another seven months as a senior ensign before promotion to lieutenant, if he gained the regimental commander's final approval.[123] Failure meant discharge from the army.[124]

In 1926, psychology-inspired testing entered the officer selection process for the first time.[125] The tests took place over a three-day period prior to acceptance as an officer-aspirant, as well as for pilots, drivers, and radio operators, and those in other technical fields. The first day included a battery of standard intellectual and mechanical tests designed to show powers of perception, among other qualities, and included in this package were observations of the aspirant's facial reactions to different stimuli including a film. Speech and handwriting analysis required a written essay on his life history—scores were less important than impressions. The second day was for "action analysis" tests concerned with observing the candidate's reactions in military settings, while tackling problems under time and physical stress. The third day emphasized special testing for pilots and other specific technical specialties. The intent of the testing was the discovery and confirmation of character and intellect.[126]

As in the British Army, the psychological portion of the tests was contentious among the senior officers of the *Heer* from the beginning:

> From the point of view of the senior officers, the German Army had always placed a special value on the evaluation of the character, quality, and the inner nature and upbringing of officer aspirants through an evaluation [process] based upon the competence of the senior officer. The duty of the senior officer was to discern these qualities. This policy in the old tradition of the Prussian-German Army had always proven successful. Little wonder that the attempt to do the same with scientific investigation and tests was viewed as an encroachment on the prerogative of the senior officers, and met with great distrust and outright rejection.[127]

The psychological tests continued until 1942. The inability to replace officer casualties and the shortage of pilots brought the use of these tests to the attention of Hitler through Hermann Göring, the head of the *Luftwaffe*. Göring demanded an end to the tests upon being told that many qualified men, possessing an *Abitur*, were so disturbed by the process of psychological testing that they did not volunteer for pilot training. Given the level of senior officer mistrust of the testing, Hitler's decision to end the tests and disband the entire testing office was little remarked on by the officer corps.[128]

On the reintroduction of conscription in 1935, the requirement for an *Abitur* became standard again. Minimum height was 5 feet 5 inches. Eligibility precluded anyone older than twenty-two years old from gaining acceptance, as well as the rejection of candidates with Jewish heritage as part of the Nazi-specific racial policy.[129] While the opportunity existed for soldiers to gain admittance to officer rank without the *Abitur* even during the Weimar Republic period, it was not common in the *Reichsheer*. The path to officer rank opened to otherwise qualified enlisted men without the *Abitur* in 1936, with specific requests for submission of candidates for officer aspirant without it.[130] The army reduced the four-year officer selection process of the years 1921–1934 to less than two years in 1935 by paring down the time required in the troop unit as a private soldier to nine months before attending his officer's school and then assignment back to a troop unit.[131]

Discipline Reform in the Reichsheer

The legal reforms of this period were of particular note, given the later racial stridency of the Nazi era, and considered at the time (1926) the most permissive in the world, placing great trust in the experience of those charged with discipline in the small units of the army. The reforms anticipated that the long-term volunteer required less strict handling than his conscripted counterparts in the old army, and that this viewpoint carried over into the tactical units of the conscript army of World War II.[132] One reform not undertaken was the elimination of the death penalty for military offenses, which, when recommended by the *Reichstag,* Seeckt specifically rejected as naive.[133] (See Appendix 3.)

The German General Officers and Adolf Hitler

The descent of the German general officers into complete abnegation and harmony with National Socialism began on the initiative of those leaders. Four officers occupied the primary leadership positions affecting army policy and its interface with the civilian, elected government. First was General Werner von Blomberg, appointed to the cabinet as *Reichswehr* minister one day before von Hindenburg chose Adolf Hitler to become chancellor in 1933. Blomberg was known to Hitler as his leading proponent among the general officers. His deputy in the *Wehrmachtsamt* (Armed Forces Office) in the ministry was his former chief of staff in *Wehrkreis* I (East Prussia), Oberst Walther von Reichenau, also known to Hitler as one of the most ardent Nazi supporters in the army.[134] Within the army itself, the chief of the Army Command Generaloberst, Werner von Fritsch, was an ardent opponent of Adolf Hitler and the Nazis. General Ludwig Beck, as chief of the Troop Office (*Truppenamt*) from 1933 to 1935, and later as chief of the General Staff from 1935 to 1938, was also an opponent. Beck became a chief conspirator from 1937 in attempts to overthrow Hitler, dying by his own hand on the day after discovery of his role in the attempt on Hitler's life on July 20, 1944.[135]

The chief architects of army acquiescence were Blomberg and Reichenau. Blomberg feared the loss of primacy of place in the defense of the nation to the Nazi Party–armed thugs, the *Sturmabteilung* of Ernst Röhm. This led Blomberg to direct the placement of the Nazi Party emblem on all *Wehrmacht* uniforms and the institution of Nazi Party policies on non-Aryans and Jews, the Arier Paragraph, during February 1934.[136] Hitler responded well, and on two occasions, February 28 and June 16, 1934, held conferences in which he gave direction affirming the primacy of the army and his intent to build the army for the war tasks he envisioned for the future. The February 28 meeting included the leadership of both organizations, the army and the *Sturmabteilung,* including all army senior general officers. Hitler outlined in specific detail the planned stages for war with the Western powers and the Soviet Union, for which he demanded an army prepared to initiate war in no less than eight years.[137]

Acquiescence in the Nazi blood purge of the *Sturmabteilung* on June 30, 1934, the "Night of the Long Knives," with the murder of at least eighty-three people, including two retired general officers, was the first significant and repugnant moral lapse for the general officers of the army, few of whom, if there were misgivings, voiced them.[138]

On August 2, 1934, on the death of von Hindenburg and Hitler's assumption of the role of both chancellor and president, Reichenau and Blomberg, on their own initiative, discarded the oath sworn to the German constitution and directed the entire *Wehrmacht* to swear a personal oath to Adolf Hitler: "I swear by God this holy oath, that I will render to Adolf Hitler, Leader of the German nation and people, Supreme Commander of the Armed Forces, unconditional obedience, and I am ready as a brave soldier to risk my life at any time for this oath."[139]

The final gift for Adolf Hitler on that day, presented by Blomberg, was the recommendation that the armed forces no longer call him "Herr Hitler," but "Mein Führer." Hitler accepted.

Prior to the Nazi *Machtübernahme* (takeover) in 1933, the army began preparation in secret during 1930 to expand to twenty-one divisions and succeeded in gaining a plan for 484 million reichsmarks beginning in 1932. The plan, known as the A-Plan or A-*Heer,* was to take place between 1932 and 1938.[140] The expansion began on October 1, 1934, with an increase in volunteers intended to double the number of infantry as well as increase motor troops, armor, and aircraft. Hitler expanded the plan on March 16, 1935, when he demanded a thirty-six-division force with twelve corps headquarters following the reintroduction of the draft on March 15, 1935.[141]

Arm-and-Arm into Hell: Army Leadership and the Nazi Party

Adolf Hitler's stated goal to overthrow the terms of the Versailles Treaty and to return international prominence to Germany was in line with the intent of the senior officers of the army. While these were identical political goals of both the late Weimar Republic leaders and of the Nazi Party, the weakness of the Weimar regime stymied most of the army's hopes for restoration. The Nazis appeared to desire what the army believed necessary for itself and for Germany. The key divergence from the military stories of the armies of the Western democracies and Germany lay in the manner in which the leadership of the German Army embraced Adolf Hitler and the Nazi dictatorship.

Should the army have acted to prevent the chancellorship of Adolf Hitler? R. J. O'Neill stated that the question rested on

> the assumption that any army has the right to interfere in the internal workings of the political system of the country which has raised it, even though none of the contending political parties has resorted to illegal means, though law and order is being maintained by the civil police, and though the Head of State or Commander-in-Chief does not desire any military assistance. Needless to say, the law of no civilized nation, including the Weimar Republic, has recognized the validity of any such principle.[142]

From this point of view, even if the army leaders disliked Hitler or his methods or governing philosophy and intent, Hitler achieved power within the legal bounds of the Weimar constitution, and ensured that each additional assumption of greater authority leading to dictatorship remained within that established legal framework. The army established no pretext for preventing the (legal) Nazi *Machtübernahme*.[143] Many of its senior officers, however, embraced the politics of National Socialism, as well as many more junior. The army's semiofficial publication the *Militär-Wochenblatt,* in its lead article on July 4, 1931, reminded its readers of the dangers of politicizing the *Wehrmacht,* as in the dark period following the opening of the Great Depression in Germany and the later unrest.[144] This view changed by the lead article of the April 25, 1933, issue, introducing Adolf Hitler to its readers as "a symbol of every nameless frontline soldier who from German manhood stood against a world of enemies. . . . [Under the chancellor], the flag of nationalism and socialism flies free . . . for the whole people."[145]

In an August 18, 1933, editorial, the magazine editors sought to define what the German "revolution" of 1933 meant for the soldier, declaring: "National Socialism is not a political program, but a world view [*Weltanschauung*] . . . [that] would bring together the state, the church, the people, culture, and justice in a new way."[146] For the first time, the editors sought to connect for the soldiers the idea of the people (*Volk*) with the soldier's sense of mission and comradeship (*Kameradschaft*).[147] The first swastika appears on the magazine cover in July 1935.[148]

From August 1934, the German general officers embraced a political viewpoint for the *Wehrmacht* and complicity with the objectives of the Nazi state and its *Weltanschauung*. While perhaps complete recognition of the dangers of this total political and personal subservience appeared after events a few years in the future, the moral lapses of the generals refuse denial or excuse. The naive and obsequious manner of the moral descent of the *Wehrmacht* must be laid at the feet of the generals. With or without reservations, they turned to the difficult task of expanding the army to achieve the stated war goals of Adolf Hitler.

Notes

1. Tallett, *War and Society in Early-Modern Europe,* 7.
2. Frost, *The Northern Wars,* 2. An early example of a regional mustering of military forces occurred in western Pomerania in 1547; Schnitter, "Die überlieferte Defensionspflicht," 31.
3. Frost, *The Northern Wars,* 79.
4. Ibid., 209; Schnitter, "Die überlieferte Defensionspflicht," 35; Tilly, *Coercion, Capital, and European States,* 20–21; Maehl, *Germany in Western Civilization,* 211.
5. Tilly, *Coercion, Capital, and European States,* 20–25; Maehl, *Germany in Western Civilization,* 211; Frost, *The Northern Wars,* 200.
6. Frost, *The Northern Wars,* 199.
7. Tallett, *War and Society in Early-Modern Europe,* 79.
8. Frost, *The Northern Wars,* 200; Tallett, *War and Society in Early-Modern Europe,* 199.

9. Stübig, "Die Verfassung Preußens in der Reformzeit," 40–41; Büsch, *Military System and Social Life in Old-Regime Prussia,* 7; Maehl, *Germany in Western Civilization,* 219.

10. Finer, "State- and Nation-Building in Europe," 142.

11. Ibid., 136; Tallett, *War and Society in Early-Modern Europe,* 199.

12. Büsch, *Military System and Social Life in Old-Regime Prussia,* 8.

13. Gagliardo, "Translator's Preface," xi.

14. Luvaas, *Frederick the Great on the Art of War,* 72.

15. Luvaas and Griess, *The Regulations of the Prussian Infantry,* 371.

16. Büsch, *Military System and Social Life in Old-Regime Prussia,* 2.

17. Paret, *Yorck and the Era of Prussian Reform,* 15–17.

18. Luvaas and Griess, *The Regulations of the Prussian Infantry,* 372.

19. Ibid., 373.

20. Luvaas, *Frederick the Great on the Art of War,* 75–76.

21. Ibid., 73.

22. Fann, "On the Infantryman's Age in Eighteenth Century Prussia," 165.

23. Ibid., 166.

24. Luvaas, *Frederick the Great on the Art of War,* 144; Simon, *The Failure of the Prussian Reform Movement,* 3–5; Dupuy, *A Genius for War* 22; Dupuy and Dupuy, *The Encyclopedia of Military History,* 753; Chandler, *The Campaigns of Napoleon,* 589; Maehl, *Germany in Western Civilization,* 291.

25. Paret, *Yorck and the Era of Prussian Reform,* 86–87; Echevarria, *After Clausewitz,* 36; Fann, "On the Infantryman's Age in Eighteenth Century Prussia," 169; Dupuy, *A Genius for War,* 34.

26. Paret, *Yorck and the Era of Prussian Reform,* 75.

27. Ibid., 117; Simon, *The Failure of the Prussian Reform Movement,* 6–8; Maehl, *Germany in Western Civilization,* 293–305; Stübig, "Die Verfassung Preußens in der Reformzeit," 42.

28. "Prussian Reform Edict of October 9, 1807," in Snyder, *Documents of German History,* 132; Büsch, *Military System and Social Life in Old-Regime Prussia,* 40; Simon, *The Failure of the Prussian Reform Movement,* 20–24.

29. Simon, *The Failure of the Prussian Reform Movement,* 154; Paret, *Yorck and the Era of Prussian Reform,* 244; Maehl, *Germany in Western Civilization,* 299.

30. Absolon, *Wehrgesetz und Wehrdienst,* 1–2; Absolon, *Die Wehrmacht im Dritten Reich: Band I,* 3; Howard, *The Franco-Prussian War,* 11.

31. Howard, *The Franco-Prussian War,* 13.

32. Simon, *The Failure of the Prussian Reform Movement,* 223–224.

33. Ibid., 101.

34. Koch, *A Constitutional History of Germany,* 164; Absolon, *Die Wehrmacht im Dritten Reich: Band I,* 1–9.

35. Craig, *Germany,* 50.

36. Craig, *The Politics of the Prussian Army,* 136–140; Craig, *The Battle of Könniggrätz,* 3–4; Stübig, "Die Verfassung Preußens in der Reformzeit," 52.

37. Howard, *The Franco-Prussian War,* 22.

38. Absolon, *Die Wehrmacht im Dritten Reich: Band I,* 2.

39. Craig, *Germany,* 51.

40. Ibid., 53.

41. Craig, *Germany,* 5, 129, 237, 279, 293–296; Absolon, *Die Wehrmacht im Dritten Reich: Band I,* 4; Förster, "Militär und staatsbürgerlicher Partizipation," 62–63.

42. Absolon, *Die Wehrmacht im Dritten Reich: Band I,* 2.
43. Ibid., 4–5.
44. Absolon, *Die Wehrmacht im Dritten Reich: Band I,* 6–9.
45. General Staff, War Office, *Handbook of the German Army in War,* 18.
46. Ibid.
47. Hardach, *The First World War,* 71–73; Groß, *Mythos und Wirklichkeit,* 135.
48. Hardach, *The First World War,* 71–73.
49. Holborn, *A History of Modern Germany,* 568; Eyck, *A History of the Weimar Republic,* 93–95; Craig, *Germany,* 424, 455.
50. Lewis, *Forgotten Legions,* 3; Dupuy, *A Genius for War,* 185.
51. Holborn, *A History of Modern Germany,* 578–591; Eyck, *A History of the Weimar Republic,* 147–160; Craig, *Germany,* 428–433.
52. Absolon, *Die Wehrmacht im Dritten Reich: Band I,* 30–31; "Text of the Versailles Treaty," in Snyder, *Documents of German History,* 376–381.
53. Absolon, *Die Wehrmacht im Dritten Reich: Band I,* 29; Müller-Hillebrand, *Das Heer bis zum Kriegsbeginn,* vol. 1, 14–15.
54. Von Rabenau, *Seeckt,* 180–185; Holborn, *A History of Modern Germany,* 574–575; Eyck, *A History of the Weimar Republic,* 65, 100–105; Dupuy, *A Genius for War,* 196; Craig, *Germany,* 425–428.
55. Deist et al., *Ursachen und Voraussetzungen der deutschen Kriegspolitik,* vol. 1, 373; Müller-Hillebrand, *Das Heer bis zum Kriegsbeginn,* vol. 1, 16.
56. Von Rabenau, *Seeckt,* 174; Craig, *The Politics of the Prussian Army,* 394–398, 415–419, 422; Dupuy, *A Genius for War,* 212; Büsch, *Military System and Social Life in Old-Regime Prussia,* 46–47.
57. Skinner, "Transformation of the German *Reichsheer,*" 27.
58. Von Seeckt, *Thoughts of a Soldier,* 80; von Seeckt, *Gedanken Eines Soldaten,* 116, emphasis added in English translation.
59. Von Rabenau, *Seeckt,* 190.
60. Ibid., 192–194 (letter to General Wetzli of July 24, 1919); Corum, *The Roots of Blitzkrieg,* 34–35.
61. Edward Davis, Military Observer, handwritten memorandum of April 26, 1920, to Division of Military Intelligence, US War Department, and monograph report, *Present Military Organization of Germany,* February 21, 1922, as reported from the Second Section of the Belgian General Staff, Records of the War Department General Staff, Military Intelligence Division correspondence, 1921–1942, Record Group 165, NACP; Skinner, "Transformation of the German Reichsheer," 24.
62. Von Rabenau, *Seeckt,* 203, 248; Eyck, *A History of the Weimar Republic,* 151–52.
63. Von Rabenau, *Seeckt,* 475, 503; Corum, *The Roots of Blitzkrieg,* 69.
64. Von Rabenau, *Seeckt,* 28; Absolon, *Die Wehrmacht im Dritten Reich: Band I,* 26.
65. Division of Military Intelligence, War Department, subject: Traditions of the Old German Army in the *Reichsheer,* monograph report, June 17, 1921; Division of Military Intelligence, War Department, subject: German Army and Navy List, June 11, 1921. A transmitted copy of the *Deutsche Wehrmachtseinteillung* (*Reichsheer und Reichsmarine*), April 1, 1921, detailed not only the locations of all units, but also the affiliations of battalions, for example the First Battalion of the new First Infantry Regiment, located in Königsberg, affiliated to the *Kaiserheer* formation; see von Rabenau, *Seeckt,* 486; Citino, *The German Way of War,* 244.

66. Von Rabenau, *Seeckt,* 201 (letter to Minister Groener of September 30, 1919), 501 (thoughts on the training of these staff officers as *Führergehilfenausbildung* in 1920); Corum, *The Roots of Blitzkrieg,* 36; Dupuy, *A Genius for War,* 216.

67. Corum, *The Roots of Blitzkrieg,* 37.

68. Ibid., 37.

69. D.V. Pl. Nr. 487, *Führung und Gefecht der verbundenen Waffen,* 29–34, 165–174; Groß, *Mythos und Wirklichkeit,* 153–154; Corum, *The Roots of Blitzkrieg,* 39; Skinner, "Transformation of the German Reichsheer," 32; Reinicke, *Das Reichsheer,* 92.

70. Heeresleitung, *Führung und Gefecht der verbundenen Waffen,* 3; Corum, *The Roots of Blitzkrieg,* 39; Skinner, "Transformation of the German Reichsheer," 32; Reinicke, *Das Reichsheer,* 92.

71. Skinner, "Transformation of the German Reichsheer," 28; Citino, *The German Way of War,* 240; Corum, *The Roots of Blitzkrieg,* 40; Lewis, *Forgotten Legions,* 17–19.

72. Skinner, "Transformation of the German Reichsheer," 21.

73. Ibid., 23.

74. BAMA RH 17/541 Heeresnachrichtenschule, 8 Offz.-Anw.-Lehrgang, Merkblatt 11, *Anhaltspunkte für Beurteilung von Lagen u. Entschlußfassung,* October 22, 1941.

75. Müller, *Das Tor Zur Weltmacht,* 66–71; Dupuy, *A Genius for War,* 215–216.

76. Müller, *Das Tor Zur Weltmacht,* 66–71.

77. Craig, *The Politics of the Prussian Army,* 410; Müller, *Das Tor Zur Weltmacht,* 143–147.

78. Habeck, *Storm of Steel,* 82–83; Eyck, *A History of the Weimar Republic,* 291, 293; Müller, *Das Tor Zur Weltmacht,* 147–148.

79. Habeck, *Storm of Steel,* 97, 166–167; Dyakov and Bushuyeva, *The Red Army and the Wehrmacht,* 20–21, 167–175.

80. Isserson, "The Evolution of Operational Art," 114–122; Guderian, "Kraftfahrkampftruppen," 75–77; Guderian, "Die Panzertruppen und ihr Zusammenwirken," 623–626.

81. Corum, *The Roots of Blitzkrieg,* 70.

82. *Verordnung über die Ergänzung des Heeres, den 4.Dezember 1920,* from Division of Military Intelligence, US War Department, memorandum transmitting the above document to War Department Assistant Chief of Staff, G-2, April 26, 1920, Records of the War Department General Staff, Military Intelligence Division Correspondence, 1921–1942, Record Group 165, NACP.

83. Division of Military Intelligence, War Department, subject: Duties of a German Soldier—Service Report, April 13, 1922, translation of *Reichswehrministerium Heeres-Verordnungsblatt,* Berlin, March 2, 1922, Records of the War Department General Staff, Military Intelligence Division correspondence, 1921–1942, Record Group 165, NACP; Skinner, "Transformation of the German Reichsheer," 44.

84. Division of Military Intelligence, US War Department, subject: The German Army—Depot Battalions, monograph, February 1, 1921, Reichswehr Order no. 1801/12.20, T 2 II, Berlin, January 12, 1921, Records of the War Department General Staff, Military Intelligence Division correspondence, 1921–1942, Record Group 165, NACP; Corum, *The Roots of Blitzkrieg,* 72–73.

85. Craig, *Germany,* 187.

86. Tessin, *Formationsgeschichte der Wehrmacht,* 21, 108; Absolon, *Die Wehrmacht im Dritten Reich: Band I,* 26.

87. Corum, *The Roots of Blitzkrieg,* 70.

88. Ibid.

89. O'Neill, *The German Army and the Nazi Party,* 84; Corum, *The Roots of Blitzkrieg,* 176.

90. General Walther von Lüttwitz (1859–1942), commander of the Berlin regional forces of the *Reichsheer,* occupied Berlin and proclaimed a new government with Wolfgang Kapp (1858–1922), an obscure right-wing politician as chancellor. The coup lasted four days, from March 13 to 17, 1920, and was defeated by the combined refusal of the rest of the army to abandon the Ebert government and by a general labor strike called by the legitimate government. General von Seeckt did not order the army to crush the coup and refused to sanction military action to cause army forces to fire upon other army forces in Berlin in support of the Ebert government. It was the national strike, and it was the failure of the civilian population to support the coup that brought it to an end. Von Rabenau, *Seeckt,* 233–235.

91. Craig, *The Politics of the Prussian Army,* 377–381, 416–417.

92. Ibid., 419–422.

93. Von Rabenau, *Seeckt,* 174

94. Deist et al., *Ursachen und Voraussetzungen der Deutschen Kriegspolitik,* 372; Citino, *The German Way of War,* 241.

95. Von Rabenau, *Seeckt,* 503–505 (Seeckt critique of 1922 maneuvers: *Das man sich von Fesseln des Stellungskrieges freimachen müsse und mehr vollig losgelöste Operationen des Bewegungskrieg üben solle*); Citino, *The German Way of War,* 242; Corum, *The Roots of Blitzkrieg,* 70; Skinner, "Transformation of the German Reichsheer," 26; Bruno Thoß, "Vom Führerheer zur Wehrmacht," 7.

96. Echevarria, *After Clausewitz,* 213.

97. Groß, *Mythos und Wirklichkeit,* 40–41. Moltke the elder's concept of leader initiative in decentralized execution of the attack into Bohemia in 1866; Hughes, *Moltke on the Art of War,* 45–46, 132–133; Echevarria, *After Clausewitz,* 38–42; English and Gudmundsson, *On Infantry,* 63.

98. Echevarria, *After Clausewitz,* 32–33; Brose, *The Kaiser's Army,* 22–23, 56–57.

99. Echevarria, *After Clausewitz,* 32–33.

100. Leistenschneider, *Auftragstaktik im preußen-deutschen Heer,* 57–123; Groß, *Mythos und Wirklichkeit,* 64–71; Echevarria, *After Clausewitz,* 32–42; Brose, *The Kaiser's Army,* 58–61; Spohn, *Ratschläge und Winke für die praktische Ausbildung und Schulung,* 28, 104–106.

101. Sonnenberger, "Mission Command During the War of Movement," 44; Groß, *Mythos und Wirklichkeit,* 81–83.

102. Hamilton and Herwig, *War Planning,* 66–73.

103. Brose, *The Kaiser's Army,* 154, 188–203.

104. Echevarria, *After Clausewitz,* 214.

105. Groß, *Mythos und Wirklichkeit,* 136–139; Lupfer, "The Dynamics of Doctrine," 43–49; Brose, *The Kaiser's Army,* 239; English and Gudmundsson, *On Infantry,* 29–31; Citino, *The German Way of War,* 240–44; Dupuy, *A Genius for War,* 212–215.

106. BAMA ZA 1/1778, Personnel and Administration (Project 2b), P-006, Teil I, *Das Rekrutierungssystem der deutschen Wehrmacht in Frieden und im Kriege;* Anlage I, *Durchführung einer Musterung und Aushebung,* 1948, 121.

107. Seeckt, *Thoughts of a Soldier,* 73–75; and Seeckt, *Gedanken Eines Soldaten,* 109–111.

108. Von Rabenau, *Seeckt,* 505. Friedrich von Rabenau was the head of the army's history archives when he wrote the biography of General Seeckt in two volumes, concluding with the second in 1940. His sources included both official and personal correspondence. Von Rabenau unabashedly sought to connect the reforms and ideas of Seeckt to the successes of the German Army between 1939 and 1940, and to tie Seeckt to support of Adolf Hitler. Thus von Rabenau often ventures into interpretations of the notes, letters, and speeches of Seeckt in order to make a political connection to the Nazi regime. With this caution in mind, and understanding that Seeckt did not disavow the Hitler regime nor voice any disapproval of National Socialism, the biography provides a unique window into the actual words written by Seeckt about the politics of his time in office, as a primary participant in the final acceptance of the Versailles Treaty, and his position as leader of the German Army between 1919 and his resignation under duress in the fall of 1926. This is especially so concerning his correspondence as the leader of the *Reichswehr* and the preparation of the German Army for the war to come. Significant references to Seeckt's correspondence from this period are also present in English translation in Carsten, *The Reichswehr and Politics.*

109. Seeckt, *Thoughts of a Soldier,* 73; Seeckt, *Gedanken Eines Soldaten,* 109.

110. Von Rabenau, *Seeckt,* 504.

111. Craig, *Germany,* 684; Dyakov and Bushuyeva, *The Red Army and the Wehrmacht,* 20–21; Corum, *The Roots of Blitzkrieg,* 67, 118–119.

112. Groehler, *Selbstmörderische Allianz,* 39–45, 50–51; Müller, *Das Tor Zur Weltmacht,* 143–145.

113. Groß, *Mythos und Wirklichkeit,* 171–172; Condell and Zabecki, *On the German Art of War,* 2–10; Lewis, *Forgotten Legions,* 34; Corum, *The Roots of Blitzkrieg,* 199.

114. H.Dv.300, Truppenführung, Teil I, 1–4, 127–134, Record Group 242, T-78, Roll 201, Frame 6144931–6144933 and Frame 6144995–6144999, NACP; D.V. Pl. Nr. 487, *Führung und Gefecht der verbundenen Waffen,* 29–30.

115. D.V. Pl. Nr. 487, *Führung und Gefecht der verbundenen Waffen,* 29.

116. USAMHI, MS # B-649; Guenther, *Training Regular Army NCOs,* 4.

117. Reinicke, *Das Reichsheer,* 311; Skinner, "Transformation of the German Reichsheer," 27.

118. BAMA ZA 1/1785, Personnel and Administration (Project 2b), Teil II: *Ausbildung und Verwendung der Unteroffiziere und Mannschaften im deutschen Heere,* 1949; Anlage I, 3.

119. Van Creveld, *Fighting Power,* 121–124.

120. Ibid.

121. BAMA ZA 1/1785, Personnel and Administration (Project 2b), Teil II: *Ausbildung und Verwendung der Unteroffiziere und Mannschaften im deutschen Heere,* 1949, 35.

122. Reinicke, *Das Reichsheer,* 310–312.

123. *Fahnenjunker* (officer-aspirant), *Fahnenjunkergefreiter* (officer-aspirant lance corporal), *Fahnenjunkerunteroffizier* (officer-aspirant corporal), *Fähnrich* (ensign), and *Oberfähnrich* (senior ensign) all fell under the title *Offizieranwärter* (officer candidate), and these were all steps in the progression of an officer-aspirant as he moved from his enlisted grade to lieutenant or second lieutenant equivalent.

124. Skinner, "Transformation of the German Reichsheer," 48–49.
125. The addition of these tests began in the British Army in 1942, just when the tests were halted in the German Army. Luvaas, *Frederick the Great on the Art of War.*
126. BAMA ZA 1/1779, Personnel and Administration (Project 2b) P-005, Teil V, Dr. Max Simoneit, *Die Anwendung psychologischer Prüfungen in der deutschen Wehrmacht, Deutsche Wehrmachtpsychologie von 1927–1942,* 1948, 30–32; Vernon, *Personnel Selection in the British Forces,* 22.
127. BAMA ZA 1/1779, Personnel and Administration (Project 2b) P-005, Teil V, Generalmajor Hellmuth Rheinhardt, *Vorbemerkungen: Die Anwendung psychologischer Prüfungen in der deutschen Wehrmacht, Deutsche Wehrmachtpsychologie von 1927–1942,* 1948, 30–32.
128. BAMA ZA 1/1779, Personnel and Administration (Project 2b) P-005, Teil V, Dr. Max Simoneit, *Die Anwendung psychologischer Prüfungen in der deutschen Wehrmacht, Deutsche Wehrmachtpsychologie von 1927–1942,* 1948, 56–71.
129. BAMA RH 26-17/96, *Merkblatt für Eintritt als Fahnenjunker in das Reichsheer, Reichsministerium—Heersleitung,* January 1935.
130. BAMA RH 26-17/96, OKH Aktz: 22 b 16 PA 1 (A), *Vorlage der Bewerberlisten für Übernahme von Nicthabiturenten aus der Truppe in die Offizierlaufbahn,* Berlin, May 8, 1936.
131. Absolon, *Die Wehrmacht im Dritten Reich: Band III,* 213.
132. Reinicke, *Das Reichsheer,* 64–65.
133. Von Rabenau, *Seeckt,* 489.
134. O'Neill, *The German Army and the Nazi Party,* 16.
135. Ibid., 21.
136. Ibid., 38.
137. O'Neill, *The German Army and the Nazi Party,* 39–40.
138. Ibid., 50.
139. Ibid., 58.
140. Müller-Hillebrand, *Das Heer bis zum Kriegsbeginn,* 18–20.
141. Tessin, *Formationsgeschichte der Wehrmacht,* 8; Müller-Hillebrand, *Das Heer bis zum Kriegsbeginn,* vol. 1, 21–25.
142. O'Neill, *The German Army and the Nazi Party,* 11–12.
143. Craig, *The Politics of the Prussian Army,* 467.
144. Von Altrock, "Politizierung der Wehrmacht."
145. Von Altrock, "Hitler," 1301.
146. Von Altrock, "Der Soldat und die nationale Revolution."
147. Ibid., 211.
148. Von Altrock, "Politizierung der Wehrmacht."

PART 2

Building Combat Divisions: From Conscription to War

5

The United Kingdom: Fighting a Defensive War

WINSTON CHURCHILL PROCLAIMED THE MUNICH AGREEment of September 29, 1938, an appeasement of Hitler and Germany, "a total and unmitigated defeat."[1] In his subsequent speech before the House of Commons on October 5, 1938, he took note of the joy of the British people in an outcome that prevented war for the moment. He challenged that "they should know the truth. They should know there has been gross neglect and deficiency in our defenses; they should know that we have sustained a defeat without a war, the consequence of which will travel far with us along our road . . . unless by a supreme recovery of our moral health and martial vigor, we arise again and take our stand for freedom as in the olden time."[2]

Churchill was among those in the United Kingdom who recognized a likelihood of renewed war in the preparations apparent in Germany following the Nazi *Machtübernahme* (takeover) by Adolf Hitler and the Nazis in 1933–1934.[3] Coming to power with a vow to restore Germany and end the humiliation of the Versailles Treaty of 1919, Hitler reintroduced conscription in 1935, unveiling a new air force, and threw out all restrictions on German military forces imposed by the Versailles Treaty, which ended the Great War. Germany appeared resurgent both politically and militarily. Events multiplied over the ensuing years to build a growing awareness and concern in the United Kingdom. These included the return of the Saar region to Germany by plebiscite in 1935, and the military reoccupation of the Rhineland in 1936. The annexation by consent of Austria in March 1938, along with the September 1938 Munich settlement revoking French and British diplomatic guarantees and dismembering an allied Czechoslovakia with the German occupation of the Sudetenland, was in Churchill's words "an awful milestone in our history, when the whole equilibrium of Europe has been deranged."[4] Totalitarian and fascist regimes appeared on the march, with the Italian aggression in Ethiopia from 1935, as well as the ongoing Japanese attacks on the Chinese mainland since 1931. German violation of the Munich Agreement and the occupation of Bohemia and Moravia on March 15–16, 1939, and the plebiscite rejoining the Memel Land to Germany on March 23, 1939, accompanied Prime Minister Neville Chamberlain's decision to institute a new conscription law in the United Kingdom.[5] Fascist success in the Spanish Civil War in April 1939 and the Italian invasion of Albania in April 1939 further awakened the British government and people to the necessity to rearm as they held themselves in readiness once again for the specter of war.[6]

Manpower Selection and Allocation

Responding to the growing alarm over the various continental and world security crises, Chamberlain issued a letter to the nation, a call to national service, in January 1939. He emphasized: "The call is to peace not to war. We have no thought of aggression; our one wish is to live at peace with all peoples. But if this wish is to be fulfilled we must be up and doing. We must make ourselves strong so that our influence for peace may be real, and we must make ourselves safe so that others cannot be tempted to thoughts of aggression against us."[7]

Voluntary military service, however, among the men whose fathers fought and died in France, weakened in the postwar years, given their reaction to the horrifying casualties of the Great War. Economic uncertainty, exacerbated by the effects of worldwide economic depression in the 1930s, continued to keep the army from meeting its manpower goals, and prevented substantial modernization.[8] Also in January 1939, in anticipation of a general war that entailed the mobilization of the manpower of the nation in both military service and industry, the Ministry of Labor issued its schedule of reserved occupations. The intent was "to prevent the indiscriminate calling up of men who were urgently needed in industry and whose loss to industry might seriously cripple the production of munitions," presuming the resumption of conscription.[9] The War Office appointed its own director of mobilization at this time, who became "the link between the War Office and the Ministry of Labour [which was] concerned with the supply distribution of manpower for the country as a whole."[10] With the schedule in place, the government sought to avoid the mistakes of World War I, when it allowed the unrestricted enlistment of men with critical skills.

While the published schedule of reserved occupations did not apply to either the existing regular or Territorial Army or new recruits until May 1939, it established the principles of national manpower management for the renamed Ministry of Labor and National Service from the beginning of any authorization for conscription. While inducting just 35,000 men prior to mobilization at the outbreak of war, the ability to test and resolve difficulties between the War Office and the Ministry of Labor for three months prior to the declaration war proved invaluable.[11]

Foreseeing an elevated need for skilled manpower, "tradesmen," in the services of the coming war, the armed forces and the ministry predicted an extraordinary reliance on the mechanization of the forces and on the technology needed to support these requirements. Skills anticipated included electrical and mechanical engineering in areas from radio communications and transport drivers to radar operators, anti-aircraft gunners, and tank crews, among many others. During the war, this list grew to more than 500 recognized trades in all arms and services.[12] In its struggle for priority against the needs of the Royal Air Force and Royal Navy, the postwar report of the army's efforts acknowledged that "the Army's search for tradesmen throughout the war may be briefly described as the insatiable in pursuit of the unobtainable. There were never enough tradesmen to meet every demand, and no subject was the source of more complaint, misunderstanding and misrepresentation."[13] The retention in industry of tradesmen further diluted the quality manpower available to an infantry arm already sharing with the civil defense forces and the other fighting services in the priority accorded the enlarged, tradesmen-heavy needs of the Royal Air Force and Royal Navy.[14]

Following his call to national service, Chamberlain proclaimed the decision to double the Territorial Army on March 29, 1939. This was a demonstration of resolve that the United Kingdom was serious about rearming. This gesture followed in time Germany's overthrow of the Munich Agreement of 1938 and the dismemberment of

Czechoslovakia with the occupation of Bohemia and Moravia on March 15–16, 1939.[15] This action of the prime minister was neither foreseen nor welcomed by the army, but deemed a political necessity. The government hoped to deter Germany with a demonstration of British resolve and seriousness. While having no discernible effect on Germany, its principal effects were unfortunate for the British Army. This political decision disrupted any attempt at a rational expansion and training that extended well into the war, as "the whole organization of the Territorial Army was violently disrupted a few months before the war began."[16]

With an increasing awareness of the unpreparedness of the United Kingdom's armed forces for possible war, the Chamberlain government passed the Military Training Act on May 26, 1939, the first peacetime conscription law in the history of the United Kingdom. This measure demonstrated a tangible effort backing up guarantees to France as well as Poland. Its endurance was for a period of three years and created a limited form of compulsory service that applied to male British subjects between ages twenty and twenty-one, with a training period of six months and status as an enlisted man (other ranks) in the militia.[17] The assignment as militia meant that those conscripted were not available for service outside the United Kingdom.[18]

The Infantryman: Characteristics and Expectations

The field service regulations of the British Army provided the guide that articulated the distinct character sought in combat soldiers. The principles articulated in these documents represented both the accumulated historical understanding of past experience and anticipation of the requirements of future war. Doctrine, it noted, "should be regarded by all ranks as authoritative," and that "in this way only will subordinates be able to act in accordance with their superior's views and decide on the right course without frequent reference to him, and even without receiving his orders."[19]

Regulations divided the soldiers into two categories: the fighting troops, or arms, and those supporting them, the services. The fighting troops were those intended to carry the fighting in close combat in the offense or defense, and included the "armoured troops, cavalry and infantry; the latter, artillery, engineers and signals."[20] The army defined a "service" as "a branch of the army whose primary duty is to supply the fighting troops with what they require, or to perform some necessary service for them."[21] These included personnel, supply, medical and veterinary, graves, and transportation, among many others.

The infantry soldier was of paramount importance. Doctrine described the essential characteristics desired of the individual infantryman, both as an individual, and for his role as a member of a group in the cohesive behavior necessary to survival in battle in the small unit. The battle experience of the Great War led the doctrine writers of the interwar period to state the requirement, in terms of the maximum economy in manpower management, that the highest-quality men, destined for action in direct contact with an enemy, men with the highest physical fitness and discipline, be reserved for a combat role. The army intended to reserve for duties outside of combat those positions demanding less physical fitness and military training.[22] Remembering the painful losses of that war, the army emphasized a future war fought with a maximum of firepower and technology and a minimum of manpower.[23] The army focused its primary organization for mobile warfare in support of offensive action in a manner reminiscent of the victorious movements in the final months of 1918.[24]

The End of Peace and Defense of the Realm

On September 3, 1939, a state of war began between the United Kingdom and Germany. Two days after the German invasion of Poland on September 1, 1939, a more far-ranging compulsory military service became law with the passage of the National Service (Armed Forces) Act and changed the Ministry of Labor to the Ministry of Labor and National Service. The act provided for enlistment for the duration of the emergency and made males from ages eighteen to forty-one eligible for service under the act. Further acts passed on September 3, 1939, included the Military and Air Forces (Prolongation of Service) Act, and the Armed Forces (Conditions of Service) Act, authorizing the transfer of soldiers to any corps, as well as the authority to call up any reservist or member of the Territorial Army for the duration. These laws provided the authority to impose the liability for coerced service on all males ages eighteen to forty-one, requiring males aged between twenty and twenty-three to register by October 31, 1939. Amendments passed in 1940, 1941, and 1942 both authorized the eligibility of women for conscription and moved the eligible age of conscripted service to eighteen for males.[25]

Together with the schedule of reserved occupations, the acts gave the government complete control over manpower. The principal feature of the management of personnel assets was the centralized nature of the British system—the needs of the national war effort overrode any commitment to a regional or regimental method of recruitment or assignment, or the wishes of any recruit, despite the continuation of voluntary enlistment alongside conscription.[26] The foremost requirement of the system was the need for an objective identification of measurable skills. Technical and trade experience were the most amenable to quantification and classification for the purpose of personnel placement, rather than the subjective personality intangibles of potential combat aptitude such as those required in infantry battle.

The Ministry of Labor and National Service was the central agency for the management of manpower for all national requirements. Within the ministry, its National Service Department was responsible for the administration of the schedule of reserved occupations, "which shows for each occupation in wartime an age of reservation from military service."[27] Its Military Recruiting Department's main functions were the registration of those liable for general service as well as selection of tradesmen not deferred in the schedule of reserved occupations and all volunteers. Thus a man denied the opportunity to volunteer early received his call-up later in the war through conscription as an outcome of changing requirements.

Control of the intakes for the army passed to the Ministry of Labor and National Service within induction and recruiting centers established across the country for this purpose. A War Office recruiting officer administered each center, with Royal Air Force and Royal Navy recruiters located in the larger centers. The Ministry of Labor's divisional controller, assisted by the chief recruiting officer, posted individuals to a corps in accordance with demands issued by the War Office for the various corps. The corps included all branches, such as the Royal Artillery, Royal Engineers, Royal Armored Corps, and Royal Army Service Corps, as well as Royal Army Ordnance Corps, among many others. Each created its own criteria for the classification, selection, and assignment of persons to its corps. These were then consolidated by the army and sent to the Ministry of Labor and National Service induction centers to fill the quotas necessary to man the corps. There was no centralized infantry corps organization to fulfill this function—that is, there was no Royal Infantry Corps.[28]

Civilian medical boards performed the entry physical examinations. Military officers conducted individual interviews with those found physically qualified and made recommendations for the service and branch within each service best suited for each individual. Selected men were then posted to the arms and services according to the quotas and prescriptions provided by the service departments as indents. Across the nation, 114 local offices carried out these activities, with medical examination boards consisting of five doctors determining medical fitness, with one or more boards associated with each office. The law allowed for the right of individuals to express their preference for naval or air service and to be granted such posting if qualified for the assignment and a vacancy existed. The boards also adjudicated all cases of conscientious objectors and hardship deferments.[29]

The male strength of the British Army (see Table 5.1) on the day of the German attack on Poland was impressive, considering its strength a few months earlier in 1939. Of the totals listed in the table, 40,000 were in India and Burma, 31,000 in the Middle East, and 10,000 in Mediterranean stations,with additional potential personnel available on September 1 on active duty.[30]

The system attempted to keep national groups together, such as Welsh, Scottish, and Northern Ireland recruits. Strict adherence to the territorial recruitment by regimental affiliation proved challenging, though recruits with a familial background with certain regiments had some likelihood of assignment.[31] Throughout World War II as in the World War I, the system functioned to send men to whatever army formations held the greatest need. This condition led to an increasing consolidation of the selection process in three stages. First, in June 1940, with the expulsion of the British Expeditionary Force from France, 120 infantry battalions were formed at home, and were assigned to regiments without equipment due to the losses incurred in the evacuation from France. Second, in July 1941, the army consolidated the regimental training depots. And third, in July 1942, the General Service Corps came into effect to the end of the war. At each stage, the army retreated from the prewar regimental system to one entirely focused on the central control of its manpower, more and more posting men regardless of any regimental or regional consideration.[32]

Table 5.1 Male British Army Assets and Potential Assets, September 1, 1939

Detail	Serving with the Colors (including those called up or embodied): Total Officers and Other Ranks	Unembodied (those still in reserves, Territorial Army, etc.): Total Officers and Other Ranks	Total Strength: Officers	Total Strength: Other Ranks	Total Strength: Total
Regular army	224,188	—	14,552	209,636	224,188
Militia	34,542	—	—	34,542	34,542
Regular army reserve	28,742	111,768	9,414	102,354	111,768
Supplementary reserve	1,096	43,465	2,666	41,895	44,561
Territorial Army	113,037	315,035	18,879	409,193	428,072
Territorial Army reserve	—	20,824	7,776	13,048	20,824
Miscellaneous officer cadets and reserves	—	—	10,640	—	10,640
Colonial and Indian troops (with British Army formations)	8,174	14	177	8,011	8,188
Total	410,041	543,262	105,882	847,421	953,303

Source: Pigott, *Manpower Problems*, 79.

First Consolidation of Recruiting and Induction

What began as a benign atmosphere for recruiting in 1939 ended with the calamitous defeat of the British Expeditionary Force in France in 1940. Until then, the priorities were adjusting to the disorganization caused by the doubling of the Territorial Army; training of new intakes; drafting, or the mobilization and transport, of whole formations and replacements for the British Expeditionary Force across the English Channel; and releasing those demanded by industry. The army processed and discharged thousands embodied during the breakneck expansion of the Territorial Army in 1939, but found unfit due to age or medical condition. The War Office and the Ministry of Labor and National Service planned the coordination of induction calls on the projected availability of equipment for new formations and the replacement of men lost through wastage.[33] Before May 1940, equipment and munitions shortages restrained the intake of personnel and were the key factors that circumscribed the British Army's war preparations in support of battle on the continent.[34] Despite earlier measures to identify and prevent the induction of men deemed essential as tradesmen for war industry, the army released 65,000 men back to industry between September 1939 and February 1941, demonstrating an ever-increasing drain on the army as the needs of essential war industry, agriculture, and coal mining combined to reduce the pool of available manpower.[35]

The army's state of training was poor. Postwar the army noted that "mere figures are illusory and the great majority of our so-called assets were sadly lacking in training and equipment. . . . In appraising military values other things must also be taken into account, and however good the human material, it cannot be claimed that Great Britain entered the war with military forces of strength, armament and equipment worthy of her prestige and responsibilities as one of the world's leading powers."[36]

The British Expeditionary Force was the primary field force of the British Army until June 1940. Five infantry divisions composed the force. There were five first-line territorial infantry divisions, while three territorial divisions came into France for labor duties and without heavy weapons or equipment.[37] South of the Somme River, a territorial infantry division landed and then evacuated in June as part of an abortive rebuilt British Expeditionary Force.[38] An armored division arrived between May 14 and 21, 1940, never fought as a complete unit, never linked up with the main body of the original British Expeditionary Force, and confined to fighting south of the Somme River, evacuated in June 1940. A highland division deployed near the Maginot Line between April 30 and May 6 was unable to reach the English Channel coast in time to avoid capture. With the exception of 4,000 men, the highland division surrendered to the German Army at Saint-Valéry-en-Caux and reconstituted in England following the battle.[39]

Operation Dynamo, the evacuation at Dunkirk, succeeded in removing over 338,000 British and French soldiers, while the lesser-known Operation Ariel evacuated an additional 215,000 from Cherbourg and St. Malo. Operation Cycle rescued another 11,000 from Le Havre. The British Expeditionary Force left behind 3,500 killed, 40,000 prisoners of war, 88 percent of its artillery, and 93 percent of its vehicles and armor. The year 1940 began a long series of defeats for the British Army not alleviated until late 1942.[40]

Upon the return of the British Expeditionary Force from France, the army removed limitations on intakes, regardless of equipment availability. Enlisting the maximum number of men for the defense of the home islands became a strategic and political imperative.[41] During the three months of June, July, and August of 1940, 324,000 men enlisted.[42] With little or no equipment, 120 additional infantry battalions formed to house these men until sufficient training facilities became available.

The expulsion of the British Expeditionary Force and the need to prepare for a feared invasion of the home islands further disorganized the British Army. With the sudden creation of so many infantry battalions, the army set aside plans to build a mobile, balanced, and mechanized force of all arms due to the immediate threat. The later wholesale conversion of infantry battalions to Royal Armored Corps, Royal Artillery, and other corps occurred as equipment became available and the war economy began to increase the output of munitions and vehicles. The army and its political leaders recognized by early 1941 that there was insufficient manpower to create an army capable of defeating the German Army on the continent, setting the initial ceiling for the British Army at 2.1 million personnel. The strength of the army rose to almost 2.9 million by war's end.[43] Thirty-five British Army divisions, fifteen deployed in some form with the British Expeditionary Force in France, contrasted with 135 of 156 German Army divisions available for combat in France by June 1940. This disparity illustrated the inability of the United Kingdom to defeat Germany without allies after the fall of France.[44]

The United Kingdom on the Defensive

From its return to the home island in June 1940, preparation for the defense of the British Isles from invasion was the army's strategic priority. Precedence went to naval and air defense. Britain found itself with few ground weapons remaining, and industry was unable to replace the losses from the British Expeditionary Force due to the higher naval and air priority. The Battle of Britain in the air and combat against the German Navy highlighted the period from June 1940 through June 1941. With the invasion of the Soviet Union on June 22, 1941, the German air offensive against the United Kingdom ended as the German Air Force concentrated on defending against the increasing Royal Air Force bomber offensive and fighting the war in the east. The destruction of German naval surface ships such as the *Graf Spee* and the *Bismarck* highlighted the importance of the Royal Navy in maintaining the sea lanes against the ever-increasing success of the German U-boat arm in contesting British control of the Atlantic and Mediterranean seaborne routes of supply.

Small British and Dominion forces in the Western Desert of North Africa, in protection of the Mediterranean and the Suez Canal, defeated the Italian Army in December 1940. With German reinforcements under General Irwin Rommel, the German *Afrika Korps* succeeded during the period from January 1941 through a series of back-and-forth ground, air, and sea battles in forcing the British to retreat in North Africa to defenses at El Alamein by July 1942. These defeats, combined with the expulsion of British Army forces from Greece and Crete in 1941, when added to the defeats in Norway and France in 1940, brought the British Army to a low ebb, further pummeled by the loss of Singapore and Malaya and the capture of the bulk of British and Dominion soldiers, over 130,000 prisoners of war, as well as the destruction of air and naval forces by the Japanese Empire in the Far East by May 1942.

The entry of the United States into the war following the December 7, 1941, Japanese attack on Pearl Harbor, followed by Adolf Hitler's declaration of war against the United States on December 10, 1941, brought renewed optimism to the British people. Public opinion of the British Army and its ability to fight was poor. With the mass of its soldiers remaining at home to repel invasion, the bulk of what appeared to be idle soldiers in perpetual training reflected on both the quality and the performance of the army in relation to the Royal Navy and Royal Air Force.[45]

Manpower Shortage

The two critical factors in the selection of infantry (other ranks) in the British Army were, first, the relative limits on the numbers of men available for all of the armed services of the United Kingdom; and second, the relative priority assigned to the army in relation to the Royal Navy and Royal Air Force. Winston Churchill acknowledged in March 1941 the diminished strategic role of the British Army following the defeat in France:

> [The size and strength of the German Army] make it impossible for the Army, except in resisting invasion, to play a primary role in the

> defeat of the enemy. That task can only be done by the staying power of the Navy, and above all by the effect of Air predominance. Very valuable and important services may be rendered overseas by the Army in operations of a secondary order, and it is for these special operations that its organization and character should be adapted.[46]

Despite a stated ceiling, as war requirements and strategy changed, so did the army's size. Prime Minister Churchill chided the army leadership for failing to economize the numbers of men sought in relation to the overall war effort, noting his belief that it took 60,000 men to support thirty-four divisions, and the poor comparison to the German enemy with 237 divisions and less than a third of the support manpower of the British forces. He remarked on the small number of fighting divisions relative to overall strength of the army by December 1940 that "we really ought to try to do better than this. It is certain our war effort cannot afford such lush disorganization. Before therefore I can ask the Cabinet to assent to any further call-up from the public, it is necessary that this whole subject be thrashed out, and that at least a million are combed out of the fluff and flummery behind the fighting troops, and made to serve effective military purposes."[47] While the German *Wehrmacht* numbered over 5 million men in June 1940, of which the army, or *Heer,* numbered well over 4 million, the actual number of German divisions stood at 156 and grew to 208 by the opening of the offensive against the Soviet Union in June 1941.[48]

Despite the strategic priority afforded the roles of the Royal Navy and Royal Air Force, the British Army was, and remained, the largest of the armed forces, growing from 930,000 under arms in October 1939 to an initial ceiling of 2.1 million in 1940 and 2.3 million by October 1941, and to 2.9 million by the end of the war. In contrast, the Royal Navy reached 857,000 men, and the Royal Air Force just over 1 million, by 1945.[49] The strategic priority at this stage of the war, however, lay outside the army and its roles and missions. This reality reflected in the quality of the infantry, as "the Army had even greater difficulties than the Royal Navy or [Royal Air Force] in meeting its skilled manpower requirements, since it attracted a smaller proportion of men of high intelligence and education, and being unable to reject any except the medically unfit at recruiting centres, it had to take in a considerable portion of very dull men."[50]

Building the Combat Divisions

From the opening of the war, shortage dominated the contest between industry, agriculture, and the armed forces for manpower. The army found itself at the end of the line for manpower in general, and quality manpower foremost. The strength of the army was a field force of thirty-three divisions. This nominal strength was limited by the inclusion of twelve second-line Territorial Army divisions from the sudden doubling of the Territorial Army by Prime Minister Chamberlin in March 1939. These formations, lacking equipment, existed on paper, and four disbanded by the end of 1940. The British Expeditionary Force went to France with fifteen divisions between September 1939 and June 1940, consisting of five regular infantry and five first-line Territorial Army divisions as its field force for combat operations. Three second-line Territorial Army divisions accomplished labor and line-of-communications duties. One additional first-line Territorial Army division and an armored division arrived too late for the fighting and evacuated France during June 1940.

Ten home defense county divisions were formed in 1940 and disbanded by the end of 1941 due to lack of mission, equipment, and personnel. The average number of divisions, through creation, loss in combat, and disbanding due to manpower and equipment shortages, was thirty-five until 1944. "By the end of 1941 the total number of British field force divisions stood at one airborne, nine armoured, and twenty-seven infantry, plus eleven independent tank brigades. In 1942 another armoured division (79th Armoured) was added, as were two further tank brigades and the 78th Infantry division."[51]

The year 1942 began the process of expedients needed to maintain nominal combat strength in the field forces. Lack of equipment caused the placement of six infantry divisions on a lower establishment basis, denying them any further combat role. By the end of 1944, the strength of the British Army in northwestern Europe was five armored divisions, seven independent armored brigades, and twenty-one infantry divisions.[52] While thirty-three infantry divisions were created in the course of the war, sixteen remained in the field force at the end of the war. Eleven armored divisions came into existence during the war and five remained at war's end due to the effect of the manpower shortages and need to keep the field force units at full strength.[53]

Testing and the Selection Interview

From the opening of the war, the interview by a single army officer held at the induction center, following a medical examination, determined the corps assignment given to each individual. The interview established the recruit's educational level, based on his level of schooling attained, as well as any claimed occupational skills or aptitudes. The interview, from March 1939 until the introduction of standard testing during 1942, remained the primary method for verification and determination of both skill or occupational trade experience, or potential, and education standard or intelligence estimate for recruits.

Yet there remained general dissatisfaction with the interview as a sound basis for the discovery of the capabilities of soldiers as

> the only two pieces of information on which much store could be set were the medical category and the type of schooling. In practice, the interview tended to be even more unsatisfactory than, in principle, it was bound to be. Neither recruiting officers, nor unit officers, were trained in the difficult art of interviewing. Nor was there any "book of words" to ensure the accurate interpretation and uniform application of judgments in a uniform manner. All the above factors, in combination, led to a considerable amount of misplacement which caused mounting concern both within and without the Army during the period 1939–1941.[54]

For those assigned to the army, the medical category meant little, as three-quarters of the military population of recruits sorted into the highest medical category, and a like number just elementary schooling.[55] The interview thus served as the principal basis, however flawed, for the classification and assignment of three-fourths of the army's soldiers entering the service prior to July 1942.

Intelligence Tests

Continued unhappiness with the interview process prompted a heightened interest in a better method of determining the likelihood of success in training for individual recruits. Early experiments began after the outbreak of the war in the Anti-Aircraft Command to devise tests that gave a more accurate assignment of soldiers to their duties and weapons. Following these tests in the spring of 1940 was the implementation of further psychological methods of selection designed to test intelligence and mechanical aptitude at selected training regiments. The Army Educational Corps, which administered a test to aid in the determination of the educational status of recruits before the war, assisted in the analysis of the results from these experiments.[56]

Criticism of this initial series of tests[57] suggested that the tests focused too much on literacy rather than intelligence and failed at "differentiating between the inferior and the markedly dull. There was some substance in the criticism, although subsequent research has indicated that, in fact, educational backwardness relates almost as closely to military ineptitude as low intelligence. . . . Indeed, since it was the dull who presented the units with their most insoluble difficulties, some units referred all men whose results were outstandingly low to the psychiatrist for his opinion."[58]

To correct this defect, the army introduced a progressive matrices test. Designed to be independent of literacy or education, it demonstrated the ability to assist in the diagnosis of mental defect and nervous disorder and was therefore superior to the initial series of tests. Called the matrix test, it was first used on a large scale, following recommendations made to the newly established Directorate of Selection of Personnel in June 1941. Its advantages were threefold:

> First, it had been devised for the specific purpose of helping to diagnose the mentally defective, and was, therefore, particularly appropriate when a great deal of the Army's manpower difficulties seemed to centre around the problem of the untrainable recruit.
>
> Secondly, it was a non-verbal test so designed as to be independent of educational standard . . . likely to be regarded as giving a fair chance to all.
>
> Thirdly, it had been found in civilian life that intelligence tests of this kind gave a good estimate of how rapidly and easily a man would acquire new knowledge.[59]

The basic principles for personnel selection in the army centered on the ability to facilitate identification of the problems of personnel through the use of applied psychology. In particular, selection procedures best identified the "definitely unsuitable" and the "eminently suitable," but focused on the rejection of those deemed most unsuitable to service. Acknowledging the value of the outcomes of the tests per se, their usefulness depended upon other subjective evidence in view of a man's personality and physicality, as well as his education and occupational experience. Of particular importance, as told in the postwar study on personnel selection, was the concept that "data obtained solely from test scores provide valid evidence on group tendencies, but not on individuals. For example, a man may not be a good soldier because he reached a high level on a test battery. A unit, however, which possesses a high proportion of men of good ability is usually a good unit. If it possesses a high proportion of low-grade men, it is still more likely to be a poor unit. Taken over large groups, then, the tests are pretty reliable."[60]

This latter point introduced into the army the concept of the "selection group." This provided a means to gain an overview of individual qualifications through comparison of differing tests with different maximums, averages, and ranges. The selection groups used by the British Army were as follows:[61]

Group 1: Extremely good (top 10 percent of the army population).
Group 2: Fairly good (next 20 percent below the top).
Group 3: Good average (the 20 percent just above average).
Group 4: Poor average (the 20 percent just below average).
Group 5: Fairly poor (the 20 percent above the sixth group).
Group 6: Extremely bad (bottom 10 percent).

Ministry of Labor and National Service began administration of the matrix test in a twenty-minute form, during August 1941, at its induction centers. This short version of the test was a compromise with the Ministry of Labor due to its reluctance to allow any interference with the procedure of the medical induction boards. The initial form sought to at least provide a preliminary intelligence score to aid in identifying the dull recruits, and allow more accurate posting by the War Office recruiting officers to an appropriate corps.

Combing and Sorting the Infantry

The manpower shortages were felt at the battalion level as the sorting process continued through 1941, with continual combing for specialists as well as to create a more balanced force for battle. The army combed the infantry battalions for tradesmen for use within the army or for return to industry. Soldiers with test results demonstrating above-average intelligence, and also the correlated physical fitness, received assignment in more technical positions, or as officers or NCOs, or for other specialized infantry units such as parachute or commando. Field Marshal Alan Brooke, the wartime chief of the

Imperial General Staff, stated after the war that "one result of this formation of specialist infantry was to reduce the quality of the infantryman, on whom everything would depend in the real trial of strength, and seriously to depress his morale."[62]

Wholesale conversions of infantry battalions to Royal Armored Corps and Royal Artillery further reduced the number of infantrymen in relation to the army as a whole as the organization of the fighting forces moved toward a more rational balance from the frenzied recruitment of infantry battalions after the return of the British Expeditionary Force. Before the war, 50 percent of the army was in the infantry, and following Dunkirk this rose to over 60 percent due to lack of equipment. While the Royal Navy and the Royal Air Force fought the war, the army sorted out its organization and role. General Sir Ronald Adam, the wartime adjutant general, related after the war that "by the middle of the war infantry units formed only 20% of the Army, and the Royal Artillery, with its many Anti-aircraft and Coast Defence units was the largest Corps with 22%."[63]

The induction of soldiers by the Ministry of Labor and National Service ensured central control of the growth of the army. This allowed the sustainment of the industrial war economy and the other services in accordance with overall war strategy. Manpower shortages by 1941 created severe strain for quotas assigned to each service "within which all commitments must be met. . . . As the war continued and the shortage of manpower became more and more acute, the Ministry of Labour and National Service were sometimes unable to supply enough men to fill the proper intakes. It then became a question for authoritative decision whether more men should be called up from industry and whether one of the Services should have its full quota at the expense of the others. . . . Sometimes the Army's full requirements . . . were supplied; at other times, and increasingly so as the war went on the supply was insufficient and the Army had perforce to be content with a smaller proportion of really fit men."[64]

The limited implementation of intelligence tests starting in late 1941, and the use of the selection group, provided a better opportunity to classify incoming recruits, and a means to sort existing units to reassign men to other units, or men with officer potential, or tradesmen. Units being disbanded or converted provided the initial opportunity to use the selection-group concept.

The 120 infantry battalions recruited after the return of the British Expeditionary Force in June 1940 were the primary test bed for transition and reorganization to a more balanced force of fighting arms. Principal conversions were to Royal Armored Corps tank regiments, Royal Artillery anti-tank regiments, anti-aircraft regiments, and field artillery regiments. With the exception of some Royal Artillery anti-aircraft regiments, the infantry battalion's War Establishment[65] provided larger numbers of other ranks of enlisted members than their counterparts' units. The other ranks strength of the British infantry battalion during this period averaged 640, dependent on the year and changes to War Establishment.[66]

The prime minister's criticisms over manpower continued in October 1941 when he again admonished the chief of the Imperial General Staff over the lavish number of support elements in the army to find economy in the rearward services, fearing the army incapable of deployment overseas, and said that "I am doing my utmost to sustain the strength of the army against a growing volume of criticism about its reputed size and enforced passivity."[67]

The army's response was a renewed emphasis on the accurate evaluation of the skills and potential of all recruits and their proper placement according to aptitude and the needs of the army. The misplacement of soldiers whose skill sets, personalities, or intelligence failed to match the units to which they received initial assignment led to significant problems in training, anger among the soldiers, and costs in transfers as reported to the Executive Council of the Army Council, including that "50% of every [Royal Armored Corps tank regiment] intake and 20% of every infantry intake had not the intelligence for full efficiency in the corps to which they had been posted. 4% of all intakes were useless for training soldiers."[68] The adjutant general recommended to the Army Council the creation of a Directorate of Selection of Personnel in June 1941. He introduced a general selection test at the intake centers as well as at various training centers and within units converting from infantry to other corps.[69]

The object was "to post people to suitable arms by observing a minimum standard for those corps which could least readily use men of low intelligence."[70] The army found in this initial stage of testing that the best measure was the establishment of a minimum intelligence level for certain corps. These received no men falling into the least-intelligent 30 percent of the population, easing their training problem. Of course, those other corps, such as the men assigned to the infantry, and gunners in the Royal Artillery, were less fortunate, receiving a disproportionate number of the less intelligent men, "and, although they could more easily employ them, the position was by no means satisfactory."[71] The impact on the infantry force was difficult. "It must be remembered that since no proper allotment of the better or worse material was ever made among the three Services, the Army had to take all of the unwilling, the least intelligent, and the least desirable, and it is remarkable what good use was made of some very unpromising material."[72]

The Second Consolidation: The Regimental System Modified

In June 1941, reacting to the increasing manpower shortage, the Executive Council of the Army Council recommended to the full Army Council the consolidation and reduction of the fifty-eight regimental infantry depots and four machine gun depots into twenty-four numbered infantry training centers and one numbered machine gun training center,[73] while retaining the guards and rifle depots. Each numbered depot represented up to three consolidated regimental depots in order to preserve their character if not their function. The goal remained to keep the larger regional affiliations intact, to the extent possible, as well as to attempt to funnel Welsh, Scottish, and Irish recruits to their national formations, dependent on the needs of the army. The Executive Council of the Army Council report acknowledged that the "regimental system will be seriously affected by the disbandment of the regimental depot with its consequent effect on sentiment and esprit de corps."[74] The adjutant general, however, "considered that an outstanding lesson of the last war was that it was impossible to retain the regimental system in time of war."[75]

The dual challenges of the manpower shortage and the anticipated casualties of a more offensive role demanded recognition, and the army moved forward with changes that hurt the narrative and expected effect of regimental tradition. The inspector general of training in September 1941, whose opinions were shared by many in the infantry regiments, noted that

> the Cavalry and Infantry regiments possess a unity unknown to any other service organization in the World. A unity which, under certain conditions, in war or peace, may be the sole means of maintaining discipline in the unit, and law and order in the Country. In war, the Infantryman's lot is, more often than not, the most exacting of any arm or service. He has not the glamour of the [Royal

> Navy]R.N., the [Royal Air Force], or the Tanks, and has often to perform the most strenuous, irksome and uninspiring tasks. In spite of the disadvantages under which our Infantry serve they usually manage to pull through, largely, I maintain, owing to our unique organization. So, I would urge that, however desirable, this is not a propitious time to break up our Infantry organization.[76]

Preparation for the General Service Corps

With these modifications, the preparations for the enlistment of men into a general service corps for testing and evaluation, before posting to a corps, were complete. The principal characteristics of the period from January 1939 to July 1942 included acknowledgment of the shortage of British manpower to defeat the German Army on the continent, the assignment of a strength ceiling, and in turn its defensive strategic role vis-à-vis the Royal Navy and the Royal Air Force. In addition, the implementation of scientifically and psychologically based selection methods gave hope for greater efficiency in the assignment of soldiers. The qualitative identification and evaluation of recruits and tasks, however, combined with the ending of the character of the prewar regimental system, diminished further the army's attention to the character of the infantry force.

While doctrine demanded that the higher-quality man be assigned to combat tasks, the low standards of the prewar era conditioned the army and the Ministry of Labor and National Service, in the face of the science of personnel selection and the greater national war effort, to accept lower-quality infantrymen against the army's own needs.

> The Army had to share the available manpower with industry, the civil defence forces and the other Fighting Services, especially the vastly expanded [Royal Air Force]. This was bound to limit the quality of men available to the Army, just as it limited quantity. This quality was further lowered, however, by the way the share-out was made. Broadly speaking, the governing principle was that, except for a quota of tradesmen, which met only a fraction of the military demand, the Army's needs should be satisfied last. Civilian reservations were fixed first. The men then called up for the Fighting Services, and who formed the bulk of their wartime strength, were allowed to state a preference, and a large number asked to be considered for the Royal Navy or the Royal Air Force. This did not affect the Army numerically, since the other Services could only accept quotas of the optants, but it told qualitatively. For the Navy and the Air Force took only what seemed to them to be the best material. Hence, the Army intakes, apart from a proportion of vol-

unteers and optants, consisted of men who had been rejected by the other Services, of men who happened to be recruited at times when the quotas for those Services had been met, and of men who had no particular preference.[77]

Scientific selection speeded and enhanced the decline of infantry quality as the measures most important to combat effectiveness were least amenable to scientific identification. The belief that the regiment itself was the catalyst for the development of cohesive fighting small units in the infantry was not possible to maintain given the downward selection of the infantrymen. The selection methodology for the bulk of the infantry battalions denied the soldiers the cohesive capability expected of the narrative assumptions for the regiment. With three-quarters of the wartime army already under arms, and as the British Army began the offensive strategic role of the final period of fighting from late 1942 to 1945, the effect of the acceptance of a low-quality infantry force reverberated in all theaters to war's end.

Notes

1. Churchill, *Never Give In!* 181–182.
2. Ibid.
3. O'Neill, *The German Army and the Nazi Party,* 11.
4. Churchill, *Never Give In!* 181–182.
5. Slack, *Being the National Service Acts,* 3.
6. Keegan, *The Second World War,* 39–42.
7. TNA PRO LAB 44/247, "National Service Pamphlet," 3.
8. Dennis, *Decision by Default,* 1–2.
9. TNA PRO WO 277/12; Pigott, *Manpower Problems,* 12–13.
10. Pigott, *Manpower Problems,* 8–9.
11. Ibid., 7; TNA WO 32/9960 E.C.A.C./P (41)111, December 6, 1941, pt. 1, 2.
12. TNA PRO WO 277/19 B; Ungerson, *Personnel Selection,* 4.
13. Pigott, *Manpower Problems,* 22.
14. Ungerson, *Personnel Selection,* 4.
15. PRO PREM I/296, H. J. [Wilson] to Hore-Belisha, March 18 1939; Pigott, *Manpower Problems,* 25; Messerschmidt, "Außenpolitik und Kriegsvorbereitung," 664.
16. TNA WO 32/9960 E.C.A.C./P (41)111, paper by Adjutant General, "Use of Manpower in the Army," December 6, 1941, pt. 1, 3; "Doubling the Territorial Army"; Pigott, *Manpower Problems,* 5, 8–9; David French, *Raising Churchill's Army,* 16.
17. Slack, *Being the National Service Acts,* 6.
18. Ibid., 3.
19. General Staff, War Office, *Field Service Regulations,* vol. 1, *Organization and Administration,* xxi–xxii.

20. General Staff, War Office, *Field Service Regulations,* vol. 2, *Operations, General. 1935,* 1; General Staff, War Office, *Field Service Regulations,* vol. 1, *Organization and Administration,* 4.

21. Ibid.

22. General Staff, War Office, *Infantry Training, Training and War,* 4.

23. Ibid., 64.

24. General Staff, War Office, *Field Service Regulations,* vol. 1, *Organization and Administration,* 3.

25. Slack, *Being the National Service Acts,* 9–12.

26. TNA PRO WO 277/12; Pigott, *Manpower Problems,* 13; TNA WO 32/9960 E.C.A.C./P (41)111, pt. 1, appendix E, para. 4.

27. TNA PRO LAB 44/297, *Ministry of Labour and National Service Handbook,* January 1941, 9.

28. LHCMA Adam 3/1-3, "Notes for Secretary of State for War Estimates," speech by Adjutant General, February 15, 1941; TNA WO 32/9960 E.C.A.C./P(41)111, pt. 1, para. 12, p. 6. Used by permission.

29. TNA PRO LAB 44/297, *Ministry of Labour and National Service Handbook,* January 1941, 10; TNA PRO LAB 29/127, Ministry of Labour and National Service, Ministry of Labour Codified Circular (M.L.C.C.), appendix 2, January 1941, 120.

30. LHCMA Adam 3/1-3, "Notes for Secretary of State for War Estimates," speech by Adjutant General, February 15, 1941; TNA WO 32/9960 E.C.A.C./P (41)111, pt. 1, appendix A, para. 2. Both of these documents were authored by Adjutant General Sir Ronald Adam, and there were some differences between the numbers provided for the estimates speech in February 1941 (as mentioned) and the Executive Council of the Army Council report of December 1941—for example, 930,000 total all corps for the February 1941 report and 892,000 for the December 1941 report. Used by permission.

31. TNA PRO WO 165/3 A.G. 1 (C.D.), War Office, September 1939–December 1941, "Army Council Instructions," November 1, 1941, para. 2149, "Posting of Officers and Other Ranks of Scottish, Welsh, and Northern Irish Nationality to Their National Units"; Pigott, *Manpower Problems,* 8; Renouf, *Black Watch,* 37.

32. TNA PRO WO 32/9846, Secretary of State for War minute sheet, July 1, 1941; TNA PRO WO 32/9960, Adjutant General memorandum to Army Council, "Use of Manpower in the Army," sec. 18, "Army Class Intakes," December 6, 1941, 8, and appendix F, "Tradesmen," December 6, 1941; TNA PRO WO 32/9846, paper no. 3, June 1942–May 1943, pt. 1, sec. 6, p. 8.

33. Pigott, *Manpower Problems,* 55–64. The army defined "wastage" as the loss to the service through a variety of means, such as men returned to industry, compassionate release, battle casualties in terms of dead and wounded, invalided, or as not physically qualified for further service through war wounds or sickness, unsuitability, and desertion.

34. Pigott, *Manpower Problems,* 14–15.

35. LHCMA Adam 3/1-3, "Notes for Secretary of State for War Estimates," speech by Adjutant General, February 15, 1941. Used by permission.

36. Ibid.; Pigott, *Manpower Problems,* 11.

37. Ellis, *The War in France and Flanders,* 19–21.

38. Ibid., 276, 297.

39. Ellis, *The War in France and Flanders,* 249.

40. Ibid., 286, 302, 305; French, *Raising Churchill's Army,* 156–159; Gilbert, *The Second World War,* 98; Operation Ariel and Operation Cycle during June 1940, http://ibiblio.org/hyperwar/UN/UK/LondonGazette/37573.html (reproduction of General Alanbrooke dispatch of June 22, 1940); http://www.naval-history.net/WW2RN05–194006.htm (summary including both operations); http://www.naval-history.net/xDKWDa-Aerial.htm (summary of Ariel shipping records from Admiralty Diary, Enclosure no. 2 in Plymouth Letter no. 347, July 10, 1940).

41. Pigott, *Manpower Problems,* 13–14.

42. TNA PRO WO 32/9960 E.C.A.C./P(41)111, December 6, 1941, pt. 1, para. 18, p. 2; LHCMA Adam 3/13, unpublished draft personal memoir written during 1949, 6–8 (used by permission); Pigott, *Manpower Problems,* 16. The number 324,000 is taken from the Executive Council of the Army Council Adjutant General report from TNA PRO WO 32/9960. In Adjutant General Sir Ronald Adam's unpublished memoir (1949) (LHCMA, Adam 3/13), the number 275,000 for June, July, and August 1940 inductions was used. This number was repeated in the postwar report (TNA PRO WO 277/12). The number 275,000 accounts for the months of June and July 1940 only, according to the Executive Council of the Army Council report, and when the month of August 1940 was added, the correct figure of 324,000 was indicated.

43. Pigott, *Manpower Problems,* 13–14.

44. French, *Raising Churchill's Army,* 187–189.

45. Keegan, *The Second World War,* 47–87, 142–172, 256–266; Gilbert, *The Second World War,* 44–116, 165–188, 272–300.

46. TNA PRO PREM 3/55/I, Churchill, Directive by Minister of Defense, March 6, 1941.

47. TNA PRO CAB 120/230, Prime Minister personal minute for Secretary of State for War, no. M.400, December 9, 1940.

48. Kroener, Müller, and Umbreit, *Organisation und Mobilisierung des deutschen Machtbereichs,* vol. 5/1, 834–845, 874–875.

49. TNA PRO CAB 108/19, War Cabinet Central Statistical Office, *Statistical Digest,* British Army strength as of May 31, 1945; Wagner et al., *The Library of Congress World War II Companion,* 276–278.

50. Vernon, *Personnel Selection in the British Forces,* 39.

51. Forty, *British Army Handbook,* 328–335; French, *Raising Churchill's Army,* 188.

52. French, *Raising Churchill's Army,* 189.

53. Ibid., 187–189.

54. Ungerson, *Personnel Selection,* 7.

55. Ibid.

56. Ibid., 9. It is unclear from the archival sources if the educational corps continued with a form of educational test following the doubling of the territorial army in March 1939 and the subsequent turnover of the induction process to Ministry of Labor and National Service control in May 1939. New test forms were introduced on a limited basis in 1940, and not until 1942 were standardized tests for all inducted men the norm for selection for the various arms of service.

57. Ungerson, *Personnel Selection,* 9. The devisors of the test were Farmer, Hotoph, and Rodger. Eric Farmer was psychologist at the Cambridge Psychological Laboratory.

58. Ungerson, *Personnel Selection,* 11.

59. Ibid., 19.

60. Ibid., 16.
61. Ibid., 20.
62. LHCMA Alanbrooke, June 2, 1916. Used by permission.
63. LHCMA Adam 3/13. Used by permission.
64. Pigott, *Manpower Problems,* 16–17.
65. The War Establishment represented the official listing of personnel by billet and number and all basic equipment assigned to a unit such as an infantry battalion or a complete division. It was the equivalent of the US Army tables of organization and equipment and German *Kriegstärkennachweisungen.*
66. TNA PRO WO 32/10400, Infantry Battalion (Higher Establishment), Notified A.C.I. [Army Council Instructions], May 19, 1943; TNA PRO WO 167/744, War Diary, Gordon Highlanders, May 19, 1940; TNA PRO WO 169/5017, War Diary, Gordon Highlanders, October 1, 1942; French, *Raising Churchill's Army,* 39
67. TNA PRO WO 32/10289, Prime Minister personal minute for Secretary of State for War and C.I.G.S. no. M.948/1, October 1, 1941.
68. TNA PRO WO 163/50, Report by the E.C.A.C., memorandum for the Army Council, "Selection Tests for the Army," Paper no. A.C./P (41)/40, June 13, 1941.
69. Ibid.
70. Ungerson, *Personnel Selection,* 21.
71. Ibid.
72. Pigott, *Manpower Problems,* 63.
73. TNA PRO WO 32/9846, extract from the minutes of the seventeenth meeting of the E.C.A.C., June 27, 1941, 2.
74. Ibid.; E.C.A.C./P (41)47, June 21, 1941, *The Future of Regimental Infantry and Machine Gun Training Centers,* 3.
75. Ibid.; extract from the minutes of the seventeenth meeting of the E.C.A.C., June 27, 1941, 1.
76. TNA PRO WO 216/66, note from Inspector-General, Training, to C.I.G.S., September 27, 1941.
77. Ungerson, *Personnel Selection,* 4.

6

The United States: Mobilizing for War

THE CONDITIONS FACED BY THE UNITED STATES IN SEPtember 1940 were the result of years of active expansion and war beginning with the Japanese seizure of Manchuria in 1931. By 1937, the dictatorships of Germany and Italy embarked on aggressive expansion and rearmament, and Japan invaded China. A civil war in Spain provided yet another dictatorship in Europe. With German moves in 1938 into Austria and the resolution of the Czechoslovakian problem in Hitler's favor the same year, Americans began to awaken to the drumbeat of coming war. The Neutrality Acts of 1935 to 1939 were the US response to the growing threat of war in Europe and the Far East, intending with these acts to prevent US involvement in any war. With this as the predominant national sentiment, the US Army received minimal enhancements beginning in 1935 that culminated in 1939 with a substantial increase across the army, with notable increases for the air corps. Deteriorating relations with the Japanese Empire, highlighted by a series of embargoes intended to cause a change in Japanese policy in Japan's war in China, further heightened US war anxieties. Against this backdrop of a world at war, the American public remained against entry into the war, while approving of measures taken to rearm and prepare. With the reintroduction of conscription in September 1940 the army began to receive the resources needed for true modernization and adequate manpower in anticipation of war that began with the attack on Pearl Harbor on December 7, 1941.[1]

Whatever the intended final quality composition of the army intended to fight the coming war, the period between September 1940 and August 1943 determined its nature and composition. This occurred through the classification, selection, and assignment policies emplaced and enforced during this period. Correcting problems noted with the quality of the infantry after this time proved difficult, and resolution of these shortcomings fell to the experience of combat itself for the US infantrymen of World War II. As the first commanding general of the Army Ground Forces (AGF), Lieutenant General Lesley J. McNair held the strong belief that "American soldiers were sustaining avoidable casualties and perhaps taking longer than necessary to win the war, because the men assigned to ground combat units did not represent a fair cross section of the nation's manpower."[2] He observed and compared the results for the AGF against the Navy Department, and his own air forces component. The AGF postwar report noted that "the Navy, the Marines, and the Army Air Forces therefore had the character of hand-picked organizations, a character preserved to a large extent by the Navy and Marine Corps even after their resort to the draft (because of differences in induction standards), and by the Army Air Forces even though a large proportion of its personnel was obtained by non-voluntary means."[3] The manner of induction, selection, and classification for military service during this period was the formative condition for the combat forces of all US armed forces and entailed the principal causes of the quality concerns of the AGF throughout the war.

Manpower Selection and Allocation

The first US peacetime military draft law, the Selective Training and Service Act, took effect in September 1940. The law tasked the citizens of each of the nation's 3,070 counties and parishes to create local draft boards, numbering 6,443 in total, and given with "utter and complete confidence, the primary responsibility of determining who was to serve the Nation by training with the armed forces, and who was to serve in industry, agriculture, government, and other civilian pursuits."[4]

Local boards accomplished the classification procedures for induction into national service. By October 1940, procedures were in effect for the registration of all male citizens and aliens between the ages of twenty-one and thirty-six. While the local board's primary function was the provision of men to serve in the armed forces through its classification, acceptance for service was the function of the armed services. The services placed quotas or "calls" for which each board received a proportion of the total national requirement. National lottery determined selection prior to the declaration of war against Japan on December 8, 1941. There were four main designations in the classification process to induct citizens into armed service.

> Class I: Men available for service.
> Class II: Men deferred because of occupational status.
> Class III: Men deferred because of dependents.
> Class IV: Men deferred specifically by law or who were unfit for service.[5]

By law, selective service operated under civilian boards within every county in every state to determine who might be released for service or deferred. Like the British system, the acceptance for induction decision rested with the War and Navy Departments once a recruit reported to the local board for testing and examination. Similar to the British experience, the period in which selective service functioned prior to the declaration of war allowed the identification and resolution of problems throughout the system. Although Congress limited the number in training at any one time to 900,000, the army received authorization to train 800,000 out of a gross manpower pool of over 17 million registrants. Prior to December 7, 1941, a small percentage of this authorized number passed through induction.[6]

The selective service system then "was primarily an agency to procure men for the military and naval forces . . . to classify registrants for their identification in the process of deferment, selection, rejection, and delivery for induction."[7] Classification for *induction* was the principal task. The armed services provided the physical examination standards and the intelligence tests for acceptance and assignment into their particular institution or component. Thus there were two systems of classification: the first was the basic selective service classification following registration for the draft that brought selected individuals into a reception center for induction and for examination by representatives of the armed services; and second was the classification of men accepted for induction into a particular service based upon the peculiar and independent methods developed by each for determining where and how and in what specialty each inducted citizen served. Once the selective service classified an individual for assignment for service induction, the individual armed force further classified these citizen-soldiers in preparation for selection for a military specialty and sent them to training.

The head of the selective service, Lewis B. Hershey, noted that unregulated volunteering created grave and critical problems in the management of manpower shortages, reducing both the effectiveness of supplying men to the armed forces and the retaining in civilian life of men in essential trades. He insisted that each man "be placed or retained wherever his services would make the greatest contribution—in the armed forces, in industry, in agriculture or in some other necessary civilian occupation."[8] He noted the experience the US Army of 1918 in halting volunteering for these reasons,[9] which resonated with British Army experience in 1916.

When combined with the pirating of manpower across US industry, the removal of workers, "essential men," through volunteering decreased the efficiency of those laborers remaining in civilian occupation. Further, volunteers stripped the local board of the numbers necessary to fulfill their national calls, or quotas for induction. This was more difficult in the case of married men. When too many single men volunteered, some local boards dipped into deferred married persons, well ahead of other local boards or states.[10]

In December 1941, eighteen- and nineteen-year-olds became eligible for registration and in November 1942 for induction. The Navy Department and the US Army Air Corps resisted the end of volunteer recruiting. The army was the first to restrict voluntary enlistments, on January 14, 1941. Eighteen- and nineteen-year-olds continued to volunteer, although ineligible to register for the draft until after the attack on Pearl Harbor. Those volunteers twenty years or older required local board certification, as not "essential" to industry, and not already ordered for induction. The Navy Department ignored the latter in regard to deferred-status men until September 1942.[11] President Roosevelt signed Executive Order 9279 on December 5, 1942, to announce the end of volunteering. He also moved the selective service system from an independent agency to the War Manpower Commission.[12]

Effective during February 1943, the executive order required the US Navy, Marine Corps, and Coast Guard to receive all of their enlistments through selective service. The system permitted men, upon reporting to the joint army-navy induction centers, to volunteer for these services as well as for the US Army Air Corps, dependent on available quotas. Men were still permitted to volunteer ahead of their call-up notice for induction and to ask for assignment to the service of choice, provided a quota existed. This remained in place until the end of the war.[13]

Selection Standards

The identification of civilian job–equivalent skills was the first cut in the classification and selection for specialty assignment, such as in the army's enormous appetite for technical skills as drivers, radio technicians, or longshoremen, as well as trades such as cook, barber, or butcher. Each civilian equivalent job received a number below 500, while those specific military specialties, such as riflemen or mortar men, received a number from 500 to 999. These military occupational specialties joined two critical assignment and quota documents: the unit tables of organization, which listed all job specialties in terms of specialty requirements in numbers and types, and the requirement and replacement rates listed for military specialists. When aggregating the totals for all unit tables-of-organization requirements, the adjutant general provided the reception centers with quotas for ordering of individuals with indicated skills to the units that needed them.

This process determined the numbers needed for each specialty per 1,000 men and guided the assignment of inductees from the selective service reception centers to the army's replacement training centers. Thus the army's intent was that any man with a matching civilian occupational specialty received assignment to that job. Specialties in the combat arms, such as the infantry, held very few definitive specialties that correlated to civilian employment skills. Those without these pre-induction qualifiers, or those too young for a trade or profession, or whose skill did not match any military specialty requirement, went into the infantry and other combat arms. The selection process favored those technical jobs for which definitive qualifications existed and matched the inductee's skills. "At one extreme was the Transportation Corps, in which, according to the Requirement and Replacement Rates, 788 out of 1,000 enlisted men filled civilian-type jobs. . . . In the infantry on the other hand only 164 men out of 1,000, mostly cooks, truck drivers, and radio operators, filled civilian type jobs. . . . Occupational classification, though not primarily adapted to the needs of the combat arms, was nevertheless the main basis of assignment."[14]

The practical difficulty experienced by the infantry in this process was that men who scored well on the intelligence test or who possessed a civilian job skill received assignment away from the combat arms. The postwar report on the quality of enlisted men in the AGF noted that "men of established callings in civilian life tended to be men of the higher intelligence levels, with a sense of responsibility and initiative, and often physically stronger and better nourished."[15] With intent, this "excessively refined" system was so successful that by 1943 a mere 17 percent of army personnel with a usable civilian trade found themselves employed outside of that skill.[16]

Intelligence: The Army General Classification Test

Developed in 1939 and implemented in 1940, the Army General Classification Test (AGCT) replaced the mental tests in effect since 1927 for all inductees and volunteers. Its purpose was to classify soldiers by mental powers. The War Department forbade the confusion of AGCT scores with concepts of "IQ," or intelligence. This test was intended to measure "a compound of native endowments with the effects of schooling and social experience, amounting to 'intelligence' in the popular and practical sense in which it was useful to the Army. Scores were so arranged that 100 represented the median of all men tested."[17] The AGCT comprised the following tests:

- Vocabulary: Required knowledge to select the correct meanings of words stored in the person's long-term memory.
- Arithmetic: Required knowledge of mathematics as well as language-based knowledge to comprehend the words in the word problems. These types of problems are heavily dependent on efficiency of information processing in working memory too.
- Block-counting: Emphasized the use of literacy as graphics display processing and required visualization to imagine the presence of obscured blocks.[18]

The army found the test was effective and validated as a classification method by comparing AGCT scores with training-course grades. The young adults who formed the vast majority of the army population demonstrated little formal training in their labor experience. The army required training on a broad set of tasks from the technical and administrative to mechanized combat. "This seemed consistent with the interpretation of the AGCT as a measure of general learning ability. Given that the AGCT so much resembled a measure of past school learning, it is not surprising that it was found to correlate reasonably well with achievement in Army schools."[19]

As in the US Army experience of the Great War and in the British Army, subtest scores combined into a total score, and from this combination came the bases for grades. These ranged from Grade I, high-scoring "rapid learners," to above-average learners in Grade II, average learners in Grade III, below-average learners in Grade IV, and slow learners in Grade V.[20]

The War Department told reception centers that all arms received adequate numbers of the rapid learners, stating the requirement that

> mental ability will be distributed proportionately to all replacement training centers and units *after* occupational specialties required by installation or unit of assignment have been supplied, *except* when specifically directed to the contrary by the War Department. Particular attention will be given to the necessity of sending to the various arms and services all men who appear to have the proper qualifications for officer candidates in the respective arms and services.[21]

The "after" clause separated those with vocational specialties already identified as a civilian skill, and the "except" clause was the cause of the disproportionate assignment of large numbers of high-intelligence men into the air corps, leaving the ground arms to contend with what was left.[22] This formed a fundamental point of contention between the AGF on the one hand, and the air forces and service forces on the other. While the intent upon the reintroduction of conscription was to separate the term "intelligence" from "learning ability," the tradition of the tests and the instructions as noted in the mobilization regulations, in particular regarding the interview, noted the importance of the tests as "an immediate and reasonably dependable classification of the men according to general intelligence."[23]

Physical Capability Standards

The first responsibility at the induction center after the decision for selective service classification was the determination of physical capability to serve based on each service's standards. This was an armed forces responsibility at induction, not a responsibility of the selective service. Wartime physical and psychiatric standards established by the army for induction were quite detailed and introduced in mobilization regulations revision of March 15, 1942.[24]

Army height standards were a minimum 5 feet, up to 6 feet 6 inches maximum, and while possible to waive weight over the standard, no one under 105 pounds or over 78 inches received such waivers.[25] Dental standards were reduced after the declaration of war and again from October 1942: "no natural teeth were required for general service registrants, but the condition must be corrected or correctible by full dentures."[26]

The overall trend, from the declaration of war until induction calls shortened in 1944, due to an overall manpower shortage,[27] was to relax the standards or their application.[28] The War Department noted the necessity of reducing standards "with respect to some defects and by adopting, on a widespread scale, a program of correcting both physical defects and illiteracy after induction."[29] The AGF commanding general believed the physical classification to be too broad and inadequate for considerations unique to the demands of ground combat. He cited how "the British and German armies recognized several grades of physical capacity, according to muscular strength, endurance, agility, coordination, etc., and assigned men to positions making corresponding demands on physique, the United States Army recognized only one category of general service and one category of limited service."[30]

Impact of the Army Classification System

On March 9, 1942, a major reorganization of army headquarters resulted in the elimination of the chiefs of the branches, such as infantry, artillery, and cavalry, and their subordination to the new Army Ground Forces. The AGF was equivalent to the Army Air Forces (AAF) and the Army Service Forces (ASF). The ASF took control and superseded the technical and supply services of the army of the former services of supply.[31]

Lieutenant General Lesley J. McNair commanded the AGF until his death on July 25, 1944, by Allied air bombardment in France while observing the launching of the US offensive to break out of the Normandy bridgehead near St. Lo.[32] McNair recognized the problem created by the army's classification system for the quality assignment of men to the ground combat arms, and its special effect on the infantry. The method in which men of higher intelligence, maturity, and physical strength were being directed away from the combat arms and into more technical pursuits was alarming. Until his death, he fought to limit the impact of this policy and to increase the percentage with higher scores on the AGCT into the combat forces and the infantry.[33] The impact of these policies meant that "the arms suffered because so many of their men had no established occupation in civilian life. Except where this lack of occupational experience was due to extreme youth it indicated a lack of ability which was reflected in high rates of 'combat exhaustion' and in the early mental and physical break up of men who entered combat."[34]

While the AGCT score was not intended to serve as an indicator of intelligence, it became, in the absence of any form of reliable measure of the leadership potential of individuals, the best available. The principal shortcoming of the scientific- and psychology-based standards and classification system was the failure to identify the qualities required of the men to perform the tasks of the ground combat arms. There remained no definitive means to evaluate a man's potential as either a fighter or a combat leader other than the AGCT. Little use was made of the physical qualification in assignment, while the extraordinary refinement of the occupational skill requirements tended to put men with the most initiative, alertness, and intelligence in the technical arms and the AAF. The AGCT was the best measure of intelligence and predictor of combat proficiency, even if this was not its primary intent. The science and psychology of selection was least prepared to identify the traits with the greatest impact on the battlefield—who was best qualified to fight and to lead in ground combat.

Physical robustness correlated to strength, aggressiveness, and emotional stability. Vocational achievement in civilian life identified a suitable man as one possessing the maturity, initiative, ambition, alertness, and self-reliance deemed essential in fighting men and battle leaders. General McNair found the evidence compelling that the most physically robust and emotionally stable men tended to fall disproportionately among those with the highest AGCT scores, sending these men to the Army Air Forces and the Army Service Forces. He proposed in 1943 that the primary classification system be weighted toward physical robustness and that the most physically healthy men be given priority in assignment to combat positions.[35]

The Army Air Forces and Infantry Quality

The exception to the rule that the principal qualification for assignment was civilian occupational skill was the AAF's reliance on the AGCT as the primary measure of mental powers for assignment. "During most of the period of rapid mobilization, from early in 1942 to the middle of 1943, the War Department ruled that a stated percentage of inductees shipped to the Air Forces should be men scoring over 100 in the AGCT. The percentage varied, but was always well above the percentage found among inductees as a whole. . . . [T]he practice greatly reduced the number of high-intelligence men available to the remainder of the Army."[36]

This was perhaps the most divisive issue of the war in the struggle over the assignment of the highest-quality individuals identified in the classification and selection process. The War Department acknowledged in March 1942 the need for proportional distribution of these high-quality men across the three major component commands, the AGF, AAF, and ASF. Each within its sphere was responsible to organize, prepare, train, and supply the forces that carried the US Army to victory. Many factors separated this wish from reality. Chief among them was the use of the AGCT to identify men of higher intelligence levels. This became the holy grail of personnel assignment and over time strengthened the correlation between physical robustness and the potential for handling the stringent physical, moral, and psychological requirements of battle and of the identification of leaders from the overall draft population.[37]

McNair believed that the disproportionate assignment of high-intelligence men to the technical services of the Army Service Forces and the Army Air Forces damaged the ability to find leaders for the infantry. The system placed those most capable in maturity, physical ruggedness, and initiative in positions other than ground combat.[38] The preference to the Army Air Forces took place throughout the period of mobilization in three stages, each distinct in its requirements and each more difficult for the AGF in effect.

The first went into effect on February 2, 1942. Under this policy, 75 percent of all white inductees going to the Army Air Forces required an AGCT score of 100 or better. Both the AGF and ASF protested this rule in June 1942 as it became more and more apparent that the effect of sending so many Class I and II (AGCT scores above 110) men to the Army Air Forces was damaging the officer candidate programs and small-unit leadership in the ground arms. The War

Department rescinded the rule on July 18, 1942. General H. H. "Hap" Arnold, the commanding general of the Army Air Forces, appealed this ruling to General Marshall, noting that the AAF was the first combat force planned for deployment. Marshall reinstated the rule in a different, more damaging form to the AGF on September 7, 1942.[39]

Under the new preference, 71 percent of those scoring over 100 on the AGCT *and* the mechanical-aptitude test went to the AAF. "Although roughly half the men tested scored 100 on each test, only about 33 percent scored 100 on both. Combining two kinds of ability, these men were exceptionally desirable. The Air Corps was now due to receive almost three-quarters of its new personnel from the top third of available manpower."[40]

This preference remained in effect until June 30, 1943, but changed downward in November 1942 from 71 percent to 55 percent upon the lowering of the induction age to eighteen. On that date, however, a new rule emerged stating that any inductee expressing a desire for flying training in the army, prior to classification—that is, before assignment to any component of the army—and "apparently" physically qualified and scoring over 100 on the AGCT, went to the AAF. This rule meant that any inductee stating such a preference did not count against the 55 percent of inductees going to the AAF who scored over 100 on both tests, and thus helped to raise the numbers of quality men going to the AAF so that "the percentage of high-grade men received by the Air Forces soared to heights never attained before. In the last half of 1943 over 60 percent of inductees shipped to the Air Forces fell in AGCT Classes I and II (with scores over 110) while only 30 percent of inductees shipped to the ground arms [both AGF and ASF] fell in these categories. . . . It was perhaps fortunate for the Ground Forces that mobilization took place chiefly under the comparatively benign influence of the 75-percent and 55-percent rules."[41] These inductees remained in the Army Air Forces even if found not physically qualified later, or failed in flight training.

In addition, War Department policy directed that any already assigned enlisted men who tested and met the basic qualifications for pilot training or officer candidate school (both restricted to men in Class I and II with AGCT scores over 110), over and above these selection preferences, received encouragement and permission to go to either the AAF or officer candidate school. The AGF leaders believed that men inducted and favored toward the AAF who failed to make selection as pilots or airmen, or for technical and mechanical tasks associated to AAF combat missions, were lost to the ground forces even if not used in accordance with their capabilities. The inspector general confirmed this belief in November 1942, noting that "over a third of these high-intelligence privates at various air bases . . . were men in AGCT Classes I and II. Over half these high-intelligence privates were acting in such jobs as messengers, warehousemen, clerks, guards, orderlies, truck-drivers, firemen, and assistant cooks."[42]

Officer Candidates and the Army Specialized Training Program

The effect of requirements to identify and transfer aspirants for officer candidate school, parachute training, and flight school was a further negative for the infantry divisions. General Marshall stressed in February 1942 the importance of the effort to find officers, causing the commanding general of the Second Army to direct this focus to his divisions: "You will personally and immediately explore your source of supply and recommend to this headquarters every noncommissioned officer and private, particularly college and high school graduates, who is qualified to take the officer candidate course no matter what duty he may be filling now. . . . Men holding key positions and all others must be encouraged to compete for officer privileges."[43]

By July 1942, the Thirtieth Division provided 900 officer candidates with quotas for 313 more and indicated that it had begun examining men with one year of high school or less. The Eighth Motorized Division[44] sent 1,232 men to officer candidate school. The Seventy-Seventh Division, activated in March 1942, sent 762 by October and received quotas for 225 more. Moreover, there was no guarantee that enlisted soldiers who accepted officer candidate school and commissions from infantry divisions would return to the infantry. "In one infantry division in 1942, of 1,200 enlisted men accepted as officer candidates, 800 elected officer training in quartermaster, medical administration, and finance. From a limited, and continuing depleted stock of men in the higher AGCT grades, the Army Ground Forces struggled to meet its own requirements for officer candidates, for men to be sent to enlisted specialist courses at the service schools, for parachute volunteers, and for cadremen for new units. Men remaining with their organizations were a very picked over lot."[45]

The Third Army experienced similar difficulties, with the Twenty-Eighth Division reporting in late July 1942 that leadership within the infantry was critical, with 334 of 706 officers present for training and most platoons commanded by noncommissioned officers while "over eight hundred noncommissioned officers . . . [had] been sent to Officers Candidate Schools; many have been sent on cadres, and those who are left are below the desired standard. . . . [I]nspectors found that 'the spirit' of the division has showed a lack of objective and the will to do."[46]

Finding, training, and retaining sufficient officers, as well as noncommissioned officers, from the limited numbers of AGCT Class I and II soldiers once again became even more difficult with the introduction in December 1942 of the Army Specialized Training Program (ASTP). General McNair opposed this program from its inception to its eventual demise in March 1944. Intended to provide military training on college campuses to selected soldiers as preparation for unique and special technical requirements demanding a higher level of education, there was no promise to become a commissioned officer upon completion of a college education program. The planned strength of the ASTP was 150,000 men. The qualifications for this on-campus training required a higher test score than for other officer candidate training, and sought soldiers younger than twenty-two and scoring a minimum of 115 on the AGCT. About 100,000 soldiers came from those already enlisted and 50,000 from new inductees.

Disappointed with the initial response to the ASTP requirement, on April 4, 1943, the commanding general of Second Army demanded that his division commanders comply without reservation with this order, directing that "nothing was permitted to interfere with the selection of eligible men. All enlisted men with the prescribed qualifications were required to appear before the Field Selection Boards. . . . Division Commanders felt, as one said, that 'it just splits the division wide open . . . it breaks down [the] training program so seriously.' . . . 750 left the 79th Division in the last week of April [1943] alone."[47]

The AGF provided 47 percent of the total quota for the ASTP. The overall share of men with a score of over 115 on the AGCT was already less than the numbers of the ASF and AAF, and the mandatory screening and assignment of soldiers of this caliber and their ultimate loss to the infantry divisions was significant.[48]

Plans for Cross-Channel Attack, 1942–1943

Exacerbating and limiting the ability to fill and train the combat divisions were the multiple and disruptive mobilization plans and deployments for Germany's invasion of the continent of Europe. These plans affected the timing and pace of the activation of AGF combat divisions and required the heightened activation and shipment of the service elements needed to prepare the ground for them in Great Britain (Operation Bolero) over and above the manning of the ground combat force. The advanced deployment of the Army Air Forces as the sole offensive capability against Germany until the arrival of the combat divisions demanded priority for manning, equipping, and deployment. These plans through 1944 in the European theater of operations alone were included first in 1942 (Operation Sledgehammer) and then in 1943 (Operation Roundup). Each required advancing the deployment and manning of divisions going overseas and entailed the stripping of divisions going later and the curtailment of the losing division's training. The combat operations beginning in 1942 increased the manning problems of activating divisions through additional stripping of trained infantrymen to go overseas as combat casualty replacements in ever greater numbers to fight Germany's invasion of North Africa (Operation Torch), Sicily (Operation Husky), and Italy (Operation Avalanche) during 1942 and 1943. All of these stressed the training and manning demands of the AGF prior to the final decision for the cross-channel attack into France (Operation Overlord) in 1944.[49]

Shipping proved the primary factor in determining the ability of the US Army to move to Great Britain in preparation for the cross-channel attack with sufficient numbers to ensure a reasonable prospect of success. Shipping proved inadequate in both 1942 and 1943, but the activations of divisions as planned by the troop basis of 1942 and 1943 continued due to the operational impetus of the invasion plans in the Mediterranean and northwestern Europe. With the eighty-nine combat divisions activated and piling up in the United States awaiting men, equipment, ammunition, and training, and transport overseas, major disruptions in manpower hindered their combat preparation. As for the British Army, the strategic priority in 1942 and 1943 was to gain naval and air supremacy. As a result, the combat forces of the army received a lower priority for manpower, equipment, and shipping.[50]

Building the Combat Divisions

The period of growth saw the activation of ninety-one divisions (see Table 6.1). Not all remained on the troop basis at the end of the war.[51] The troop basis document was an outline of the kind of army authorized to exist. It looked to the future in anticipation of the general war strategy and operational requirements through estimates of the manpower and equipment available and length of time necessary to train. The troop basis of July 1, 1943, projected the army of December 31, 1943, in numbers of authorized units, and served as the basis for either mobilization of future units or deactivations. These unit strengths were based upon the tables of organization and therefore represented authorized rather than actual strength.

The largest actual number of divisions active at any one time was ninety on December 31, 1943, including sixty-four infantry, three light infantry, two cavalry, sixteen armored, and five airborne divisions. The number of divisions, set at ninety, was the approved War Department troop basis of October 4, 1943. The eighty-nine divisions active on June 30, 1944, became the authorized troop basis of July 1, 1944 (indicating also the planned strength of the army on June 30, 1945) and included sixty-seven infantry, one cavalry, sixteen armored, and five airborne divisions.

Compounding the personnel problems throughout the period of growth was the shortage of equipment, arms, and ammunition. Weapons were in such short supply in the Ninety-Fourth Division

Table 6.1 Division Activation During the Period of Growth, 1939–1943

	To 1939	1940	1941	1942	1943	Total
Infantry	6	14	9	27	11	67
Cavalry	1		1		1	3
Armored		2	3	9	2	16
Airborne				2	3	5
Total	7	16	13	38	17	
Grand total						91

Source: US Army Ground Forces, *Study no. 12,* 2–6.

Note: The Second Cavalry Division was activated and deactivated twice, bringing the total number of divisions to the January 1944 Troop Basis authorization of ninety. Of the ninety divisions authorized, eighty-nine were active on June 30, 1944 (sixty-seven infantry [including one mountain division], one cavalry, sixteen armored, and five airborne), and this number remained constant until the war's end.

that rifles were issued on a 25 percent basis for rotation among soldiers in training, many not serviceable and with insufficient ammunition to fire them.[52] In March 1943 these difficulties were further complicated by the War Department's decision to supply a 250,000-man French army in North Africa with US weapons, further reducing the ability to field US divisions.[53] As late as July 1943, inspectors reported insufficient arms and equipment to train the divisions activated, regardless of the shortage of manpower to fill their tables-of-organization strength.[54] The effect of the training shortages appeared in combat. The Forty-Fifth division listed the lack of ammunition available for anti-tank training, with the bazooka as a causative factor in the division's disastrous introduction to combat against German tanks at Persano, Italy, on September 13, 1943.[55]

The Decline of Infantry Quality

Manpower to fill activated units was never available in sufficient numbers in 1942. The army's strategic priority was to the AAF and the ASF. The ASF, in particular, received priority for expansion because of its responsibility to prepare the ground for the arrival both of the aircraft wings of the AAF and then of the AGF combat divisions in England, first for the anticipated cross-channel Operations Sledgehammer in 1942 and then Roundup for 1943. This meant fewer men for the AGF,[56] with the effect that "by 30 June 1942 the Army Ground Forces were short 162,505 men . . . [causing AGF to recommend] deferment of the 97th Division . . . to refill the depleted older divisions. . . . [The] 2nd Cavalry Division was inactivated. . . . New units were formed, but men failed to appear and shortages mounted. . . . By September 1942 the Ground Forces were short 330,000 men, or over 30 percent. . . . Shortages in the Ground Forces threatened to make proper training impossible."[57]

In October 1942, President Roosevelt began to reduce the planned manpower of the army and other services. While it was then unclear that this reduction was the permanent ceiling on the strength of the army, the troop basis for July 1, 1943, came in at 500,000 fewer than the original planned strength. The savings occurred in the reduction of AGF ground combat divisions from 125 divisions to 90, while the ASF and AAF were augmented in strength.[58]

The twin problems of quantity and quality became paramount for the infantry. The AGF divisions received 42 percent of their soldiers from the bottom two categories of test score, with corresponding reductions in Class I through III. Forty-two percent of a 14,000-man infantry division meant that 5,880 soldiers were in this lowest category, signaling the placement of soldiers into the infantry squads lacking identifiable skills for any other assignment (see Table 6.2).

Divisions formed early in the process were subject to providing cadres for later divisions. The Eighth Motorized Division, activated in July 1940, provided three different division cadres in 1942. After these losses and the loss of AGCT Class I and II enlisted men to officer candidate schools, the infantry of the division saw its AGCT averages decline.

In October 1941, the Eighth Motorized Division held 3,388 men in three regiments classified by AGCT in Classes I and II for a combined 38.9 percent. After two rounds of giving up cadres for the Seventy-Seventh Division (activated March 1942) and the Eightieth Division (activated July 1942), the numbers fell to 2,489 or a combined 29.9 percent of AGCT Classes I and II, with 41.7 percent of the infantry in Classes IV and V. The Second Army noted that "many of the ablest and most intelligent men had been taken from the field forces."[59] The trend of assigning those with lower scores for combat organizations received further evidence in statistics gathered between July 1, and December 31, 1943, wherein data indicated an inverse ratio of the AGCT scores of inductees going from reception centers to their assignments in either combat or noncombat arms (see Table 6.3).

The AGF Replacement Center at Fort Meade, Maryland, compared the characteristics of the infantry in comparison to other arms (see Table 6.4), showing "twelve thousand combat soldiers at the AGF overseas replacement depot at Ft. Meade were examined in November 1943. They were general service men from units and

Table 6.2 AGCT Class Assigned to Newly Activated Divisions, March–September 1942 (percentage)

	I	II	III	IV and V
National average	7.0	27.4	32.4	33.2
11 divisions, March–July 1942	5.3	24.3	33.1	37.4
5 divisions, September 1942	3.8	23.0	31.2	42.0

Source: Palmer, *Study no. 5,* 8.

Table 6.3 AGCT Classes Assigned to Combat vs. Noncombat Organizations, July 1–December 31, 1943 (percentage)

	Combat	Noncombat	Number of Men
I	24.2	75.8	62,000
II	34.9	65.1	247,000
III	51.5	48.9	229,000
IV and V	64.3	35.7	174,000
All classes	46.4	53.6	713,000

Source: Palmer, *Study no. 5,* 41.

Table 6.4 Characteristics of Enlisted Replacements, November 1943

Number and Branch	Age (years)	Height (inches)	Weight (pounds)	AGCT[a]	Education (years)
6,000 infantry	23.29	67.74	145.40	95.10	9.52
2,800 FA and TD	22.07	67.78	147.48	96.40	9.69
1,500 CAC and AA	22.20	68.13	148.50	100.52	10.40
1,300 armored and cavalry	22.40	68.21	146.62	94.67	9.79

Source: Headquarters, AGF Replacement Depot no. 1, Fort Meade, Maryland, Subject: Characteristics of Enlisted Replacements by Arm, WD, AGF (1916–1954), Ref. Files L. J. McNair (1940–44), RG 337.3, NACP, enclosure 6.

Note: a. AGCT = Army General Classification Test (US); this was similar to an IQ and aptitude test. 100 represented the median of all men tested.

replacement training centers in all parts of the country and so offered a representative cross-section of personnel in the ground arms. They were inferior in height, weight, AGCT grade, and education [years of school] to the average for the Army."[60]

With the indefinite postponement of the cross-channel attack in 1943, the movement of divisions overseas slowed. Divisions activated in early 1943 began to compensate for the shortages of 1942 and to train effectively. Inductions failed to meet stated goals, however, and shortages began to reappear. The Forty-Second Infantry Division, activated in July 1943, did not receive required personnel for basic training until September 1943. The Sixty-Fifth Infantry Division, activated in August 1943 as the last US Army division mobilized, waited for sufficient personnel to begin its combat preparation until January 1944.[61]

The Transition to Offensive War

The period of growth ended with the activation of the Sixty-Fifth Division in August 1943. The principal characteristics of the period from September 1940 to August 1943 lay in the enormous rapidity of the buildup of manpower. From a force of just over 1.6 million soldiers in December 1941, the army grew to almost 7.5 million men and ninety-one divisions in August 1943. These numbers did not translate into higher priority for ground combat units, and the number of divisions remained constant as "with the development of air power and with Russian victories, there was no certainty that United States ground forces would be needed in large numbers. Among the many demands for military manpower those of the Army Ground Forces were judged to be of low priority in 1943."[62]

As noted, during the interwar period the quality of the infantryman was not a stated priority, despite the acknowledgment that the ultimate price for victory lay upon the lives of the infantrymen. Low priority, plus a classification system that put a premium on civilian job skills and AGCT scores, speeded and enhanced the decline of infantry quality, as the measures most important to combat effectiveness were least amenable to scientific identification. The surprise effect of this methodology was the negation of the US metanarrative of the characteristics of the infantry soldier. Unlike the mass infantry armies of the American past, the US Army of World War II created infantrymen without the assumed nature of expected physicality and intelligence deemed essential and assumed in abundance in prior wars. As army combat divisions began their offensive strategic role in the final period of fighting, beginning in November 1942, following the end of the period of growth from August 1943 to 1945, the acceptance of a low-quality infantry force, as in the British Army, resounded in all theaters to war's end.

Notes

1. Iriye, *The Globalizing of America,* 138–164; Conn, *American Military History,* 415–418; Stoler, *Allies and Adversaries,* 15–18.
2. Palmer, Wiley, and Keast, *The Procurement and Training of Ground Combat Troops,* 4.
3. Ibid., 4.
4. Hershey, *Special Monograph no. 5,* 35.
5. Ibid., 37; Hershey, *Special Monograph no. 16,* 32.

6. Hershey, *Special Monograph no. 16,* 65.
7. Ibid., 168.
8. Ibid., 54.
9. Ibid., 48.
10. Ibid., 68–69.
11. Ibid., 56–59.
12. Ibid., 60–61.
13. Ibid., 61–62; Commandant US Marine Corps, subject: Induction and Enlistment of Men Received Through the Selective Service System, December 1, 1943, US Marine Corps, Office of the Commandant, General Correspondence (1939–1950), Record Group 127, NACP, 4; Headquarters US Marine Corps, Letter of Instruction no. 280, December 10, 1942, USMC, Office of the Commandant, General Correspondence (1939–1950), RG 127, NACP.
14. Palmer, *Study no. 5,* 4; War Department, Classification Memorandum no. 6, February 18, 1942, 1; War Department, memorandum for Chief of Staff, subject: Assignment of Enlisted Men from Reception Centers, June 6, 1942, WD, AGF (1916–1954), Ref. Files L. J. McNair (1940–1944), RG 337.3, NACP.
15. Palmer, *Study no. 5,* 5.
16. Ibid., 6.
17. Armstrong and Sticht, *Adult Literacy in the United States,* 36–37.
18. Ibid.
19. Ibid.
20. Ibid.
21. Palmer, *Study no. 5,* 6; War Department, Classification Memorandum no. 6, January 17, 1942, WD, AGF, Ground Hist. Sec. 1942–1947, RG 337.6.1, NACP; US Army, Army Regulation no. 615-25, *Enlisted Men: Initial Classification,* 1–2; US Army Air Forces, *The History of the US Army Air Forces in World War II,* 538, emphasis added.
22 Palmer, *Study no. 5,* 6.
23. US Army, Mobilization Regulation 1-9, *Reception of Selective Service Men,* 10.
24. Palmer, *Study no. 5,* 3.
25. Forster et al., *Physical Standards in World War II,* 132.
26. Hershey, *Special Monograph no. 15,* 133.
27. Hershey, *Special Monograph no. 16,* 36.
28. Hershey, *Special Monograph no. 5,* 128.
29. Ibid., 142.
30. Palmer, *Study no. 5,* 3.
31. Conn, *American Military History,* 431; Keast, *Study no. 32,* 2.
32. Weigley, *History of the United States Army,* 170.
33. War Department, memorandum for the Chief of Staff, subject: Assignment of Enlisted Men from Reception Centers, June 6, 1942; War Department, memorandum for the Assistant Chief of Staff (G-1), subject: Assignment of Enlisted Men for the Army Air Forces from Reception Centers, September 2, 1942; War Department, memorandum for the Chief of Staff, subject: Preferential Treatment of Army Air Forces in Assignment of Enlisted Personnel, November 13 and 17, 1942; War Department, memorandum for the Chief of Staff, subject: Distribution of Manpower, December 17, 1943; War Department, memorandum for the Chief of Staff, subject: Utilization of Available Manpower Based upon Physical Capability, December 21, 1943; War Department, memorandum for the

Chief of Staff, subject: Distribution of Manpower, February 8, 1944, WD, AGF (1916–1954), Ref. Files L. J. McNair (1940–1944), RG 337.3, NACP.

34. Lerwill, *The Personnel Replacement System in the United States Army,* 317.

35. War Department, memorandum for the Chief of Staff, subject: Utilization of Available Manpower Based upon Physical Capability, December 21, 1943, WD, AGF (1916–1954), Ref. Files L. J. McNair (1940–1944), RG 337.3, NACP; Palmer, Wiley, and Keast, *The Procurement and Training of Ground Combat Troops,* 65.

36. Palmer, *Study no. 5,* 5; US Army Air Forces, *The History of the US Army Air Forces in World War II,* 540.

37. Palmer, Wiley, and Keast, *The Procurement and Training of Ground Combat Troops,* 65.

38. US Army Ground Forces, *Study no. 11,* 42.

39. Palmer, *Study no. 5,* 10; Headquarters Army Ground Forces, staff routing sheet (memo slip), "Notes on Air Corps 75%," October 27, 1942, WD, AGF, Ground Hist. Sec. 1942–1947, RG 337.6.1, NACP.

40. Palmer, *Study no. 5,* 11; Headquarters Army Ground Forces, staff routing sheet (memo slip), "Notes on Air Corps 71%," November 6, 1942, WD, AGF (1916–1954), Ref. Files L. J. McNair (1940–1944), RG 337.3, NACP.

41. Palmer, *Study no. 5,* 13; War Department, memorandum for the Adjutant General, subject: Recruits for the Air Corps, November 29, 1942; Headquarters Army Ground Forces, staff routing sheet (memo slip), subject: Reception Center Procedures in Assignment of Personnel to Army Air Forces, November 10, 1943, WD, AGF (1916–1954), Ref. Files L. J. McNair (1940–1944), RG 337.3, NACP.

42. Palmer, *Study no. 5,* 11.

43. Wiley and Govan, *Study no. 16,* 94.

44. US Army Ground Forces, *Study no. 8,* 42. The motorized division was an infantry division equipped to move entirely and simultaneously by motor transport. It was designed for use in conjunction with the armored division. The Fourth, Sixth, Seventh, Eighth, Ninth, and Ninetieth divisions were converted to motorized divisions in April 1942. All motorized divisions were reconverted to standard infantry in March 1943.

45. Palmer, *Study no. 5,* 8–9.

46. US Army Ground Forces, *Study no. 12,* 13.

47. Wiley and Govan, *Study no. 16,* 95.

48. Palmer, *Study no. 5,* 14–17.

49. Palmer, Wiley, and Keast, *The Procurement and Training of Ground Combat Troops,* 592.

50. Stoler, *Allies and Adversaries,* 84–90; Conn, *American Military History,* 439–446.

51. Palmer, *Study no. 3,* 1–3.

52. US Army Ground Forces, *Study no. 12,* 20.

53. Palmer, *Study no. 4,* 13.

54. US Army Ground Forces, *Study no. 12,* 20.

55. Ibid., 20–21.

56. US Army Ground Forces, *Study no. 11,* 11–12.

57. Palmer, *Study no. 4,* 8; Headquarters Army Ground Forces, memorandum for the Chief of Staff (attn: G-3 Division), subject: Shortage of Personnel in Army Ground Forces Units, WD, AGF (1916–1954), Ref. Files L. J. McNair (1940–1944), RG 337.3, NACP; Secretary of War, War Department, memorandum for

the Commanding General, Army Ground Forces, subject: Shortage of Personnel in Army Ground Forces, December 4, 1943, WD, AGF (1916–1954), Ref. Files L. J. McNair (1940–1944), RG 337.3, NACP.

58. Palmer, *Study no. 4,* 16–17.

59. Wiley and Govan, *Study no. 16,* 84.

60. Palmer, *Study no. 5,* 24; Headquarters Army Ground Forces, memorandum for the Chief of Staff, US Army, subject: Utilization of Available Manpower Based upon Physical Capability, tab D, "Characteristics of Infantry," December 21, 1943, WD, AGF (1916–1954), Ref. Files L. J. McNair (1940–1944), RG 337.3, NACP.

61. Palmer, *Study no. 4,* 17.

62. Ibid., 29; War Department, memorandum for the Chief of Staff, subject: Revision of the Current Military Program, June 1, 1943, WD, AGF (1916–1954), Ref. Files L. J. McNair (1940–1944), RG 337.3, NACP, 2.

7

Germany: Fighting an Offensive War

THE EUROPEAN OPENING OF WORLD WAR II BEGAN WITH the rearmament and mobilization of Germany. Hitler left the League of Nations in 1933 and declared Germany's abrogation of its Versailles Treaty obligations in 1935. He reinstituted conscription in 1935, expanding the planned twenty-one-division increase to one of thirty-six divisions. The military reoccupation of the Rhineland in 1936 was the first action in Hitler's goal to redress the territorial losses of the Great War. He benefited from the hesitation of the guarantors of the Versailles Treaty in France and a reluctant United Kingdom. He counted on the European powers doing little or nothing to stop him, and he exaggerated the preparation of the army and air force to give them pause.[1] The differences between the projection of force for international consumption and the reality of the preparation of the army on the ground spoke to the profound conflict between the soldiers tasked to fight if Hitler's bluff failed and opponents attacked an unprepared Germany.

On November 5, 1937, Hitler held a meeting with his minister of war, Werner von Blomberg; and the leaders of the army, Werner Freiherr von Fritsch; the air force, Hermann Göring; and navy, Admiral Erich Raeder; as well as the foreign minister, Konstantin von Neurath. Hitler laid out his intent to relieve Germany's problems through war, if needed, no later than the period 1943–1945. When asked for comments, these leaders of the armed forces and the foreign office voiced opposition. This was the final break, and "within a few days of this speech, Hitler had dismissed both his War Minister and the Commander-in-Chief of the Army [and von Neurath], and was planning a reorganization of the upper ranks of the Army. . . . [T]here is little doubt that the consistent refusal of Fritsch, and to a lesser extent, Blomberg, to enter whole-heartedly into Hitler's plans for aggression played a big part in their dismissal. . . . The crisis in confidence between Hitler and the High Command had increased to the point at which the dictator felt that a sweeping purge of the higher ranks was justified, so that the German Army entered the period of international tension of early 1938 in a state of disunity and confusion."[2]

Hitler dismissed von Blomberg, discrediting him with the past of his new young bride, with Wilhelm Keitel as the leader of the Supreme Command of the German Armed Forces from 1938 until the end of the war. Keitel's chief of operations was Alfred Jodl. The victorious Allies later executed both Keitel and Jodl in 1946 for war crimes. Hitler also cashiered Werner von Fritsch following trumped-up charges of being a homosexual, replacing him with Werner von Brauchitsch. After Blomberg, Hitler exercised personal command of the entire *Wehrmacht* with Keitel in charge of the new Supreme Command of the German Armed Forces as his military staff, cementing the humiliation and subordination of the army, "and these indignities, in which the Army acquiesced, were the forerunners of worse indignities to come."[3]

General Ludwig Beck, chief of the General Staff, encouraged von Brauchitsch, the new commander in chief of the army, to express the collective caution of senior officers to Hitler's intent to attack Czechoslovakia, and the threat of mass resignations if their recommendations fell on deaf ears. When von Brauchitsch refused to go along, Beck offered his resignation and departed the army on August 27, 1938. He offered these words to von Brauchitsch on July 16, 1938:

> What are at stake here are decisions which in the last analysis affect the existence of the nation. History will burden these leaders with blood-guilt if they do not act in accord with their specialized political knowledge and conscience. Their military obedience has a limit where their knowledge, their conscience and their sense of responsibility forbid the execution of a command. . . . It is a lack of greatness and of recognition of the task if a soldier in the highest position in such times regards his duties and tasks only within the limited framework of his military instructions without being aware of the highest responsibilities towards the nation as a whole. Extraordinary times demand extraordinary measures.[4]

The crises of 1936–1939 provided the German Army with the opportunity to test and adjust its mobilization plans. Unlike the lack of flexibility inherent in the mobilization of 1914, which denied any political room for adjustment once initiated, the army tread very lightly in terms of mobilization in response to Hitler's demands for offensive action until 1939. The army's plans provided for minimal mobilization beyond the existing peacetime army formations, fearing the responses of Germany's opponents if a more muscular mobilization become known. The forces used in the reoccupation of the Rhineland came from the active forces, while a partial mobilization of reserves came from two regions, *Wehrkreise* VII and XIII, bordering Austria in early 1938, in preparations prior to the *Anschluss*. A larger, though not general, mobilization occurred for the possible invasion of Czechoslovakia late in 1938.[5]

The accession of Austria brought the immediate benefit of additional infantry, Panzer, and light divisions. The occupation of the remainder of the industrial heart of Czechoslovakia, Bohemia, Moravia, and Slovakia, in March 1939, brought the full production capability of Czech heavy industry into the *Reich*. General Franz Halder replaced Ludwig Beck as the chief of the General Staff. In this capacity, he continued with secret efforts to depose Hitler in the event of war over the Sudetenland. The capitulation of the Western powers at Munich in late 1938 deprived the conspirators of the opportunity to strike, and Hitler's soaring popularity after another bloodless coup ended the perceived chance that the German people might support a military action to depose Hitler in either 1938 or 1939. The final violation of Czech territory and the reoccupation of Memel, in the north of East Prussia in March 1939, prompted the British government to pass its first peacetime conscription act in that same month and its guarantee of Polish territorial integrity in August 1939.[6]

Manpower Selection and Allocation

The law of March 1935 introduced conscription and war service requirements including a universal mandate for registration of all males eighteen to forty-five years of age. German organization for recruitment was in place in the form of the regional headquarters, the *Wehrkreise,* which numbered seven in 1933, and rose to fifteen with the absorption of Austrian territory into the *Reich* in 1938. The

Wehrkreise were under the direction of army headquarters, *Oberkommando des Heeres,* in peacetime. Fromm, from his position in the General Army Office, *Allgemeines Heeresamt,* oversaw standards and inspections and mobilization until August 31, 1939, when, on the cusp of the European war, the operational *Feldheer* and the *Ersatzheer* separated, and Fromm became the commander in chief of the replacement army, retaining the office of the *Allgemeines Heeresamt.* This responsibility included jurisdiction over the replacements for the entire *Wehrmacht* (army, navy, and air force).[7]

The local registration offices (*Wehrmeldeamt*) throughout Germany delivered the notices for inductees to appear for registration and examination (*Musterung*) and, in 1935, immediate enlistment into active service (*Aushebung/Einberufung*). Throughout peace and war, the *Wehrmeldeamten* were the permanent home of each recruit's service record book (*Wehrstammbuch*), holding the entire history of service from entry to end of service.[8] The law defined the various levels of service based upon age and training as:

- Active service of two years' length for draftees at age twenty.
- Deferred service:
  - *Der Reserve:* For those under thirty-five and trained (first reserve).
  - *Der Ersatzreserve:* For those not yet trained and under thirty-five (second reserve).
  - *Der Landwehr:* For those between thirty-five and forty-five regardless of training.
  - *Der Landsturm I:* For those over forty-five.
  - *Der Landsturm II:* For those over forty-five but not organized.[9]

The Interview

The German system used a collective interview by a panel of military, civilian, police, medical, and labor officials, following the doctor's examination and recommendation. This traditional method intended to provide sufficient information to formulate a sound picture of the capabilities, and that elusive quality—character—of the soldiers, ensuring the best likelihood for successful employment in combat, before all other characteristics.

> The characteristic of the German solution was widely decentralized and . . . avoided any directed scheme. This [method] was well-

> founded in the German Army principle of "*Auftragstaktik*." Subordinate offices were energized with particular responsibility. Only what was categorically needed would be ordered hard and fast from above.
>
> A further basis for the German solution was the strong emphasis on those things which were not measurable in men. . . . Tradition, regional affiliation, bonds of family and friends, personal inclinations, in short all emotional and mental qualities, cannot be measured and thus not brought into a formula. Greater value was placed directly upon these [characteristics]. They can only be assessed man to man, i.e., through personal, individual handling. . . . [T]he great benefit of this method had been established again and again, which was reflected at the end of the day in the bearing of the troops.[10]

The key to German standards for combat infantry physicality and intelligence, as for the other combat arms, was the infantry's tie to the nearness of the fighting. The infantry had the largest requirement for combat-capable soldiers. During the mustering process, 50 percent of those qualified for combat duty, regardless of other qualifications, received assignment to an infantry unit. Given the relationship between physicality and intelligence, the effect of the assignment of 50 percent of combat-capable soldiers to the infantry ensured a disproportionate arrival of the men most qualified physically and mentally for the most exacting form of battle—face-to-face combat in the infantry small unit.[11]

The army also sought its share of men with experience in the technical fields such as drivers, mechanics, armaments artificers, communications, and the like. Every unit wanted "the right man for the right job," but there were never enough of these men to go around. Industries claimed exemption for many of the most technically qualified, further reducing the pool. As the German infantry divisions possessed very little in motorized or mechanized assets, the problem was not as acute as with the levels of motorized and mechanized assets within the infantry divisions of the US and British Armies. The assignment of 50 percent of the qualified recruits to the infantry likely ensured that a fair share of men with civilian technical or other specialist qualifications filled infantry requirements.[12]

Selection

The induction board met at the local recruiting office (*Wehrmeldeamt*). The members of the induction board consisted of a combination of local civilian administrative, labor, school, and police representatives, with the military officer presiding, and included two doctors as well as staff recorders and medical aid men to assist the doctors. The head of the local labor office verified work history or deferment of an individual as essential to industry or agriculture. In the German method, the interview was collective. The induction board made its decisions for combat service using the multiple perspectives of the members of the board. The German induction board combined both the form and the function of the Anglo-American induction system with the military determination for physical capability to serve and the service branch. This was in contrast to the Anglo-American induction method, which separated the induction classification from the military judgment that placed an inductee into a service branch after the selective service induction physical examination.[13] The *Wehrmeldeamten* examined up to eighty recruits each day. The induction process was the responsibility of the local military subregion commander (*Wehrbezirkskommando*).[14]

Recruits reported for induction with a birth certificate, proof of ancestry, school certificate or proof of work skill, as well as any evidence of membership in Nazi organizations, as well as the driver, flying, or radio groups. In addition to these, evidence of awards for sports was important. Using all these categories, the mustering built upon the doctor's recommendation at the joint interview process to determine which service and arm of service and branch to assign a recruit.[15]

The *Wehrkreise* created the tactical organizations, such as infantry and Panzer divisions, within their boundaries and recruited their members within their own areas. An example of this process was that in *Wehrkreis VI,* Rheinland-Westphalia, which prior to the expansion of the army after 1935 held a single infantry division.[16] As the expansion advanced, the army added more to each *Wehrkreis,* as well as the responsibility for organization of air force units and recruiting, and for the coastal *Wehrkreise* (II and X), the navy. Individual infantry companies from within the *Wehrkreis* VI boundaries consisted of up to 97 percent of men born within the region.[17]

Twenty-one-year-old members of year-group 1914 were the first conscripted, during June 1935, for one year of active service followed by first reserve. Beginning with the 1915 year-group in 1936, twenty-year-old inductees served for two years in active service followed by first reserve. The 1916 year-group followed in 1937, overlapping its service with the 1915 year-group, followed by the 1917 year-group in 1938 and the mustering out into the first reserve of the 1915 year-group. Thus each year-group overlapped the previous two, growing the army and first reserve until the mustering of the 1918 year-group in 1939 and the advancement of the 1919 year-group and the mustering into active duty of the first and second reserves and some *Landwehr* members for the opening war offensive against Poland on September 1, 1939. The gap between the last conscription class of World War I, year-group 1900, and the 1914 year-group left thirteen year-groups with no training, also known as the "white" year-groups (*die weißen Jahrgänge*). These became known by the official term *Ergänzungseinheiten (E-Einheiten).* Upon the 1908–1913 year-groups receiving short training courses of two and later three months during the years 1936–1939, they entered as second reserve, or for those receiving a minimum of nine months training, first reserve.[18]

Selection Standards

The physical standards for induction on May 29, 1935, are shown in Table 7.1. The initial minimum height restriction was 1.6 meters (5 feet 2 inches), but allowed an exception if the doctor noted in particular an inductee's good health and robustness, and those judged as *Tauglich* 1 or 2 (fit for service) went into the combat arms.[19] The doctor conducted an examination with the intent to determine the recruit's capability to withstand the rigors of combat, both physically and mentally. While the terms changed during the course of the war, with some categories combined and others added or eliminated, the basic intent of the medical qualification levels remained the same: to indicate fitness for battle duty within a combat organization. The following seven degrees of physical qualifications existed in 1935:[20]

Tauglich 1: Fit, highest category.
Tauglich 2: Fit.
Bedingt tauglich: Provisionally fit.
Zeitlich untauglich: Temporarily unfit.
Beschränkt untauglich: Restricted fitness.
Untauglich (für Wehrdienst): Unfit for war service.
Völlig untauglich: Completely unfit.

Guidance to examining physicians included the following examples for deciding on an arm of service in relation to physicality:

Table 7.1 Minimum Standards for German Army Recruits, January 26, 1936

Arms of the Service *(Waffengattungen)*	Desired Characteristics: General	Height	Eyesight	Assignment Not Permitted
Infantry			For communications and anti-tank, no color-blindness	
• Rifle and machine gun	Robust, persevering, able to cope with the strain of marches and battle			
• Communication platoon	Same as for communications			
• Mortars	Same as for light artillery			
• Anti-tank	Same as for Panzer units			
• Horse platoon (riders)	Same as for cavalry			
Mountain troops *(Gebirgstruppe)*	Robust, not overweight, strong bones, taut ligaments well-formed chest, wiry long limbs			Recruited only from mountain regions
Cavalry	Short upper body, long legs			
Artillery		For heavy artillery, preferably not under 5 feet 7 inches	For observers, no color-blindness	
• Light	Body weight not excessive			
• Heavy	Especially robust, muscular			
• Observers	Especially good eyesight and hearing			
Mechanized, tracked, and battle units (Panzer) *(Kraftfahrkampftruppe)*	Good eyesight and hearing, clear nasal passages, free from eye disease, no indication of skin disease	Not over 5 feet 9 inches	No color-blindness, normal vision without glasses	No glasses, not over 5 feet 9 inches
Engineers (*Pioniere*)	Athletic, muscular, or complete body form	Not under 5 feet 5 inches		
Communications *(Nachrichtentruppe)*	Good eyesight and hearing		No color-blindness	
Motorized *(Kraftfahrtruppe)*	Same as for Panzer units			
Combat medical *(Sanitätstruppe)*	Same as for hospital units			

Sources: BAMA RH 26-17/104, *Der Reichskriegsminister* 49n 11.10; AHA/S Jn(IV), Nr. 374.12.35.II; Ang, *Änderung der Anleitung zur Untersuchung Wehrpflichtiger und Freiwilliger für die Wehrmacht* 26 (January 1936), 3.

- *Tauglich* 1: Completely physically and mentally sound and a minimum of 160 centimeters (5 feet 2 inches) in height and of robust build.
- *Tauglich* 2: A man with a greater number of physical faults from within the same table as *Tauglich* 1, but not sufficient in the view of the doctor to impair his unrestricted performance of duty and a minimum of 154 centimeters (5 feet) in height.[21]

The first determination at mustering was the recruit's physical capability. The doctor recommended the service (*Wehrmachtteile*), army, navy, or air force, and the arm of service (*Waffengattungen*) for those found qualified for combat duty in *Der Reserve, Der Ersatzreserve,* and *Der Landwehr*, following the physical examination.[22] The basis for the physical determination lay in the nearness to physical combat expected for the soldier, with those bearing the highest physical capability intended for the *fechtende Truppe,* or front, or fighting soldiers. Within this first category, German regulations separated the arms listed in Table 7.1 as those requiring the most physically and mentally fit with the anticipation that their position and task was nearest the actual combat environment. The regulation provided two examples to the mustering board for arm of service based upon the medical faults tables (*Fehlertabelle*) for *Tauglich* 1 and *Tauglich* 2:[23]

- Height 180 centimeters (5 feet 11 inches), weight 75 kilograms (165 pounds), no noteworthy defects. *Tauglich* 1: Qualified for all arms, preferably heavy artillery.
- Wears glasses, noteworthy faults: one eye below minimum, and color-blind. *Tauglich* 2: Qualified for all arms except infantry rifle and machine gun, communication, Panzer units, and *Luftwaffe* flak artillery.

In addition, insufficient teeth or teeth requiring extensive dental work disqualified a man from combat duty before 1941.

There was no standard intelligence test and no psychological testing for the vast majority of German recruits. Psychological tests, introduced in 1926,[24] became a requirement limited only to officer candidates, pilots, aircrew, drivers, radio operators, and a few other specialties. The army discontinued these tests in 1942.[25] Illiteracy was not a concern within the borders of the German state, and there was no program as in the US and British Armies to train illiterate soldiers to a minimum acceptable literacy for service. Complaints of illiteracy focused on the German-language capabilities of the *Volksdeutsche,* or ethnic Germans, from countries or annexed regions outside the 1918 borders of the German *Reich.*[26]

During the course of the war, this first category opened in 1942, and allowed other specialties to seek soldiers from the highest physical category. The priority for fighting or front soldiers remained in effect:

- Railroad artillery and railroad pioneers.
- Technicians (petrol, electric, water).
- Propaganda/correspondents.
- Railroad maintenance troops.
- Map and survey troops.
- Fortress troops.
- Military prison guards.
- Military police.

Men with the highest physical qualifications received no assignments to the following specialties for the support services (*Versorgungstruppen*) of those higher tactical formations such as divisions, regiments, and battalions/squadrons, including:

- Combat unit supply (within divisions, regiments, and battalions/batteries/squadrons).
- Ordnance/ammunition supply.
- Division and regimental medical personnel for aid stations and field hospitals.
- Veterinarian.
- Tank and tracked vehicle maintenance at division or higher level.
- Water supply.
- Traffic control.
- Field post/mail.

Further toward the rear of tactical formations were the supply and logistics services (*Nachschubtruppen*), including higher-level transport for tanks and tracked vehicles, army and army group–level services, and corps and division logistic support. These were the transport columns for division and higher-command echelons for supply, ordnance, maintenance, petrol, veterinary, and medical battalions, organized into company-level units assigned within the rear-area boundaries of the higher tactical formations, and supporting the army groups, armies, and corps fighting units.

A final grouping with the lowest physical category that might be assigned to positions near the front were the security troops (*Sicherungstruppen*). Members ranged from the actual security companies and battalions responsible for rear-area security behind the armies, to the militia and home guard units (*Landesschützen*) in the homeland, to the units of the security police (*Geheime Polizei*), and the prisoner-of-war organizations, camps, and guards.[27]

Building the Combat Divisions

By the beginning of 1939 the peacetime army had grown from 100,000 in 1932 to 3.8 million. Conscription enabled the buildup of the army to forty-five divisions by the end of 1938 (up from ten in 1932).[28] Despite its focus on building a mechanized army, the thirty-five infantry divisions remained foot-mobile, dependent on horse-drawn supplies and artillery.[29] The army built and trained the divisions, but experienced a continuing weakness in its ability to equip, arm, and build ammunition reserves. This left the army short of four months' supply of ammunition of all types, and the *Luftwaffe* with three months' supply, in 1939.[30]

Mobilization increased during the summer of 1939 with the addition of divisions in three waves (*Welle*), increasing the total divisions expected as available for war to 103:[31]

- Eighty-six infantry divisions (including thirty-five active first-wave and sixteen second-wave).
- Three mountain divisions (active).
- One cavalry brigade (active).
- Four infantry divisions (motorized) (active).
- Four light divisions (one active).
- Five Panzer divisions (active).

The eighty-six infantry (foot-mobile) divisions consisted of the following:[32]

- Thirty-five active infantry divisions from peacetime: first-wave (for offensive combat).

- Sixteen infantry divisions: second-wave (for offensive combat).
- Twenty infantry divisions: third-wave (for border and fortress duty).
- Fourteen infantry divisions: fourth-wave (reserve duty for occupation and defensive only).
- One *Landwehr* division (activated for border and fortress duty).

Each wave represented a unique combination of soldiers on active service and the various reserves as well as differing strengths, based on what equipment and support formations were available upon activation.

The fifty-one infantry divisions of the first and second waves constituted the primary infantry formations prepared for combat in 1939 and 1940. Given the focus on the 1908–1913 year-groups for first and second reserves and the return to civilian life in the first reserve of the 1915–1916 year-groups by fall 1938, members of the first- and second-wave divisions were in the optimal age groups for infantry, with the longest and freshest periods of recent training.

The third-wave divisions received the majority of their soldiers from the first and second reserves under age thirty-five and with minimal training during the period after the reestablishment of conscription in 1935. These divisions also activated *Landwehr* soldiers, between the ages of thirty-five and forty-five with either no training or limited World War I experience. The army placed these divisions in static positions in the west, recognizing the need for extensive combat training, and replaced their older soldiers in preparation for the assault into the Soviet Union in 1941.

The fourth-wave divisions activated in August and September 1939. The army considered them unprepared for offensive operations in the west in 1940, and so they entered France for occupation duty after the cessation of combat operations. Their *E-Einheiten* soldiers from the first and second reserves and active soldiers manned the combat arms positions, with the *Landwehr* soldiers in support specialties. The thirty-four divisions of the third and fourth waves and the *Landwehr* division did not participate in combat operations in 1939–1940.[33] By the spring of 1939 the active army had, either in active units or arriving units, the three year-group classes of 1916–1918 available in active divisions (conscripted in 1937–1938), and the trained first reserve manpower of year-groups 1914 and 1915 (conscripted in 1935–1936).

Each division by wave had a different organization to account for its differing strength totals. These changes occurred in the areas of combat support weapons and reconnaissance units, as well as the support and supply columns. The first-wave formations achieved the highest level of organization and personnel as they were the first line combat infantry divisions of the German Army. For example, a second-wave division organized without a motorized heavy machine gun company in its anti-tank battalion when compared to a first-wave division in 1939–1940. Various small changes throughout these formations represented the state of munitions and vehicle and weapon system availability as each wave activated.[34] Prior to the attack on the Soviet Union, the infantry divisions achieved a more standard appearance, exemplified by the 253rd Infantry Division, activated on August 28, 1939 (fourth wave).[35]

An example of the growth of the replacement army within a *Wehrkreis* was the process of activation in *Wehrkreis* VI. This region was the home to a single division, the active Sixth Infantry Division of the *Reichswehr.* This increased to ten divisions by the opening of the war, as each wave assigned a new set of divisions and new replacement units to support them.[36]

Through the course of the war, thirty-five waves of mobilized infantry divisions came into service. Many divisions, including all Panzer/armored divisions, organized outside this structure of waves, but each retained its regional affiliation as all other divisions upon activation.[37]

The Field Army and the Replacement and Training Army

At the opening of the offensive phase of the European war in 1939, the German Army was a single organization with two main parts: the field army (*Feldheer*) and the replacement and training army (*Ersatzheer*), both under the leadership of the commander in chief of the army. The *Feldheer* consisted of all active fighting formations engaged in combat operations. The *Ersatzheer* owned responsibility for the entry and initial training of all men of the German *Wehrmacht,* which included the air force and navy. The *Ersatzheer* linked each soldier at his induction to a replacement army training formation connected to a *Feldheer* unit. For all of a soldier's time in the army, the connection to his original induction and training regiment or battalion endured.[38]

His field army unit updated his record book through the office (*Wehrmeldeamt*) of his induction. If wounded and requiring more than eight weeks to recover as combat-capable, he returned to a hospital in his own region. His original local *Ersatzheer* unit took responsibility for his recovery and eventual return to his original field army formation. In addition to these responsibilities, the commander-in-chief of the *Ersatzheer* was also responsible for ensuring uniformity of training, weapons and equipment distribution, and the inspectors for training in each arm of service.[39]

The responsibility for induction, training, and equipping incoming soldiers and moving them to their active army formations, as well as the inspectorates for all combat and support specialties, lay with the original *Allgemeines Heeresamt* with Friedrich Fromm as its head. Fromm began in the *Allgemeines Heeresamt* as a lieutenant colonel in 1934, and on August 31, 1939, one day prior to the opening of hostilities with Poland, he assumed the position of commander in chief of the replacement army.[40] He continued with responsibility for the *Allgemeines Heeresamt* in his prior role. In order to resolve ongoing differences between the office charged with the weapons and ammunition procurement for the army, under the general of artillery Karl Becker, Fromm gained the combination of the two offices in one title with himself as the chief of army armaments and commander in chief of the replacement army on November 16, 1939.[41] The essential responsibilities of the new office were the training, replenishment, and armament of all personnel (less officers) and the materiel, and financial dealings for weapons, munitions, and supply management for the army.[42]

The Supreme Command of the German Armed Forces, the staff supporting the War Ministry, also exercised direct access to activities within the *Ersatzheer,* often bypassing the army command. There was an unavoidable interface between the Supreme Command, the army headquarters, and the civilian economic entities charged to supply manpower, armaments, and ammunition to the field and replacement armies. The inevitable conflicts between the soldiers and the civilian economy placed great strain on the relationships between the Supreme Command and army headquarters staffs. The bizarre workings of the Nazi hierarchy further allowed individuals, such as General Fromm, to gain direct access to Hitler if he became someone he trusted to carry out his desires. The ad hoc command relationships of the Nazi bureaucracy led to satrapies within government, whose members and influence waxed and waned with the fortunes of the war, to the ultimate detriment of the soldier and the war effort.[43]

The Field Army

Poland. On April 3, 1939, following the Anglo-French guarantee against German aggression toward Poland, Hitler advised the army to prepare for war against Poland. With an initial preparation date for invasion of August 20, 1939, the attack awaited the August 21 conclusion of the German-Soviet nonaggression pact.[44] The attack on Poland, assured of no Soviet interference, went forward. Germany organized 103 divisions by September 1, 1939. Of these, fifty-two saw combat in Poland. A further forty-three divisions received assignment to the Western Front to guard against possible French attack, with the remainder (eight) held in army headquarters reserve. Within eighteen days, the campaign was over. The collapse of Polish resistance in the face of German combined arms superiority was dramatic. The Soviet Army entered Poland after its defeat became obvious in accordance with the 1939 treaty.[45]

While the attack on Poland fulfilled the doctrinal expectations for a war of fast-moving, mechanized forces, supported by tactical air power, many problems came to light by the speed and depth of maneuver. The army found itself unprepared for the quick turnaround and attack on the West that Hitler ordered for mid-November 1939, and prevailed over time in postponing that attack until the spring to await the replenishment of arms, equipment, and training. After-action reviews noted, despite the spectacular success, that the preparation of officers and NCOs was seen as not up to the standards of the *Kaiserheer,* due to the expansion of the small cadre of officers of the old *Reichsheer* across the new army. The army set about a thorough reorganization, training, rearmament, and expansion program, as Hitler demanded an immediate attack on the West.[46]

Denmark and Norway. Frustrated at his generals' success in delaying the attack in the West, given the extraordinary success and rapid victory over Poland, Hitler reacted to the threat of possible British assistance to Norway by directing the Supreme Command staff to deploy the navy and air force with a few army divisions to occupy Denmark and attack and occupy Norway on April 9, 1940. This offensive operation was again a complete success, although it entailed significant and irreplaceable German naval force losses. The British evacuated their counter-landings at various Norwegian ports by sea by the end of June 1940, and "there is no doubt that the German demonstration of effective combined operations reinforced the psychological impact of the Polish campaign and contributed to the growing defeatism of the French and Belgian governments."[47] Hitler's decision to use the Supreme Command staff to prepare and direct operations against Norway and Denmark, as well as his overriding of army headquarters resistance to the immediate attack against the West, "signaled his intention to take over command of the Army itself. . . . [T]hus did the [Supreme Command] begin to supplant the [army headquarters]."[48]

France, the British Expeditionary Force, and the Low Countries. Of 135 divisions formed at the time of offensive operations in the west on May 10, 1940, 93 engaged in combat operations. The remaining divisions were not considered ready for the offensive or were held in reserve in the east. The collapse of the French Army and expulsion of the British Army concluded the campaign in six weeks to the armistice of June 25, 1940. Events caught the generals and Hitler by surprise. Superior combined arms maneuver, and the accompanying destruction and surrender of surrounded Allied formations, contributed to panic and paralyzed uncommitted Allied commands. The concentric operational maneuver of army groups to achieve a battle of annihilation confirmed the interwar doctrinal training and expectations for *Bewegungskrieg*.[49]

The Replacement Army

The German Army suffered over 43,000 casualties in the attack on Poland, including over 11,000 killed and 3,400 missing.[50] The primary concern of the army following combat operations in Poland lay in the area of training. From 1934 to 1940, the army expanded from 100,000 men and 10 divisions to 3.7 million men in 156 divisions in the field army and the replacement army, the *Ersatzheer*. The additions in 1940 stressed the capacity to find and train the numbers of officers and NCOs required. Critical after-action reports noted that infantry small units demonstrated weakness in fire discipline, use of supporting fires, proper reconnaissance, and dispersal during the attack. These defects appeared to stem from inadequate leader training from regimental commanders to squad leaders.[51] Training of battalion and regimental commanders became the responsibility of the General Staff, and brought together groups of 300 commanders in classes of three to four weeks each until May 1940. An entire infantry division and training area were placed at the disposal of the officers in training.[52]

Training of NCOs and soldiers was the responsibility of the divisions down to the company commanders. The army guided uniformity in the application of doctrine with common training directives. The *Ersatzheer* ensured compliance with its combat arms inspectorates. For infantry training, this guidance lay in a capstone document that encompassed all infantry unit guidance for battle preparation, from formations and orders, to the employment of machine guns and mortars, and the control of supporting arms.[53] Nevertheless, the breakneck pace of activations and limitations on the number and quality of remaining NCOs in the replacement pool, many of whom proved inadequate in previous NCO training, dictated that thirty-seven infantry divisions were not available and were declared fit for defensive duties.[54]

The addition of waves of infantry divisions of the fifth through tenth further complicated the task of preparation for offensive combat. The tenth-wave divisions, activated during May 1940, received ten to fourteen days of training. For these tenth-wave divisions, officer aspirants filled some officer positions, and NCOs occupied only 50 percent of their positions. Recognizing these limitations, the army disbanded these formations following the end of the campaign in the west, finding them useful only for fortress duty, due to the age and number of the second reserve and *Landwehr* I constituting the majority of their soldiers.[55] The army also activated the first four local, stationary defense (*Bodenständige Infanterie*) divisions. The army intended these weak divisions to relieve active divisions for the tasks of defense of locations behind the front, or in coastal defense, and they were not trained or equipped for offensive battle tasks.[56]

The General Staff's success in delaying the attack in the west until May allowed for the acceptable leader preparation of those divisions envisioned to participate in offensive combat operations. To improve the munitions and weapons status of the leading divisions, the army lent thousands of soldiers to the armaments industry until just prior to the attack in the west.[57]

The main tactical effort of the army was the training and preparation of the "fast" units, *die Schnellen Truppen,* consisting of the Panzer, cavalry, and infantry (motorized) divisions and brigades. Of the 156 divisions of the *Heer,* seventeen divisions and one brigade of mechanized or motorized Panzer and light divisions formed the spearhead of the attack, or 10 percent of the army. A total of 135 divisions of the overall organization of 156 divisions mobilized for the *Westfeldzug,* the attack in the west of May 10, 1940, comprising the ground forces of the German *Wehrmacht* for that attack.[58] A process of de-motorization took place prior to the attack against France and its allies in 1940, due to limitations in truck production. To ensure the fullest motorization of the mobile divisions, the German army made the infantry ever more dependent on horse-drawn artillery and supplies. These divisions became eerily similar to their World War I forebears. Of the total formations available for the attack, a maximum of 93 divisions, of the 135 made available, participated in the attack, with 42 held in reserve or in the east and deemed not ready for offensive combat.[59]

As the active forces expanded for the attack on France, so the replacement army grew. Absorbing these greater numbers of units, the *Ersatzheer* created reserve divisions to consolidate the multiplying organizations for the training and replacement of the fighting divisions. Prior to the attack in the west, each *Wehrkreis* formed two of these divisions for this purpose.[60] For example, within *Wehrkreis* XIII, *Nürnberg,* the 173rd Reserve Division in 1940 incorporated the 17th Infantry Division's (first wave) replacement regiment, as well as that for the 73rd Infantry Division (second wave), and the 231st Infantry Division (third wave). The reserve division's structure included the replacement and training units for all artillery, cavalry, anti-tank, and supply and communications units for these field divisions, allowing for greater control and responsiveness.[61]

Quality Standards

The quality standards of the German infantry and combat arms soldier did not change with the beginning of the war, but achieved greater granularity in the medical gradations for combat and other services. The September 10, 1939, change combined the two *Tauglich* grades for greater efficiency. The physical qualification standards of 1939–1940 were as follows:[62]

- k.v. (*Kriegsverwendungsfähig*): War/combat service–capable overall and suitable for any assignment (*Ersatz* first reserve).
- g.v. *Feld* (*Garnisonverwendungsfähig Feld*): Field garrison service–capable, only for limited assignment with front units except combat medical units, and namely for office, cooks, and the like. Fully suitable for rearward organizations, construction units, homeland defense units (*Landesschützen*), homeland flak units, and units in the rear of operational areas (*Ersatz* first reserve).
- g.v. *Heimat* (*Garnisonverwendungsfähig Heimat*): Homeland garrison service–capable, only in the homeland not in an operations area; suitable as a trainer, in watch units, cooks, workshops/garages, in an orderly room/office, or administrative office, and so forth, but not as fully fit for service (*Ersatz* second reserve).
- a.v. (*Arbeitsverwendungsfähig*): Labor service–capable, suitable as for *g.v. Heimat,* but not suitable as a trainer or in watch units (*Ersatz* second reserve).
- v.a.u./a.v.u. (*Arbeitsverwendungsunfähig*): Not labor service–capable, suitable for no beneficial, regular labor tasks (mustered out as unfit for war service).

Replacements, Training, and Rearmament, June 1940–June 1941

Total killed of the German Army in the war from September 1939 to the end of August 1940 was 76,938.[63] By June 1941 there were some 3.8 million men in the field army and 1.2 million men in the replacement army. The increased size of the replacement army came with the induction of year-groups 1919 and 1920 in 1940, and year-group 1921 by early 1941; year-group 1922 soldiers were not available for enlistment until after October 1941 following their labor office service.[64]

With Hitler's directive to prepare to attack the Soviet Union, already decided in August 1940, the army once again set out to reorganize, retrain, and rearm, expanding to 208 divisions, with 149 divisions earmarked for the attack. Doubling of the Panzer divisions occurred through the expedient of halving the number of tanks in each division, giving the illusion of an expanded Panzer capability. The air force continued attacks on Great Britain into the spring of 1941, unable to recover its losses by June 1941, as it turned to support the attack against the Soviet Union. The losses included transport aircraft and airborne soldiers in the attack on Crete to the extent that Germany never again conducted a major parachute assault. Attacks into the Balkans and Greece secured the southern flank of the attack and brought contingents from Bulgaria, Hungary, and Italy to help with the war in the east. Unlike preparations for the attacks against Poland and France, the anticipation of Hitler and his generals centered on planning for a short war and the decisiveness of German tactical forces to end the war in Russia within six to eight weeks. Strategic diversions to North Africa to assist the Italians were increasing, moving men and armor from the east already in 1941.[65]

As in the aftermath of the Polish campaign, the army sought to capture the principal strengths and weaknesses of the infantry after the end of combat operations in France. In December 1940, the army headquarters released a summary of experiences for the improvement of training for the infantry in the attack. The document emphasized the shortcomings of the small-unit leader initiative. Poor reconnaissance led to high casualties. The French were very good at camouflage and often let patrols go past them before opening fire. The critique stressed the importance of concentrating infantry heavy weapons for effect on the *Schwerpunkt,* or main point of the attack. Revised training mandates highlighted the importance of reconnaissance patrols (*Spähtrupps*), focusing special attention on finding the most aggressive and knowledgeable men to lead these patrols.[66]

The manpower projections after final mobilization for the attack on the Soviet Union revealed that even with the activation of year-groups 1919 through 1922, the reserve and replacement situation for the army was such that 46,600 men appeared available as casualty replacements after September 1941.[67] Arrogance and ideology drove a war plan that assumed a lightning victory on the model of 1939 and 1940.[68]

The tug-of-war between industry and the armed forces continued. As a result of the ongoing struggle with Great Britain, the priority for allocation of raw materials for armaments lay with the navy and air force. The army received orders to expand the number of divisions and returned these same soldiers back to industry to build the ammunition needed for the coming campaign. The soldiers of eighteen divisions went into the factories in 1940, with a promise of their return by February 1941.[69]

The army rated the divisions going into the Soviet Union according to battle-worthiness on June 20, 1941, as follows:[70]

- One hundred thirty-four divisions and three motorized infantry brigades as fully combat-worthy in the attack (Panzer, infantry, infantry [motorized], mountain, and cavalry divisions).
- Eight infantry and mountain divisions with somewhat limited attack capability.
- Nineteen infantry divisions with low attack capability and mobility.
- Twenty-two infantry divisions with limited attack capability.
- Fifteen infantry divisions for area security only.
- Nine security divisions.

Building these divisions required sending cadres of combat-experienced officers, NCOs, and soldiers to the new formations. This process, while unwelcome in the losing divisions, created the basis for the cohesive behavior of the gaining small infantry units. The morale of the victorious army extended into the new formations and infused their preparation for battle. Based upon a common expectation for quality leadership and a common understanding of its tactical precepts, the infantry achieved perhaps its highest capability on the eve of the attack into the Soviet Union.[71]

The Anvil Forged

The battle for France in 1940 confirmed the ability of armies to break the stalemate of position warfare that characterized World War I. Germany's concentric operational maneuver of army groups—penetration, encirclement, annihilation—to achieve the destruction of the French and British mobile armies demonstrated the possibilities of such warfare to achieve absolute political aims in the destruction and final victory over France. The German Army that Hitler now turned to destroy the Soviet Union remained a functional foot-mobile infantry force with a small percentage of mechanized and motorized divisions leading the attack. The highest expression of both the quality of the German infantry, and its combined arms application, occurred in the concentric movement of army groups in France. The attack on the Soviet Union featured planning for an eccentric operational maneuver of the army groups to achieve separate geographic and strategic objectives in the Soviet Union: Leningrad for Army Group North, Moscow for Army Group Center, and Kiev for Army Group South.

These eccentric movements denied the army groups a common focus on the enemy in favor of geographic objectives. The grand concentric encirclement battle on an operational scale as accomplished against France and the British Expeditionary Force devolved in June 1941 into a tactical method for each army group to achieve within its own boundaries and with its own forces. This devolution, coupled with Hitler's fixation on territory rather than the destruction of the Soviet Army, denied the strategic effect that the German method of war achieved against the Western Allies in France.[72]

The Infantryman and the Division

"A soldier relied upon the division until his death or capture, or injury which might have led to change to another organization or his release from war service or other conditional change from the division. Between his entry and exit he found himself in a steady cycle between field and replacement units as well as within the compass of its individual components. This is of special meaning for the analysis of the members of the division as a social group. Because irrespective of their membership in a field or replacement troop unit at a definite time, the members of a division always constitute only *one* social organization, which can only be captured through an analysis of the complete connectivity of the field and replacement units."[73]

The 253rd Infantry Division provided an example of the process of mustering, enlistment, training, and movement from the *Ersatzheer* to the *Feldheer.* Inducted German soldiers joined the infantry regiment formed in their local area, basing all recruitment on the creation and sustainment of divisions and their regiments within the region.[74] The 253rd Infantry Division organized with a replacement and training regiment in the *Ersatzheer,* with the same number—the 253rd Infantry Replacement Regiment. The replacement organization was a functional part of its parent division, and combined within the regiment each sub-element of the division with a corresponding replacement unit. The *Ersatzheer* replicated this throughout Germany for each division regardless of type.

Once mustered (*Musterung),* a recruit for the 253rd Infantry Division spent up to six months in labor service with the *Reichsarbeitsdienst.* Enlistment (*Einberufung*) into the army proper followed *Reichsarbeitsdienst* service. The infantry recruit then entered one of the three infantry replacement battalions for basic and combat specialty training. These battalions' designations, 453rd, 464th, and 473rd, matched the infantry regiments of the 253rd Infantry Division.

Upon completion of his initial and combat training, he joined other trained soldiers of all military specialties for movement to the division in the field. The organization for movement was a march unit of battalion or company size (*Marschbataillon/Marschkompanie*). From April 1941 to January 1945, seventy-six such movements arrived at the 253rd Infantry Division in the field, comprising 18,405 men, both new recruits and returning recovered wounded, as well as 4,613 replacements originating from other units from the *Wehrkreis* VI region.[75] Among these seventy-six movements the term *Genesenenmarschkompanie,* or "recovered wounded march company," predominated. Within these march units recruits combined with the recovered wounded and moved as a unit to the front under the leadership of their own officers and NCOs. These leaders, recovered from wounds and returned to physical capability as war/combat service–capable, served as trainers before assignment back to combat.[76] On arrival, members of the march unit received training in the division's field replacement/training battalion (*Feldersatzbataillon*) for combat preparation and familiarization with the division's battle area before final assignment to a company, platoon, and squad within the division. Up to 97 percent of the men in the unit were from the home area and spoke the same dialect of German.[77] The German Army was both the field army and the replacement army. The basis of their juncture was the localization of recruitment. The German soldier entered in or near his hometown and remained so connected throughout his service.[78]

The 253rd Division was not immune to the changing fortunes of the personnel situation during the war. Twice in 1940, the division lost the personnel of one infantry battalion and one regimental staff to form the basis for the new establishment of another division in *Wehrkreis* VI, the 298th Infantry Division (eighth wave), and later the 126th Infantry Division (eleventh wave), *Wehrkreis* VI.[79]

Entry Training

The entry of a soldier into the German Army, with its roots in the local community, when combined with realistic and difficult initial training, provided the strongest initial basis for the development of cohesion in combat. "The remarkable cohesion and fighting performance of the German army, its ability time and again to cobble units together out of broken fragments and use them effectively, owed much to the extensive, realistic and continual training given the *Landser*."[80] Seeckt stated in his *Observations of the Chief to the Army Command* in 1925 that "the principal thing now is to increase the responsibilities of the individual man, particularly his independence of action, and thereby to increase the efficiency of the entire army. . . . The limitations imposed by exterior circumstances causes us to give the mind more freedom of activity, with the profitable result of increasing the ability of the individual."[81]

This is the essence of the embodiment of the term *Auftragstaktik,* often translated as "mission orders"; the term characterized an entire range of expectations for the soldier and leader. It was the responsibility of both subordinate and leader to ask "Why?" and to understand the purpose of one's mission.[82] The *Truppenführung* stated: "Every man, from the youngest soldier upward, must be required at all times and in all situations to commit his whole mental, spiritual, and physical strength. Only in this way will the full force of a unit be brought to bear in a decisive action."[83]

The intent of the organization for the instruction of the German recruit was to build a close connection to the unit and the men with whom he served. In peacetime this lasted sixteen weeks.[84] During the various phases of the war, the length of this initial period for infantry lasted from eight weeks in 1940, to twelve to fourteen weeks in 1944,[85] and settled on six to eight weeks during April 1945.[86]

In both peacetime and war, the foundation of the induction, selection, and assignment process was to foster the connection of the soldier to his unit, such that "the inner bonding of the soldier was materially raised when the young soldier was placed from his first days of service in *his* company, *his* battalion, etc. He gained from the outset a sense of home [*Heimatgefühl*] in his unit. These psychological factors were very highly stressed."[87]

Before conscription, all volunteer recruits went into a company during the fall of each year. The company commander was responsible for preparing his recruits. The recruit became familiar with both the unit and the older soldiers. The *Principles for Education and Training in the Army* noted that "each leader bears the duty for the training of troops entrusted to him. . . . [T]he primary responsibility for the education and training task in the unit is the Company Commander. All superiors have the duty to support him in this solemn duty, without restricting him."[88]

The responsibility for the preparation of the army for battle lay with the commander in chief. The chief of the General Staff directed the combat operations of the army, and oversaw all combat preparation of the army. The commander of the replacement army owned the responsibility for training, equipping, and arming in matters of replacement personnel and material. Through his General Army Office, he executed his specific responsibility for all troop officer and soldier schools and for all instructional directives, owning each of the inspectorates of the combat arms, such as infantry, artillery, mechanized and motorized troops (*die Schnellen Truppen*), and others of the fighting troops (*Fechtende Truppe*), as well as the supporting arms and services. The oversight of school curriculum and tactical training directives came from the Training Detachment, located in the headquarters of the chief of the General Staff. Thus all combat preparation in the army came under the final oversight of the commander in chief of the army and the chief of the General Staff through its combat specialty inspectorates.[89]

The army sought demanding and realistic training. The term *Härteübung* ("hardness training") conveys some sense of the goal of basic and advanced preparation in the *Kaserne* (barracks). Live fire, sleep deprivation, and constant physical exertion, while common in degrees to all armies, were the normal fare of basic and advanced infantry preparation for the typical young German soldier. As Karl Fuchs noted: "The intensity of training is tremendous and there is not rest for anyone . . . [as] all of us are eager to make progress and no one complains. . . . I've become such an integral part of my company that I couldn't leave it ever again."[90] And as Hans Woltersdorf noted: "My unit was my home, my family, which I had to protect."[91]

Once conscription began and the divisions received their training regiments, the administration of combat preparation changed to the replacement regiments (*Infanterieersatzregiment*) associated with each field division. The 253rd Infantry Division followed this pattern. The division's replacement organization provided both basic and combat training for recruits. During the attack on France, the division and its replacement units moved to the vicinity of Thorn and Graudenz near East Prussia. During this stay, the division's replacement echelons joined the 156th Division, along with similar formations for four other divisions of *Wehrkreis* VI.[92] With an intervening command element now entered between the division and its replacement regiment, "this naturally diluted some of the idea of the soldier's home, but did not fully break it. Every field unit maintained not only for each individual soldier his homeland entity [mustering place], but also his appropriate 'replacement unit,' which was entered into his personnel record."[93]

Pre-Induction Training

A January 12, 1936, law required the membership in the *Hitlerjugend* organization of all German youth aged ten to eighteen years.[94] This law ended all other similar organizations, such as Boy Scouts, or Catholic and Protestant faith groups. The intent was the "war education of the German youth and their training in the beginning of soldierly qualities through field and shooting sports."[95] On April 6, 1939, in an agreement with the Supreme Command, young men designated as leaders (*HJ-Führer*) received assignment to infantry units upon their enlistment.[96] During 1940, a special three-week course of physical hardening intended to strengthen the youth in preparation for their war service was introduced.[97]

The *Reichsarbeitsdienst* was a required labor obligation following mustering for all civilians conducted prior to their enlistment (*Einberufung*). This was a twenty-six-week commitment of labor service for the *Reich.* At the first induction in March 1935, year-group 1915 received exemption for this service, and all subsequent year-groups went into the labor service for six months, shortened to three months for volunteer recruits.[98] The *Reichsarbeitsdienst* sought to inculcate a National Socialist ideal of the national community (*Volksgemeinschaft*) in a setting that discouraged religious discussion and permitted no class or social barriers.[99] Siegfried Knappe described his experience in the *Reichsarbeitsdienst* between April and September 1936:

> We spent the first four weeks learning military drill and routine. An important function of Labor Service was to free the Army from having to do this very basic type of training. Everyone who went into Labor Service would also be drafted in to the Army, and we would enter the Army already partially trained. . . . On our first work day following our training period, we marched forty minutes to a strip mine . . . [and] we were then issued working spades. . . . [T]he work we did consisted of removing layers of dirt from veins of coal and loading the dirt onto lorries.[100]

The primary benefit to the army of the *Hitlerjugend* and the *Reichsarbeitsdienst* in the view of author Stephen Fritz was that "the daily routine of the Labor Service had a clear paramilitary content whose aims were to instill the rudiments of military training and physical stamina but also to foster character, comradeship, and cohesion within the Nazi world view. Not surprisingly, given the degree of hard work, intense physical training, and military drill to which they had already been subjected, when these thousands of graduates of [*Hitlerjugend*] and [*Reichsarbeitsdienst*] flooded into *Wehrmacht* training centers, they were generally better prepared for what lay ahead than were their counterparts in Great Britain and the United States."[101]

Small-Unit Combat Training and Tactical Organization

> The soldier of the rifle company is the actual attacker and the one who carries the real attack onto the enemy with his light weapons.[102]
>
> The success of the attack depends on the aggressive spirit of the infantry.[103]

The basic regulations for the battle preparation of the infantry provided the basis for instruction of all combat arms.[104] Infantry combat training separated this experience into two parts. Both basic and combat training occurred within the same organization and prepared the soldier to take his place on the line in his infantry division. As Martin Pöppel noted:

> Our training was unbelievably hard, but basically fair. . . . The marches, exercises, night alerts, the shooting and radio practices were all even worse than before [in basic training]. Every day we fell into our beds completely exhausted. In action later on, we realized time and again how valuable this training had been for us. Sweat saves blood, that was a truism we confirmed later. We didn't know it yet, though, so we cursed and swore at everything and everyone.[105]

The key characteristic of the German Army squad was its simplicity. *Die Kriegsstärkenachweisungen des Heeres* was the equivalent of the British Army War Establishment and the US Army's table of organization and equipment. The *Kriegsstärkenachweisungen* defined the rifle company (*Schützenkompanie*) of a standard foot-mobile infantry division; the vast majority of the infantrymen of the German Army were in these divisions. The rifle company headquarters included the horse-drawn supply, armorers, and cooks, along with a machine gun platoon or half-platoon, a mortar platoon, and three rifle platoons (*Züge*). The German rifle platoon (*Zug*) consisted of three rifle squads (*Gruppen*). The squad of a 1939 rifle company was thirteen men organized into a machine gun element of four men, and a rifle element of seven riflemen, along with an NCO squad leader (*Gruppenführer*) and an assistant squad leader (*Truppführer*). This remained intact through the 1940 campaign in France.[106]

For the invasion of the Soviet Union in 1941, the army changed the squad again to one NCO and ten men. The machine gun team was now a three-man fire element, and the remaining six men constituted the movement element, all led by a single NCO squad leader. While the new organization reduced the size of the rifle squad, the addition of a fourth rifle squad, along with a light mortar section in one or two of the three platoons, mitigated the reduction. Other variations to the number of infantrymen existed in several different types of organizations within divisions. Reconnaissance companies, motorized rifle companies, and armored infantry companies organized with twelve-man squads to include a driver for the car or half-track carrier, and a minimum of two machine guns per squad.[107]

The rifle squad and platoon in which the German soldier found himself between 1939 and 1945 maintained its form and function throughout the war. If a soldier belonged to a rifle squad that began in 1939 with thirteen men and later transitioned in 1941 to ten men for the invasion of Russia, an additional squad, the fourth, absorbed the additional men and so he remained in the same platoon in a squad built around a single machine gun, comprising a fire element and a movement element. In 1943 the squad changed again with a reduction to nine men with one NCO and one machine gun, and this basic organization carried through the final course of the war, again organized with three rifle squads per platoon. An officer commanded one platoon in the new company organization, while NCOs commanded the other two platoons. Platoon firepower increased by the addition of two machine guns, not counting the one carried by each squad for a new total of five per platoon, and the company increased to twenty-seven MP43-type automatic rifles, with each squad leader authorized to carry one of these weapons.[108] The reduction of the squad to nine men in 1943 still retained the concept of a machine gun fire element and rifle movement element.[109]

It was not the intent of the army to keep divisions in the line, if capable of moving them out for periods of time to maintain their fighting edge. This was done more often with smaller units than the entire division as the war progressed. The rotation allowed the infantry to absorb its losses, accept new or returning members, refresh its equipment and training, and return to the front capable again of providing a formidable resistance, even as their overall numbers and condition diminished as the army devolved.[110]

The combination of adherence to quality manpower, regional manning of regiments within divisions, and a standard organizational and training concept for the employment of the infantryman within his squad, platoon, and company, created a coherence and opportunity to bond across these small units unequalled in other armies. With these factors in the background of his tactical organization, the small unit became the home to the German soldier and was a primary facet in the creation of the development of the personal bonding and staying power of the German infantry small unit.

Junior Leaders

The army reduced the four-year process of the years 1921–1934 to less than two years in 1935 by paring down the time required in the troop unit as a private soldier to nine months before attending his officer school and then assignment back to a troop unit.[111] Siegfried Knappe provided a typical example of this path to becoming on officer after his induction into an artillery unit in October 1936. He possessed the *Abitur,* and began officer training six months after the beginning of his service. Upon requesting the path to officer rank, he found himself in the military academy in Potsdam one year later. He graduated with the equivalent rank of an ensign after ten months in the military academy and was promoted to lieutenant two months later.[112]

The army selected NCOs of the former *Reichsheer* for officer-aspirants on the basis of an exam and recommendation by their regimental commander. Georg Grossjohan was an example of such a promotion, when after eleven years in service, he received promotion in 1940 and found himself as pioneer platoon commander in the attack against France. He followed this with assignment as an infantry lieutenant in June 1941 with the 198th Infantry Division, *Wehrkreis* V, Stuttgart. He later received the award of the German Cross in Gold and the Knight's Cross to the Iron Cross during his wartime service, encompassing the entire period of the war.[113]

Inducted into the 132nd Infantry Division, *Wehrkreis* XII, Wiesbaden, on its formation in 1940, Gottlob Bidermann entered the Soviet Union in late summer 1941, in combat on the Crimean Peninsula, participating in the taking of Sevastopol before transfer in summer 1942 to the Leningrad front at Lake Ladoga. Wounded in action several times, he received his notification of promotion to officer-aspirant while in hospital and transferred in September 1942 to infantry officer school near Prague. His time in officer preparation was three months before promotion and return to combat as an infantry officer for duty with the same division and regiment he served with as an NCO, the 437th Regiment, in January 1943.[114]

These three officers provide examples of the various methods of promotion to officer rank prior to and during World War II. The key characteristic was the requirement for service as a private soldier for all officer-aspirants prior to consideration and recommendation by their regimental commander for officer rank, both before and during the war. With the expansion of the army, it reduced the four-year program of the *Reichsheer* to less than two years. For NCOs whose service overlapped the *Reichsheer* and the *Heer,* the time was further shortened due to their experience. For recommendation toward advancement to officer rank during the war, combat service while leading soldiers was the first prerequisite.[115]

The intent of the *Heer* was to elevate the status, role, and capability of the NCO corps. Their success was a major factor in the cohesion of the German infantry. Well-trained, well-selected, veteran, and respected squad and platoon leaders held the infantry together in the face of an ever-increasing likelihood of their defeat and the end of the Germany for which they fought. In the words of the postwar study, the role of the German *Unteroffizier* was one of the highest importance and the highest impact.

> The noncommissioned officer assists the officer and must replace him if necessary. Upon his steadiness rests the inner cohesion of the troops. . . . [All] training of the noncommissioned officers . . . must emanate from standards . . . for the noncommissioned officer in the field [of battle]. *Correspondingly they must be the very same basic principles as for the training of officers*. . . . The assistants of the officers in the education, training and leadership of the troops are the noncommissioned officers. As well, their example in their professional calling and lifestyle, their mastery of the soldier's art, and their teaching competence are of decisive impact for the spirit and proficiency of the troops. The officer must be able to depend without fail on his noncommissioned officers.[116]

Discipline and Courts-Martial

As with the British system, soldiers lived under the restrictions of two sets of law, subject to both, while providing a strong dividing line between military and civil. An extensive taxonomy of serious civil crimes was held under the jurisdiction of civil authority, restricting the military to the category of discipline most associated to military offenses against good order. Thus murder, for example, fell within the German civilian legal code.[117] There were, from 1934, three types of court-martial for which to refer a German soldier or officer for military, or in exceptional circumstances, civil offenses such as treason. In garrison during wartime, the primary form was the *Kriegsgericht* convened by a division commander of general officer rank or equivalent, including the *Wehrkreis* commanders in the replacement army (*Ersatzheer*). In the field during wartime, this was the *Feldkriegsgericht,* convened by the division commander. In exceptional cases, the division commander possessed the authority in wartime to delegate this court to a regiment, although in the *Wehrkreis* VI database there are no instances of regimental level courts-martial. These courts were the equivalent of the British Army general court-martial and field general court-martial and the US Army general court-martial. (See Appendix 3 for the court-martial systems in the German, US, and British Army systems.)

Penalties levied by a court-martial included death, confinement to a civilian prison and hard labor, confinement to a military fortress prison, or confinement to a military field prison or confinement facility. Awards for specified offenses were as varied as the organizations and units and the individuals themselves, and reviewed based upon the severity of the offense. For instance, the penalties for desertion or AWOL (absence without leave) ranged from death to fourteen days under close arrest.[118] Punishments for assaulting a subordinate included death, two to ten years in a penitentiary, incarceration in a fortress prison for six months, or less than fourteen days' confinement.[119] It required a court-martial judgment to reduce an NCO (*Unteroffizier*) in rank, or award confinement greater than four weeks.[120]

As in the US and British Armies, soldiers were at risk for a variety of opportunities for incarceration short of prison or firing squad, ranging from simple restriction to quarters or extra duty. The examples presented for the British Army of a field prison camp and detention barracks, and the US Army correctional facilities for bread-and-water discipline or the field (straight time) brig, were examples of these less-than-prison-level sentencing locations intended to deal with varying lower levels of soldier missteps.

Three types of punishment battalions existed in the German Army, short of civilian prison, field prison, or fortress prison incarceration, and all were located within the replacement army. Two, a probationary punishment battalion and a special field punishment battalion, were designated for soldiers sentenced by the field army. One, a special punishment training battalion, was for soldiers sentenced from within units of the replacement army.

Commanding officer discipline. Unit-level discipline, as in the British and US Armies, was the most prevalent form. The *Disziplinarstrafordnung,* or disciplinary regimen, for the German soldier read much like that for both the British and US soldier. The authority to sanction lay with the position of the office held, not on the rank of the individual. Thus, if a lieutenant was the company commander, he held the authority of a company commander; however, if the acting company commander was a sergeant, that power for the company was held by the battalion commander. Typical to the authority of the battalion- and company-level command were penalties for offenses ranging from unauthorized absence or overstaying leave up to seven days, or in the field for one day, escape from confinement, violating orders, lying to a senior on official business, and insulting a senior, unless the result of gossip or mudslinging against the offending individual.[121]

There were two caveats to the German concept of disobedience or insubordination as an infraction, allowing the soldier to disobey an order if, in his judgment, such an order constituted a significant threat to human life or property, the safety of others, or the alertness of the troops; and if the order intended to deprive a soldier of his ability to defend himself by laying down or giving up his arms or his service. Other offenses included encouraging disobedience in a subordinate or insulting a subordinate, willfully issuing a false statement to cause friction in a unit, abandoning a post or march without permission, drunk on duty, as well as actions to interfere with the duties of a sentry.

The penalties available at the unit level consisted of reprimands, extra duty, withholding of pay, and confinement to barracks for up to four weeks through duty restrictions requiring return to barracks before, at, or after taps. Disciplinary exercise was allowable for soldiers with less than four years' service, but for no more than one-half hour at a time and supervised by an officer. The authority for confinement included three forms of barracks or quarters arrest. The first was simple restriction to barracks or quarters allowing no movement outside of barracks except for duty and meals. Mild arrest was in solitary confinement with meals taken with the unit, and bed, books, and writing materials were permitted, as was outside exercise, after one week under supervision. Severe arrest was also awarded in solitary confinement, but with only hard bread and water, stopping on the fourth day of severe arrest and every three days thereafter.[122]

The attitude toward "chickenshit" of the British and US soldiers in the barracks was a common theme with the German soldier, as the following examples of offenses and punishments demonstrate:

- Twelve days' severe arrest for failure to have a proper haircut and being one hour and forty minutes late for taps.
- Three days' severe arrest and fourteen days' restriction to barracks for being twenty minutes late from liberty at taps.
- Ten days' severe arrest as an NCO for failing to have a button on the coat and an improper necktie during a uniform inspection.
- Ten days' severe arrest for being drunk and telling an NCO to "kiss my ass."
- Ten days' severe arrest for not returning to unit for one day after following an official trip into Paris.
- Ten days' severe arrest for not going to bed when ordered several times to do so.[123]

The soldiers of each of these armies shared in an extraordinary distaste for the type of discipline inflicted while in garrison. Soldiers seemed to share Paul Fussell's conclusion that "chickenshit" was a positive incentive to deploy into combat where the measure of the small unit leader became less about minor discipline and more about survival of the small group.

Strategic Direction and Nazi Leadership

Adolf Hitler was sensitive to the blow to home front morale and the collapse of the German standard of living and industrial capacity during World War I, and he was determined to keep the German people and labor happy despite war. Nazi ideological beliefs saw the number of women in the workplace decrease during the war, unlike either the United Kingdom or the United States. The byzantine Nazi leadership and management methods that pitted industries, factories, and armed services all against each other in competition for resources hampered central planning of the economy and the management of essential manpower resources for labor and warfighting. Hermann Göring refused any attempts to gain central control over the resources or industries supporting his air force. Mismanagement of manpower resources in the intersection between the labor needs of industry and the armed forces, with up to 5.5 million men deferred in reserved consumer occupations through 1941, received no adequate resolution as no central management occurred until too late, in a November 1943 order to mobilize the economy for war.[124]

For the army, the lack of rational planning led to the key characteristic of the Nazi regime—improvisation—in all things regarding armament development, ammunition and supplies, and manpower. The army's tactical successes to 1940 did not mask the ultimate and undeniable deficiencies in the management of the war economy. The army's own studies indicated the inability of the nation to sustain a long war, with severe shortages of all forms of ammunition going into the initial combat operations in 1939, with the expectation of greater shortfalls, if combat were prolonged, or if Germany entered again into a multiple front war. Hitler based his insistence on the early use of the army due to his own understanding of the limited window of opportunity to act before his opponents became too strong to defeat.[125]

Above the level of tactical execution, other disturbing trends foreshadowed challenges ahead for the army. Hitler bypassed the General Staff and army headquarters to use the Supreme Command staff for the planning and execution of the attacks on Norway and Denmark in 1940. Hitler's direct interference into tactical decisions in 1940 was unique in the experience of the General Staff. Hitler did not trust the General Staff and the army headquarters, substituting emotion and intuition for military judgment. He moved the army ever further away from the very qualities most prized by the professional General Staff.[126] The abnegation and acquiescence of the leaders of the army to the Nazi worldview left them precious little moral, and no legal, room to object to any order for strategic or tactical maneuver or coordination of the army in its operational tasks.

Conflicts with industry over manpower continued without resolution during this period. Following the victory in France, Hitler decreed the return of production levels for consumer goods at the expense of the armed forces. Even after deciding that priority must go to the navy and air force, despite the army's needs for the coming invasion of the Soviet Union, industry continued to find ways to exempt itself from Hitler's decrees by playing on its access to Hitler or his chief party leaders. Central management in an environment that pitted all players against each other became an impossibility at this stage of the war and loomed as catastrophic in a long and drawn-out struggle of Germany against the greatest industrial powers of the West and the Soviet Union.[127]

The moral slide begun by the generals as early as 1933 continued with the what became known as the illegal orders of 1941, those racially charged and immoral orders issued prior to the attack on the Soviet Union. With these directives, Hitler declared a war of annihilation against the Bolshevik-Jewish order of the Soviet Union. The most disruptive to discipline in the German Army and in strengthening the resistance of the Red Army was the *Kommissarbefehl,* authorizing the immediate execution of Communist Party political commissars when captured. Hitler allowed the order to lapse in 1942 at the insistence of the army over its disciplinary implications, not its moral basis, nor its violation of the traditional laws of war. Its withdrawal was too late to prevent the initial impression, and to magnify the fighting spirit of the Russian soldier, as well as the expansion of the definition of acceptable moral conduct in its own soldiers in the east. Nazi ideology told the soldiers they were superior to their enemy. The issuance of these orders changed the tenor of the campaign to an ideological and racial total war.[128]

Demanding the harshest measures against the Red Army and Soviet people, these orders compromised the moral boundaries of the German soldier beyond any of the armies studied here. Some officers refused to issue the orders.[129] Many, including leading generals such as Walter von Reichenau, commander of the Sixth Army; Erich von Manstein,[130] commander of the Eleventh Army; and Gerd von Rundstedt, commander in chief of Army Group South in 1941, passed them on and issued clarifying orders of their own in the dissemination of these measures. Some historians cautioned against the assumption that such orders translated into widespread army participation in support of the killing units, noting that "it is difficult to determine the exact extent to which army units supported individual massacres or to make generalizations about the willingness of the Wehrmacht to participate in the 'work' of the Action Groups."[131] These groups included not just the *SS-Einsatzgruppen,* but also the police battalions that followed in the wake of the army groups.

The hubris of the army leaders after their lightning victories, coupled with Nazi propaganda of the debased Soviet Union, blinded the army to its own personnel and arms weakness if the Russians proved more resilient than expected or planned. The "catastrophe" of victory clouded the true inability of Germany to sustain a long war on multiple fronts. The incompetent intelligence picture of the Russian Army and economy failed to identify the mass of personnel and materiel available to replace losses if the *Heer* failed to destroy the Red Army in the opening weeks of the campaign. These clouds loomed just over the horizon in the early morning of June 22, 1941, and the opening of Operation Barbarossa, the campaign against the Soviet Union.

Notes

1. Wilhelm Deist et al., *Ursachen und Voraussetzungen der Deutschen Kriegspolitik,* 425; O'Neill, *The German Army and the Nazi Party,* 131.
2. O'Neill, *The German Army and the Nazi Party,* 138.
3. Craig, *The Politics of the Prussian Army,* 488–490, 492–496.
4. Müller, *General Ludwig Beck,* 551–554; Müller, *Generaloberst Ludwig Beck,* 342–343; O'Neill, *The German Army and the Nazi Party,* 157–158; Craig, *The Politics of the Prussian Army,* 499.
5. Müller-Hillebrand, *Das Heer bis zum Kriegsbeginn,* 60–64.
6. Deist et al., *Ursachen und Voraussetzungen der Deutschen Kriegspolitik,* 525–529.
7. Kroener, *Generaloberst Friedrich Fromm,* 351; Absolon, *Die Wehrmacht im Dritten Reich: Band IV,* 395.
8. BAMA RH 26 17/96, *Wehrbezirkseinteilung für das Deutsche Reich, Reichsgesetzblatt,* no. 52, May 22, 1935; Müller-Hillebrand, *Das Heer bis zum Kriegsbeginn,* 31–32.
9. Absolon, *Die Wehrmacht im Dritten Reich: Band III,* 83–85, 347–348; Müller-Hillebrand, *Das Heer bis zum Kriegsbeginn,* 31.
10. BAMA ZA 1/1778, Personnel and Administration (Project 2b), P-006, Teil I, *Das Rekrutierungssystem der deutschen Wehrmacht in Frieden und im Kriege;* Anlage I, *Durchführung einer Musterung und Aushebung,* 1948, 73–75, 97.
11. Ibid., 98.
12. Ibid.
13. TNA PRO LAB 44/297, *Ministry of Labour and National Service Handbook,* January 1941, 10; TNA PRO LAB 29/127, Ministry of Labour and National Service, Ministry of Labour Codified Circular (M.L.C.C.), appendix 2, January 1941, 120; Hershey, *Special Monograph no. 5,* 37; Hershey, *Special Monograph no. 16,* 32.
14. BAMA RH 26 17/96, *Der Reichskriegsminister, Reichsgesetzblatt,* Teil I, no. 56, sec. 37, *Musterung stab,* May 29, 1935, 706; BAMA RH 15/417a, OKW, Az. 12.29 AHA/Ag/E (Ia), no. 502/39 geh., *Mob. Vorarbeiten für Einrichtung Freiwilligenmeldestellen,* Berlin, February 9, 1939, 1; Absolon, *Die Wehrmacht im Dritten Reich: Band III,* 92.
15. BAMA RH 26 17/96, *Der Reichskriegsminister, Reichsgesetzblatt,* Teil I, sec. 40, *Personalpapiere,* May 29, 1935, 707.
16. Müller-Hillebrand, *Das Heer bis zum Kriegsbeginn,* 14. Between 1935 and 1939, the army created four waves of infantry divisions, and *Wehrkreis* VI recruited nine of these divisions: 6th, 16th, 26th, 69th, 86th, 211th, 227th, 253rd, and 254th. The 1st (*Leichte*) Light Division, a motorized formation, also came from this region during this period.
17. Rass and Rohrkamp, *Deutsche Soldaten.* Data extracted by author from analysis of eight infantry and reconnaissance companies from the Rass database.
18. BAMA RH 15/208, *Der Chef der Heeresleitung,* no. 700/35 g.Kdos., Allg. E. I., *Musterung 1935,* March 13, 1935, 98–101; Müller-Hillebrand, *Das Heer bis zum Kriegsbeginn,* 32–34.

19. BAMA RH 15/208, *Der Chef der Heeresleitung,* no. 700/35 g.Kdos., Allg. E. I., *Musterung 1935,* March 13, 1935, 1, 30, 42, 50. (The date of this order for mustering was three days before Hitler's announcement.) Müller-Hillebrand, *Das Heer bis zum Kriegsbeginn,* 32–34.

20. BAMA RH 15/208, *Der Chef der Heeresleitung,* no. 700/35 g.Kdos., Allg. E. I., *Musterung 1935,* March 13, 1935, 50–51.

21. BAMA RH 26 17/96, *Der Reichskriegsminister, Reichsgesetzblatt,* Teil I, sec. 40, *Personalpapiere,* May 29, 1935, 710.

22. BAMA RH 26 17/96, *Der Reichskriegsminister, Reichsgesetzblatt,* Teil I, sec. 47, *Untersuchung auf Wehrtauglichkeit,* May 29, 1935, 710.

23. BAMA RH 26 17/104, *Der Reichskriegsminister* 49n 11.10, AHA/S Jn (IV), no. 374.12.35.II. Ang, *Änderung der Anleitung zur Untersuchung Wehrpflichtiger und Freiwilliger für die Wehrmacht,* January 26, 1936, 3.

24. Von Rabenau, *Seeckt,* 490.

25. BAMA ZA 1/1778, Personnel and Administration (Project 2b), P-006, Teil I, *Das Rekrutierungssystem der deutschen Wehrmacht in Frieden und im Kriege;* Anlage I, *Durchführung einer Musterung und Aushebung,* 1948, 98; BAMA ZA 1/1778 ,Personnel and Administration (Project 2b) P-005, Teil V, Dr. Max Simoneit, *Die Anwendung psychologischer Prüfungen in der deutschen Wehrmacht, Deutsche Wehrmachtpsychologie von 1927–1942,* 1948, 30–32; Vernon, *Personnel Selection in the British Forces,* 22.

26. BAMA RH 15/369, *Infanterie Ersatz Regiment 212,* IIb no. 1296/41 geh., *Lagebericht der 212.Inf.Div. vom 19.6.1941,* September 8, 1941, 1.

27. Absolon, *Die Wehrmacht im Dritten Reich: Band V,* 84–86.

28. Müller-Hillebrand, *Das Heer bis zum Kriegsbeginn,* 25.

29. Frieser, *The Blitzkrieg Legend,* 29.

30. Müller-Hillebrand, *Das Heer bis zum Kriegsbeginn,* 128; Kroener, *Generaloberst Friedrich Fromm,* 342, 346.

31. Müller-Hillebrand, *Das Heer bis zum Kriegsbeginn,* 68.

32. Ibid.

33. Müller-Hillebrand, *Das Heer bis zum Kriegsbeginn,* 68–70.

34. Ibid., 158–161.

35. Rass, *"Menschenmaterial,"* 447.

36. Ibid., 82.

37. Müller-Hillebrand, *Der Zweifrontenkrieg,* 236.

38. Rass, *"Menschenmaterial,"* 61–62.

39. Absolon, *Die Wehrmacht im Dritten Reich: Band V,* 61.

40. Ibid.

41. Kroener, *Generaloberst Friedrich Fromm,* 352; Absolon, *Die Wehrmacht im Dritten Reich: Band V,* 61–64, 388; Kroener, *Generaloberst Friedrich Fromm,* 319. Fromm and Becker collaborated in 1938 to present a solution to the problems of weapons and munitions procurement that included giving this responsibility to the army rather than the staff of the Supreme Command.

42. Absolon, *Die Wehrmacht im Dritten Reich: Band V,* 61.

43. Kroener, *Generaloberst Friedrich Fromm,* 352–356.

44. Craig, *Germany,* 710–712.

45. Müller-Hillebrand, *Die Blitzfeldzüge,* 17.

46. Groß, *Mythos und Wirklichkeit,* 209–210.

47. Craig, *Germany,* 717–718.

48. Kroener, Müller, and Umbreit, *Organisation und Mobilisierung des deutschen Machtbereichs,* 915–916; Boog et al., *Der Angriff auf die Sowjetunion,*

190; Maier et al., *Die Errichtung der Hegemonie auf dem europäischen Kontinent,* 197; Cooper, *The German Army,* 188–194.

49. Frieser, *The Blitzkrieg Legend,* 36.

50. Maier et al., *Die Errichtung der Hegemonie auf dem europäischen Kontinent,* 133; Overmans, *Deutsche militärische Verluste im Zweiten Weltkrieg,* 54. Overmans noted that the casualty figures for Germans killed in action in Poland ranged from 10,244 to 14,188 at the time of the battle, and up to 15,450 from a reckoning in 1944.

51. Groß, *Mythos und Wirklichkeit,* 209–210.

52. BAMA RH 14/42 OKH, Ia(I) Nr. 118/40 geh., *Umbildung der 164.Inf.Div. zu einer Lehr-Division,* January 8, 1940; Lewis, *Forgotten Legions,* 97.

53. BAMA ZA 1/1778, Personnel and Administration (Project 2b) P-005, Teil II, *Ausbildung und Verwendung der Unteroffiziere und Mannschaften im deutschen Heere,* 1948, 22.

54. BAMA ZA 1/1778, Personnel and Administration (Project 2b) P-005, Teil II, *Ausbildung und Verwendung der Unteroffiziere und Mannschaften im deutschen Heere,* 1948, 22.

55. Kroener, Müller, and Umbreit, *Organisation und Mobilisierung des deutschen Machtbereichs,* 825–830; Lewis, *Forgotten Legions,* 92–93.

56. Müller-Hillebrand, *Die Blitzfeldzüge,* 136; Müller-Hillebrand, *Der Zweifrontenkrieg,* 153.

57. Kroener, Müller, and Umbreit, *Organisation und Mobilisierung des deutschen Machtbereichs,* 948.

58. Müller-Hillebrand, *Die Blitzfeldzüge,* 44.

59. Frieser, *The Blitzkrieg Legend,* 35–36; Müller-Hillebrand, *Die Blitzfeldzüge,* 45–46.

60. Kroener, Müller, and Umbreit, *Organisation und Mobilisierung des deutschen Machtbereichs,* 827.

61. BAMA RH 14/42, Wehrkreiskommando XIII, Anlage, no. 8707/40 geh. Ia, *Ersatzheer W.K.XIII, Stand am (10.6.40),* June 12, 1940.

62. Absolon, *Die Wehrmacht im Dritten Reich: Band V,* 116–117.

63. Kroener, Müller, and Umbreit, *Organisation und Mobilisierung des deutschen Machtbereichs,* 829; Overmans, *Deutsche militärische Verluste im Zweiten Weltkrieg,* 54. Overmans noted that the casualty figures for German killed in action to the end of the campaign against France were 62,848 men and 3,367 officers.

64. Kroener, Müller, and Umbreit, *Organisation und Mobilisierung des deutschen Machtbereichs,* 853.

65. Ibid., 984–985; Cooper, *The German Army,* 250–251, 280–281.

66. Boog et al., *Der Angriff auf die Sowjetunion,* 270.

67. Ibid., 856.

68. Frieser, 350–351.

69. Kroener, Müller, and Umbreit, *Organisation und Mobilisierung des deutschen Machtbereichs,* 851.

70. Müller-Hillebrand, *Die Blitzfeldzüge,* 111; Kroener, Müller, and Umbreit, *Organisation und Mobilisierung des deutschen Machtbereichs,* 874–875; Boog et al., *Der Angriff auf die Sowjetunion,* 270.

71. BAMA RH 11 I/129, Oberkommando des Heeres, Gen.St.d.H./Gen.d.Inf. /no. 1900/40 geh., *Auswertung der taktischen Erfahrungen der Infanterie im Westen,* H.Qu. O.K.H., December 18, 1940.

72. Groß, *Mythos und Wirklichkeit,* 222–227; Glantz, *When Titans Clashed,* 32–33.

73. Rass, *"Menschenmaterial,"* 50, emphasis in original.
74. BAMA ZA 1/1778, Personnel and Administration (Project 2b) P-006, *Das Rekrutierungssystem der deutschen Wehrmacht im Frieden und im Krieg,* September 15, 1948, 69.
75. Rass, *"Menschenmaterial,"* 420–421.
76. BAMA ZA 1/1778, Personnel and Administration (Project 2b) P-006, *Das Rekrutierungssystem der deutschen Wehrmacht im Frieden und im Krieg,* September 15, 1948, 39–40.
77. Rass and Rohrkamp, *Deutsche SoldatenS,* 197–218.
78. Rass, "Das Sozialprofil von Kampfverbänden," 659; Rass, *"Menschenmaterial,"* 50.
79. Rass, *"Menschenmaterial,"* 51.
80. Fritz, *Frontsoldaten,* 13.
81. Condell and Zabecki, *On the German Art of War,* 4.
82. Ibid.
83. H.Dv.300, *Truppenführung,* Teil I, 3–4, Record Group 242, T-78, Roll 201, Frame 6144995-6144999, NACP.
84. BAMA ZA 1/1778, Personnel and Administration (Project 2b) P-005, Teil II, *Ausbildung und Verwendung der Unteroffiziere und Mannschaften im deutschen Heere,* 1948, 22.
85. Rass, *"Menschenmaterial,"* 147, n. 241.
86. *Ausbildungsplan für die kurzfristige Ausbildung (6 Wochen) . . . der Schützenkompanie,* April 18, 1945, Record Group 242, T-78, Roll 201, Frame 6144914-6144924, NACP; BAMA RW 15/126 OKW Abschrift aus no. 3053/45 g.K. AHA/Ia an Nachr. Ob.d.E./AHA Stab Ia, Betr., *Unterstellung der Ausbildungseinheiten des Ersatzheeres unter das Feldheer,* April 27, 1945, 2 (directed that persons with less than eight weeks training and without weapons could not be directed to the front).
87. BAMA ZA 1/1778, Personnel and Administration (Project 2b) P-005, Teil II, *Ausbildung und Verwendung der Unteroffiziere und Mannschaften im deutschen Heere,* 1948, 15, emphasis in original.
88. Ibid., 15, 17.
89. Absolon, *Die Wehrmacht im Dritten Reich: Band V,* 61–65; Müller-Hillebrand, *Das Heer bis zum Kriegsbeginn,* 116–117.
90. Fritz, *Frontsoldaten,* 15–19.
91. Ibid.
92. Rass, *"Menschenmaterial,"* 59.
93. BAMA ZA 1/1778, Personnel and Administration (Project 2b) P-005, Teil II, *Ausbildung und Verwendung der Unteroffiziere und Mannschaften im deutschen Heere,* 26.
94. Absolon, *Die Wehrmacht im Dritten Reich: Band III,* 469.
95. Absolon, *Die Wehrmacht im Dritten Reich: Band IV,* 49–54.
96. Ibid., 52.
97. Absolon, *Die Wehrmacht im Dritten Reich: Band V,* 14.
98. Ibid., 34.
99. Absolon, *Die Wehrmacht im Dritten Reich: Band IV,* 98.
100. Knappe and Brusaw, *Soldat,* 82.
101. Fritz, *Frontsoldaten,* 15.
102. BAMA RH 15/138, *Vortrag der Waffenkommission der Infanterie vor dem Herrn Befehlshaber, Chef H Rüst u BdE, Stab Rüst Iia,* no. 123/41 g.Kdos.II.Ang., January 23, 1941, 2.

103. BAMA RH 14/541, Merkblatt 2, *Taktische Grundbegriffe, Heeresnachrichtenschule, 6, Offz-Anw-Lehrgang v. Inspektion,* Leipzig, October 1, 1941, 5.

104. BAMA ZA 1/1778, Personnel and Administration (Project 2b) P-005, Teil II, *Ausbildung und Verwendung der Unteroffiziere und Mannschaften im deutschen Heer,* 22; Knappe and Brusaw, *Soldat,* 92. (Siegfried Knappe entered the artillery as a basic recruit in 1936 and noted his first six weeks in the artillery was spent in infantry training.)

105. Fritz, *Frontsoldaten,* 16.

106. BAMA RH 2/1331, *Kriegstärkenachweisungen (Heer), KStN 131a/c and 131n;* Tessin, *Verbände und Truppen der deutschen Wehrmacht und Waffen-SS,* vol. 15, n. 151, sketch A1.

107. BAMA RH 2/1331, *Kriegstärkenachweisungen (Heer), KStN 131a/c and 131n.*

108. Ibid.

109. English and Gudmundsson, *On Infantry,* 130.

110. USAMHI, MS # B-068, Walter Denkert, Third Panzer Grenadier Division (Ardennes), Historical Division, Headquarters, US Army, Europe, 1947, 4–5.

111. Absolon, *Die Wehrmacht im Dritten Reich: Band III,* 213.

112. Knappe and Brusaw, *Soldat,* 101–110.

113. Grossjohan, *Five Years, Four Fronts,* 44.

114. Bidermann, *In Deadly Combat,* 151–155.

115. Absolon, *Die Wehrmacht im Dritten Reich: Band VI,* 373–374.

116. BAMA ZA 1/1785, Personnel and Administration (Project 2b) Teil II, *Ausbildung und Verwendung der Unteroffiziere und Mannschaften im deutschen Heere,* 1949, Anlage I, p. 2, emphasis in original.

117. Absolon, *Das Wehrmachtstrafrecht,* xiv–xv, 242.

118. Ibid., sec. 64, p. 49.

119. Ibid., sec. 122, p. 35.

120. Altrichter, *Der Reserveoffizier,* 82–83.

121. Ibid., 73–74.

122. Ibid., 82–83.

123. Rass and Rohrkamp, *Deutsche Soldaten,* "Punishment Tables."

124. Kroener, Müller, and Umbreit, *Organisation und Mobilisierung des deutschen Machtbereichs,* vol. 5/1, 873; Diest, *The Wehrmacht and German Rearmament,* 90–94.

125. Kroener et al., *Organisation und Mobilisierung des deutschen Machtbereichs,* 857; Müller-Hillebrand, *Das Heer bis zum Kriegsbeginn,* 128; Kroener, *Generaloberst Friedrich Fromm,* 346.

126. Frieser, *The Blitzkrieg Legend,* 257–259; Cooper, *The German Army,* 244–245.

127. Frieser, *The Blitzkrieg Legend,* 24–27.

128. Müller and Ueberschär, *Hitler's War in the East,* 245.

129. Glantz, *The Initial Period of the War on the Eastern Front,* vol. 2, 227. Noted by retired lieutenant general Heinz-Georg Lemm while a company commander, Second Company, Fusilier Regiment no. 27, Twelfth Infantry Division, June 21, 1941.

130. Excerpt from Erich von Manstein's supplemental order dated November 20, 1941, to the Eleventh Army in the Crimea; http://www.wehrmacht-history.com/personnel/m/manstein-erich-von-heer-personnel-file.htm.

131. Müller and Ueberschär, *Hitler's War in the East,* 229.

PART 3

Facing Infantry Crises: To Final Victory and Defeat

8

The United Kingdom: Consolidating Recruitment and Induction

IN JULY 1942 THE BRITISH EIGHTH ARMY STOOD ON THE defensive at El Alamein. The stalemate ended in November 1942 when the Eighth Army went over to the offensive and drove the Germans and Italians back to Tunisia, and along with the US Army took their surrender in May 1943. From the initial offensive at El Alamein in November 1942 to the end of the war in Europe, the British Army remained on the strategic offensive. Together with the US Army, and later the reconstituted Free French Army, through Sicily, Italy, and northwestern Europe, despite the continuing difficulty with the quality of the infantry combat force, the British Army as part of the Allies achieved victory over Germany in May 1945 and over Japan in August 1945.[1]

The introduction of the general service corps in July 1942 was the final method for the selection and assignment of other-ranks infantry. The manpower situation for the British Army became more acute as the offensive phase of the war presented greater numbers of battle casualties requiring ever greater numbers of replacements. Tactical methods took into account the imperative to lessen casualties, while the difficulties of maintaining the strength and number of fighting divisions grew more and more difficult as the war drew to its conclusion. The army disbanded divisions with increasing frequency in order to maintain the battle strength of those fewer divisions remaining in the front lines.[2]

The army never solved the questions surrounding the quality of the infantry force. The transfer of high-quality men from the Royal Air Force and Royal Navy in 1944 and 1945 highlighted the existing problem. British and Allied victory came despite the quality standard of the infantry force, and was a testament to the moral strength of the British cause and the fortitude and perseverance of the infantry soldier. The massive supporting air and artillery fires that characterized British infantry battle succeeded despite the decisions of the political and military leaders of the army that denied the need for quality against the battle tasks assigned to the men of the British infantry.[3]

The British Infantry Crisis

In November 1941, the Executive Council of the Army Council discussed the basic outline of what became the general service corps. The corps involved the enlistment of men into the army in basic training centers for six weeks of initial training prior to assignment to a corps for follow-on combat or skills training. The corps allowed for the testing, evaluation, and scientific selection of men for the most appropriate role in the army. The regimental system, as well as the prior consolidations for selection, seemed wasteful and unscientific, and although never abandoned, required fundamental transformation for the betterment of the army. Following the second report of the Beveridge Committee on the use of skilled men in the services, the Army Council accepted its recommendation and moved to implement the scheme in July 1942.[4]

The official announcement to the press on May 28, 1942, highlighted the benefits of the new system upon the traditional regimental affiliations:

> This departure from the traditional British Army system of direct enlistment into a particular Corps or Regiment is being introduced solely in the interests of efficiency; its aim is to get men into their right place early in their Army career and to avoid unnecessary waste of time and instructors in training men, particularly specialists, for an Arm for which they are unsuited.
>
> The regimental system, particularly as it now applies to Infantry, is not to be abolished or interfered with in any way. In fact, the new scheme will tend to foster *esprit de corps* by avoiding the incorrect posting of men, who, later on, would have in any case to be posted elsewhere.[5]

The General Service Corps Method

The methodology behind the general service corps was the objective and quantitative evaluation and scientific selection of men into particular job categories, based in large measure on psychological and behavioral principles.[6] Formal testing of recruits occurred in the first two of the six weeks training at the twenty-four primary training centers or wings located at the consolidated regimental training centers. Before assignment for service, an inductee received first a decision on classification by the Ministry of Labor and National Service as a tradesman, if possible, and was either denied induction as essential to industry or assigned to the military for his recognized civilian trade.[7]

> Although the National Service Acts gave no guarantee that any man would necessarily be called up to the Service of his choice, they did allow a man to express a choice. If this concession were to be of any value, it must mean that when a man expressed a choice for one particular Service, he must be offered to that Service first. Throughout the war there were always more men who preferred the Royal Navy and the Royal Air Force than either of those Services could absorb, with the natural result that those Services chose the men whom they wanted and left the rest to be turned over to the Army.[8]

This design provided up to four weeks for the appropriate corps record offices to find and post those men fitting their requirements. A trained personnel selection officer made the determination. Using a battery of appropriate texts and phrases for use in classification, the officer clarified a recruit's capabilities during an interview that included his medical determination, his education level, potential combat temperament, leadership, and test results to determine his selection group. The term *summed selection group* was later applied to ground the total of various testable results for the interview. In particular, the personnel selection officer noted the educational standard for each individual as follows:[9]

11 and 12: University degree or comparable professional education.

21 and 22: Higher school certificate or a high standard of professional or technical qualification.

31 and 32: School certificate or matriculation, or technical qualification such as advanced Royal Society of Arts or final city and guilds, and the like.

41 and 42: Education until the age of sixteen but without the qualifications for advanced Royal Society of the Arts or final city and guilds, but intermediate Royal Society of the Arts, intermediate city and guilds, and the like.
51 and 52: Good standard of elementary education or evening classes for at least two sessions.
61: Average standard of elementary education.
71: Educationally retarded.
81: Illiterate.

Illiteracy. During the war, illiterate or partially literate British inductees received the designation as "dull and backward personnel," and included "men who are so mentally backward that they are unable to benefit from military training and though unfit to bear arms can be employed in routine manual work." These men found assignment in the unarmed pioneer corps if medically qualified, and served there alongside refugee aliens resident in the United Kingdom, Poles, Czechs, and Jewish men from many countries, and if not otherwise medically qualified, were discharged. The pioneer corps grew from 112,000 to 150,000 men between October 1941 and May 1943. The British Army instituted literacy training for illiterate inducted men in March 1944.[10]

Selection-group criteria. The tests shown in Table 8.1, when summed, provided one component of the selection-group criteria. The combined scores from these tests, along with the indications from the interview concerning leadership, employment, and education level, and the assigned medical category, as well as a combatant temperament score (1, 2, or 3), became the basis for the selection group assigned to each individual. The combatant temperament translated as "guts," and a 1 rating was given to the small numbers of inductees who appeared to have this quality well above the average. In contrast, a 3 rating marked a man as well below average in this quality and was not given unless an army psychiatrist saw the man and confirmed the judgment of the personnel selection officer.[11]

These sought a measure of objective certainty for the range of training recommendations that accompanied the assessment of each recruit sent to the War Office. This became the basis for final selection and assignment into a corps. Within the criteria for each corps, and keeping in mind the overarching demand for tradesmen, the

Table 8.1 Battery of General Service Corps Tests

Test	Purpose	Required
Matrix	General intelligence and probable ability to acquire new knowledge easily and rapidly	All Recruits
Bennett	Mechanical aptitude, interest, and knowledge	All Recruits
Arithmetic	Elementary ability and knowledge of more advanced mathematics	All Recruits
Squares	Ability to perceive spatial relationships (important for mechanical trades)	All Recruits
Verbal	Verbal fluency	All Recruits
Agility	Speed and accuracy in movement involving coordination of hand, eye, and foot	All Recruits
Hearing	Acuteness of hearing	Only for training recommendation 3
Morse	Ability to differentiate like and unlike sounds and remember Morse combinations	Only for training recommendation 3
Clerical	Ability to perform quickly and accurately such clerical jobs as listing, checking, and codifying	Only for training recommendation 5
Assembly	Mechanical aptitude and manual dexterity	Only for training recommendation 2 or other mechanical trade

Source: Ungerson, *Personnel Selection,* 36.

interviewing officer had nine possible training recommendations to select from for recruit assignments:[12]

- 0: All tradesmen for whom the broad categorization system was likely to prove inadequate.
- 1: Drivers, all types of vehicles in all arms.
- 2: Maintenance men—that is, nearly all mechanical jobs, other than high-grade trades.
- 3: Signalers—that is, soldiers in all arms whose primary task was the operation of wireless sets.
- 4: Layers and fine operators—that is, gun layers and predictor layers.
- 5: Clerks and storemen.
- 6: Other combatant duties—that is, the majority of nonspecialists in field units, such as infantry.
- 7: Other administrative and domestic duties; mostly low-grade employment in field or static units.
- 8: Unarmed pioneers—that is, men the psychiatrist considered unfit to bear arms.

9: Armed pioneers—that is, men the psychiatrist considered might carry arms in self-defense, but were fit for simple laboring duties only.[13]

The object rested on the identification of technical and clerical skills associated with higher intelligence scores, and occupational experience or potential, and the identification of the least-fit. Recruits who scored low on the tests and were unimpressive in their interview tended to receive a lower selection group, and thus became eligible for placement into the sixth training recommendation (other combatant duties: combat arms assignments such as the infantry) if A1 medically, or the seventh training recommendation (other administrative duties) if less than A1 physically. Prior to the development of the general service corps, "the only two pieces of information on which much store could be set were the medical category and the type of schooling. These were of only limited use because about three-quarters of the military population were medically A1, and as many had had only elementary schooling."[14] A poor test score did not incur an automatic definition of a man as poor soldier material, but it did provide the opportunity to label him as dull or illiterate, requiring evaluation by the psychiatrist for possible rejection or assignment to an army educational corps school.[15] As the war progressed, the average age of the intakes grew younger and the test scores declined, indicating a reduction in available quality in the general population of young men. One reason given was the possibility that the disruption of regular schooling by the war itself caused the lower scores.[16]

To speed the production of orders posting recruits to their eventual corps, the army used a Hollerith mechanical data-sorting machine, named after its inventor and one of the original founders of IBM Corporation, Herman Hollerith.[17] Punch cards tabulated information according to the various minimum requirements for each job description, and the data were sorted against corps requirements such as "the lowest medical category it can take, proportion of each aptitude group [first, second, third-plus, third-minus, fourth, or fifth selection groups], highest age, lowest height, any selection grade restrictions, and the number of men to be taken."[18]

Job analysis through the Directorate of Selection of Personnel provided a range of selection groups with the minimum indicator for successful performance for that job. For a cook, batman, or waiter the minimum was the fourth group, while for a tank driver, driver operator, or clerk the minimum was the third-plus group. Any corps with requirements for cooks or batmen received a proportion of men of the fourth group, whereas a Royal Armored Corps unit with a quota for tank drivers accepted no one below the third-plus designation for that function.[19] An example of this method was the job analysis sheet for the reconnaissance battalion, which demonstrated the ideal distribution and demand for a high selection group. This army formation required 65 percent as third-plus or higher group.[20]

The demands by the personnel branches of each corps were requisitions for those jobs requiring training on a regular basis and therefore came with a definitive selection-group requirement. The effect was to call for higher proportions of intelligent men than were available. Clerks (fifth training recommendation), a skilled position, required an third-plus or higher selection group. Those performing other administrative tasks (seventh training recommendation) also performed some clerical work, though without requiring additional training, allowing a minimum of the fourth selection group. By specifying the need for clerical work that required training, the selection group allowed the identification of more of the lower-quality men for the less demanding jobs. Its effect was to reduce the availability of higher-quality men for the combat arms employments without such specific selection-group and training recommendation criteria. The returns "meant more satisfaction to the Personnel Branches, but inevitably caused a decline in the quality of men available for such employments as Infantry Riflemen and Gunners."[21] The general service corps accomplished its primary aim. It provided a scientific and psychologically sound basis for the selection of soldiers. It focused on the identification of those not suited to service and those suited best for particular tasks.

General Service Corps Statistics

Numbers: Nearly 710,000 recruits passed through the general service corps between July 2, 1942, and the end of June 1945, or less than one-quarter of the final number of 2.9 million in the army by that date.

Intelligence: Nearly one-half were above the median (third-minus selection group) and 10 percent were in the first group; 7 percent were in the fifth group. This low percentage was due to weeding out at the medical examination stage, but it was known that the less fit men tended to be somewhat less intelligent.[22]

Education: One-quarter were above normal elementary school standard (51 and above), with 6 percent having a school certificate or its equivalent (31 and 32 standard):

- One-half were up to normal elementary school standard (61).
- One-quarter were below normal elementary school level (71 or lower).

Combat temperament: Just over 5 percent (36,000 men) received a rating of high combatant temperament and just over 3 percent (22,000 men) a rating of poor combatant temperament.

Age: General service corps inductees grew younger year over year. In the first year about 40 percent were nineteen years of age or less; in the second year the proportion rose to 50 percent and in the third year to 70 percent.[23] The average age of the infantry (other ranks) in July 1940 was just over twenty-five years old.[24]

Infantry minimum height and weight standard: Height of 5 feet and weight of 119 pounds.[25]

> As the war progressed the intellectual and educational standards of recruits steadily decreased. The earliest intakes to primary training units contained recruits of whom over 60 per cent were in selection group 1, 2, or 3 plus. The quality steadily declined until November 1943, only 40.3 per cent of recruits were in these three upper selection groups.[26]

The inspector general of training perhaps summed up the effect best in August 1942:

> The present system of posting recruits so that the square pegs fall into square holes appears to leave the infantry a long way down the list of preferences. The Ministry of Labour, in allotting manpower to the Services, appears to allot to the [Royal Air Force] first, [Royal Navy] second and the Army third. In allotting within the Army, the Technical Corps come first, Armoured units, second and the Infantry almost last. As a result the Infantry receive the bulk of their men from the lowest grade of the lowest Ministry of Labour allocation. It is rare to find in any Infantry intake more than 5% of potential Officers and N.C.O.s. The constant drain of supplying the best personnel for new commitments such as Commandos, Paratroops, Reconnaissance Corps, etc. The result of all the above which I recognize may have been inevitable, is that the Infantry

> today is not as efficient as it should be after two years of training, and in view of the low grade of present intakes it is difficult to see how even the present standard can be maintained.[27]

The Infantry Crisis and the Rifleman

By November 30, 1944, the strength of the army had grown to over 2.7 million soldiers—officers, NCOs, and other ranks.[28] Beginning in 1940, however, the army experienced grave difficulty in maintaining the strength of its primary fighting divisions—infantry, armored, and airborne. It managed to sustain a combination of thirty-six divisions from 1939, but by the end of 1944 was down to twenty-eight. With the renewed emphasis on the preparation of the army for offensive operations, the shortage of manpower and the quality of the infantry continued to trouble the War Office in 1943. While the adjutant general was optimistic that the general service corps and the prior conversions from infantry meant a better intake of recruits for the infantry, this hope failed of realization. "There was at present [January 1943] a deficiency of 16,000. When infantry battalions were converted to other arms the cream of personnel were taken and the residue left as infantry. The general quality of the infantry was, therefore, deteriorating."[29]

It was the combat infantryman whose casualties mounted in this period. Divisions disbanded through cannibalization kept other divisions up to strength. The number of military specialties within the infantry, while not large, defined the range of tasks associated with the role of meeting the enemy face-to-face. These tasks included riflemen, machine gunners, mortar men, and projector infantry antitank men, in general. These tasks lay within the rifle company of the infantry battalion.

Given the offensive nature of the war in this period, the principal task of the rifleman was to cross no-man's land to face the enemy in close combat. He received support from the infantry specialties within his company, and the full range of battalion, brigade, division, corps, army support units and specialties, armored units, artillery, and air support. No matter how much support was available to him, the rifleman retained the primary mission to endure the enemy's fire, and to move forward to kill, wound, capture, or cause the enemy to abandon his position. The numbers of these riflemen demonstrate the small numbers of men within the army required to move into no-man's land (see Table 8.2).

Table 8.2 Total Riflemen in Rifle Units by 1944 War Establishment

Rifle Section	Rifle Platoon	Rifle Company	Infantry Battalion	Infantry Brigade	Infantry Division
10 riflemen, other ranks, and NCO	30 riflemen, other ranks, and NCOs in three rifle sections	90 riflemen, other ranks, and NCOs in three rifle platoons	360 riflemen, other ranks, and NCOs in four rifle companies	1,080 riflemen, other ranks, and NCOs in three rifle battalions	3,240 riflemen, other ranks, and NCOs in three infantry brigades

Sources: TNA: PRO WO 365/203 W.E. (War Establishment), *Analysis by Arms*; A.G. Stats (War Office) S/243/43, *Analysis of Arms of W.E. of Standard Formations,* March 26, 1943.

The actual number of riflemen in an infantry division during 1943–1944 demonstrates the low numbers of men tasked to take ground from German or Japanese soldiers, hold that ground, and then defend it against counterattack. It was this series of tasks that made the infantryman's job so physically and psychologically difficult. The War Establishment strength of an infantry division in 1943 and 1944 ranged from 19,436 to 18,347 other ranks respectively. Winston Churchill criticized the rising numbers of rear echelon troops compared to the number of infantrymen in an infantry division.[30] These rear echelon troops reflected the motorization of the infantry divisions and the large numbers of technicians required to operate, maintain, and supply the machines. In contrast, if 56 percent of the infantry division were in the median and below, the third-minus, fourth, and fifth selection groups, then there were more than 10,000 men at the median and below, and more than 7,000 of these men were of the fourth and fifth selection groups. The rifle sections of the infantry, with no mandated selection-group requirement, was the location for these soldiers.

The numbers in Table 8.2 reflect the riflemen in sections within their rifle platoons and rifle companies within the infantry battalions of the twenty-eight combat divisions and seven independent armored brigades.[31] While all other combat support units, such as machine guns, mortars, and artillery, had the task to help get these men to their objectives, it fell to just 3,240 infantry riflemen to move across and into the enemy's positions to achieve an infantry division's mission. The numbers become even starker for the entire picture of the British Army, less the Dominions, when adding together the existing formations as of December 1944. Out of the total of over 2.7 million soldiers serving in the British Army as of November 30, 1944, the other ranks assigned a military specialty in the Royal Artillery numbered over 562,000, while the other ranks assigned an infantry specialty numbered 500,000. Of the total of 500,000 soldiers assigned the infantry specialty, 81,000[32] served as riflemen.[33]

On the occasion of the disbandment of the Fiftieth (Northumberland) Infantry Division in December 1944 to cannibalize its existing infantry as replacements for other divisions, Prime Minister Churchill remarked on the continuing reduction of the fighting formations of the British Army: "I greatly regret destruction of 50th Division as a fighting force, but as you have gone so far, I fear the process must be completed. I greatly fear the dwindling of the British Army as a factor in France as it will affect our right to express our opinion upon strategic and other matters."[34] The army disbanded a total of three armored and four infantry divisions in 1944 alone.[35] On December 11, 1944, Field Marshal Bernard Law Montgomery warned that the "infantry reinforcement situation is bad and with this plan [the disbanding of the Fiftieth Division] we shall just and only just pull through."[36] The army partially resolved the acuteness of the infantry shortage through various plans to disband and retrain thousands of soldiers from Royal Artillery anti-aircraft and anti-tank units as infantry replacements. This process began in the Allied Armies Italy and carried into the Twenty-First Army Group in northwestern Europe.[37]

In addition, the War Cabinet agreed to the transfer of men serving with the Royal Navy and the Royal Air Force, as well as the Royal Air Force deferred list, for duty as infantrymen in 1944 and 1945. The salient feature of these last transfers, a total of 39,000 men, was the general quality of these men above the normal intake of the infantry, with 92 percent of the deferred-list men above the median in the first, second, or third-plus selection groups. The normal army intake was 44 percent above the median, with 56 percent at the median, third-minus selection group, and below. Potential officers represented 20 percent of the deferred-list men, against the normal army intake of 5.8 percent.[38] As welcome as these men were, there were too few to make a large difference in the infantry shortage of 1944 and 1945.

Commanders across the spectrum of battlefields remarked both upon the shortage of infantry and the impact of low quality on combat operations. Officers in the Burma-India theater summarized the deterioration of the infantry situation in June 1944:

- Junior leaders, W.O.s [Warrant Officers] and N.C.O.s are inexperienced and not up to standard either in tactics or discipline;
- The Regimental spirit is so diluted as to become almost non-existent;
- Efficiency declines and unfortunate incidents begin to occur in action;
- The Bn [battalion] becomes unfit for war.[39]

Other lessons from the Sicilian and Italian campaigns further stressed the impact of low-quality infantry on the ability to find leaders from within their ranks. "The root cause of the matter would appear to be the low standard of the infantry intake as compared with other arms. . . . [A] proportion of the infantry intake should be of the highest quality, equal to that required by the airborne forces."[40]

The Airborne Divisions of the British Army

For the army, the airborne was a concept created by outside directive, accepted with a great deal of skepticism, and begun with no immediate intent of creating elite combat infantry. On June 22, 1940, the prime minister directed the War Office to investigate the creation of "a Corps of at least 5,000 parachute troops."[41] This nascent airborne force faced formidable difficulties in becoming operational in the defensive period of the war, competing for suitable men from the army, and with no capable aircraft or gliders without crews from the Royal Air Force. The largest difficulty was a lack of policy about what to do with an airborne unit and how to get it into combat. Neither the army nor the air force "owned" the mission or the concept of parachute infantry employment, equipment, selection of personnel, or development of aircraft or glider standards and production.

As the airborne units sorted through these problems, the army decided upon a twofold tactical organization, one parachute and one glider-borne. The questions of manpower and air force airframes remained paramount and unresolved. Early recruiting for volunteers received a mixed reception. These volunteers were of uneven quality, in the words of the postwar report on airborne forces, "with the fact that units supplying volunteers used the new airborne units as an excuse for getting rid of their 'bad types,' and many new recruits to parachuting thought they were coming to an 'Eldorado' where discipline did not exist. The process of disillusionment took up much of a parachute commanding officer's time."[42]

While parachutists were volunteers, the initial glider-borne components in November 1941 came from existing infantry battalions "no better trained or equipped than the majority of the British Army at home at that time. . . . [C]ommanding officers were allowed to get rid of whom they wished and a considerable turnover took place."[43] The First Airborne Division formed at this time along with the army air corps and within it the glider pilot regiment.[44]

The key to the quality of the airborne force was the decision by the commander of its first division-sized unit to establish itself as a *corps d'élite* within the army and the parachutists as the elite of the *corps d'élite,* achieving this status through a selection process emphasizing high physical and intelligence standards. One man in three met the standard.[45]

Through August 1943, parachute interview boards accepted men to parachute duty rather than through selection testing. The army suspended the boards when it found during research tests that one-half of those rejected by the boards passed their parachute training.[46] In the end, the methods used to find and train volunteers eliminated those without sufficient levels of physical stamina and mental acuity. These elite forces impacted the regular infantry by drawing away soldiers of high quality in terms of physical fitness and leadership. It was another claim on men that the infantry divisions, with their already quality-starved ranks, were ill-prepared to accept.

The Effect of the General Service Corps

The general service corps provided the British Army with the culmination of the scientific/psychological basis for personnel selection in its commitment to a war of materiel over men. The process emphasized testing and interview techniques designed to establish the intelligence level, mechanical and technical aptitude, and psychological profiles for the purpose of selecting the best qualified for the most technical requirements as represented by the selection group. Those army jobs for which specific scores equated to individual tasks, for gun layers, surveyors, or ammunition artificers, for example, allowed the greatest discrimination in selection. Those tasks for which specific job descriptions identified no definitive (by test score or job analysis) task qualifiers against a particular selection group, such as the tasks associated to the infantry, allowed for the least discrimination. Hence, assignment to the infantry meant that the sorting process arrived at this assignment after pulling all others qualified by test score and interview to more technical needs. In essence, one did not find himself in the infantry unless one had no qualification for almost anything else. This paralleled in close fashion the experience of the US Army and the disproportionate assignment of those with the least mental capability to withstand the stress of face-to-face combat, combined with those least physically capable for that form of battle.

By the time the general service corps came into being, more than three-fourths of the ending total of over 2.9 million soldiers serving on V-J Day were already in the army. The interview was the primary method for classification, selection, and posting of individuals from 1939 to 1942. This final form of scientific selection affected less than one-quarter of the final British Army strength. The quality available to the army, by its own estimation, declined from the time of the inception of the general service corps to the conclusion of the war in intelligence as well as in the induction of younger and younger soldiers. As the overall quality of the induction population declined, so did that portion of the intakes left for assignment to the infantry.

The fears of the inspector general of training, voiced in August 1942 at the very beginning of the general service corps, concluded that

> the fact remains that the Infantry have previously been given a very indifferent intake of Recruits compared with the other Arms, and they have been continually milked of all their best instructors by perpetual calls for volunteers for Commandos, Paratroops, Reconnaissance Corps, etc., and the demand for potential officers. . . . [I]t is essential that the Infantry should not only be well-trained, but should be well selected in the first place. The Infantry of today needs more than ever, the sympathetic and understanding selection of Recruits, and it will take considerable time to recover from the damage already done in recruiting the less suitable material since the commencement of the war. The need is urgent if Infantry are efficiently to take their essential role in the final stages of the war.[47]

The effect of the downward selection of the infantryman in intelligence and physicality and the diminishment of the role of the regiment in the development of the army's intent for cohesive, and thus effective, combat behavior, continued unabated through the war. Postwar commentary on the failures in selection and development of cohesion in battle in the small unit were clear. In their turn, the general officers commanding in the campaigns of World War II adjusted their tactics and expectations for the maneuver of the infantry based on this reality. The often-quoted trope of the lesser quality of the Anglo-American officers must be modified, by noting that these generals adjusted and accommodated their tactics to the actual infantry force they created, rather than some ideal in doctrine. Through the use of their greatest advantage, mass artillery and air fires, they saved the lives of these infantrymen over time, generating a small fraction of the casualties in victory compared to the Great War. Their generals reacted to fight with the infantry they had, used the fire advantages they owned over the German enemy, destroyed that enemy, and found victory in that common sense.

Cohesion and Battle

From the initiation of combat experience in the defensive phase of the war, and into its offensive phase, the full range of the army's measures intended to create cohesive behavior in the small infantry unit came into expression in battle. The critical assumption, that cohesive behavior would develop as a matter of course, lay in the culture of the infantry regiment and its historical and narrative power to modify the behavior and expectations of the men intended for face-to-face battle. The diminishment of the regiment and its place in this development was continuous from the beginning of the war. There were three distinct reductions in the influence of the regiment, each known, each discussed, and dismissed as essential for the national and priority management of the nation's manpower.

Infantry Battle and Training

Battle drill is the term most associated to a new way of preparing the British infantry to fight during World War II. Soldiers learned battle drills from the experience of those who survived their combat encounters with the real German enemy. The army considered this the best form of combat indoctrination.[48] The role of these minor tactical evolutions comes into play at the final moments of the attack, when crossing no-man's land into the face-to-face combat necessary to seize an objective during the offensive. The planning and execution methodology of the British Army sought control above other factors. Artillery could be controlled. The fight in no-man's land could not. There existed a fear that the morale of the troops was fragile and that this mandated the provision of overwhelming, powerful fires in support of the soldier.[49]

Battle-drill concepts were not panaceas for increasing the aggressive spirit and performance of British infantry. Teaching these techniques late in the war further confused the issue of which technique to use by establishing a difference between the types of drills essential for the set-piece, artillery-heavy, supported attack, and one in which the infantry had to advance on its own merits to clear an objective. The former allowed for a more formal, complex, but slower orchestrated battle drill in the advance, nullifying the need for the type of initiative and fire and movement of the latter form.[50] "In practical experience against the German enemy in Normandy and beyond, fire and movement was necessary against almost any objective. Many Germans survived the artillery and air preparation to resist any set-piece approach."[51]

Where infantry training went awry was in the growing tendency to suppose that minor tactics were unnecessary in attacks that were based tactically around an artillery barrage. Infantry minor tactics came to be regarded as an alternative to be applied when the blunt artillery instrument was no longer suitable. The notion that two types of fighting might be necessary in the same battle passed away. It left the expectation that an infantry set-piece attack behind artillery support required little if any actual infantry fighting. When in battle, "that expectation was falsified, too few infantrymen knew what to do."[52]

In March 1944 the discussion of battle drill entered into the official lexicon of the army with the publication of a training manual.[53] The army intended battle drill as the basis of the course for officer and NCO instructors at the new infantry school formed from the existing battle school of general headquarters in late 1942. Its gradual dissemination by word of mouth was also considered a large factor in the improvement of army morale, among that infantry exposed to it, for its focus on hard, realistic combat preparedness.[54] The opportunity for the infantryman to become versed in an understanding and practice of the battle drill remained very much the purview of his immediate commanders, and was not enforced as a doctrinal standard.[55] The basis of all tactical training from the interwar period was that tactical training was the responsibility of the unit.

The army created a director of infantry in 1943 to highlight the role of the infantry, to improve morale of the force as a whole, and to advocate for increased recognition, including pay. The director had no infantry tactical training responsibility.[56] The general officer in charge of the infantry directorate noted in late 1944 that "the School of Infantry is *not* a Battle School; nor does it teach Battle Drill."[57] The purpose of the infantry school was to train instructors in the proper techniques—the how—for teaching in a classroom or field setting, not battle drill—the what—itself. Over 4,500 officers and 10,000 warrant officers and NCOs, or nearly 750 officers and 1,700 warrant officers and NCOs per year, received these instructional techniques, in the full variety of military lessons. Lacking a common tactical focus, the impact on the opportunity of the other-ranks infantry to learn battle drill or other combat-tested techniques did not improve.[58] "During the war, when there were so many men to be trained and so few instructors of experience, the tendency was to produce drills for all tactical situations. This policy served its purpose in that it enabled instructors to teach students, when confronted by a problem, to do something reasonable instead of nothing. At the same time, however, these drills tended to become masters and not servants and commanders were inclined to act blindly and without first appreciating the situation."[59]

The need to husband the number of soldiers at all costs imposed on British tactical doctrine the imperative of overwhelming firepower in support of the infantry, as the secretary of state for war, Sir P. J. Grigg, noted to Prime Minister Churchill in April 1943: "The infantry cannot perform their task without firepower, and it is just as easy to fail to provide sufficient firepower as it is to provide sufficient infantry. It is indeed more economical in infantry lives to have them well-supported by firepower (our greatest life-saving element) than it is to have a larger number of men and inadequate support."[60]

Senior commanders were aware of the limitations the overall manpower shortage placed on any encouragement of small-unit tactical aggressiveness, and the importance to the soldier of adequate support. Lieutenant General Sir Sidney Kirkman, commander of the British Thirteenth Corps in Italy, noted in November 1944 that "we may at times be lavish with artillery expenditure, but the British soldier has come to expect a certain measure of support and if at this stage support appears inadequate, I consider that attacks will be launched in so half-hearted a manner that we shall incur heavy casualties without success."[61]

The development and expectations for the British infantry did not anticipate the use of individual initiative to carry forward to objectives without the requisite massive fire support. Once engaged in battle at the small-unit level, and significant resistance encountered, British tactical preparation underscored the need for the reintroduction of massive fires to restart the forward movement of the infantry. When the German enemy tightly engaged the attackers in close terrain, such as existed in the mountains of Italy, the hedgerows of Normandy in 1944, and in the densely packed forest growth of the *Reichswald* in 1945, the effects of the heavy supporting fires were less than expected, casualties increased, and the attack ground to a halt.[62]

In Italy and northwestern Europe, the British infantryman demonstrated his ability to sustain casualties as his forefathers had, when required. Casualties among the infantry in Normandy far exceeded predictions.[63] The priority to preserving British lives was a paramount factor in all plans of attack. When confronted with the need to press an attack that ground to a halt due to heavy enemy resistance, the commonsense answer was to save the lives of the soldiers and plan to fight another day.[64]

The 1943 and 1944 War Establishment increased the section from the prewar number of eight men to that of ten men. The new organization, a response to close combat difficulty with the German squad (*Gruppe*), now included a section leader with Sten gun, six riflemen (each carrying two additional magazines for the Bren gun), and a Bren gun unit of three men, each with four additional Bren magazines.[65] Although designated a light machine gun, the Bren had the distinct disadvantage of requiring a magazine change after each twenty-nine rounds of fire. (See Appendix 2.)

The rifle section received significant fire support from a variety of sources within its division structure. Three sections together constituted a rifle platoon of one officer and thirty-six other ranks, with the addition of one 2-inch mortar. A rifle company consisted of three platoons, with five officers and 122 enlisted other ranks. The company also added 2-inch mortars and three projector infantry anti-tank launchers for support of its small units. Each infantry battalion held four rifle companies with a support company of six heavier mortars (3-inch), six 6-pounder anti-tank weapons, and a headquarters company for a strength of thirty-six officers and 809 other ranks.[66] The infantry division consisted of three brigades with three infantry battalions in each, along with 25-pounder artillery, engineers, and a machine gun battalion with additional heavy (4.2-inch) mortars and .303-caliber medium machine guns.[67]

The infantry section, platoon, and company possessed so little firepower that the requirement for artillery and other fires from battalion and above was near absolute. The absence of an organic weapons platoon within the rifle company set the British Army rifle company apart from both the German and US organizations, placing the British infantry at a disadvantage against the multiple machine guns and mortars of their German small-unit counterparts.[68] The British Army, in this as in higher organizations, sacrificed infantry small-unit firepower for a lighter, more mobile squad, platoon, and company.[69] The focus on external firepower reduced the need for enhanced cohesive behavior in the infantry primary groups, whether attack or defense, nor did it encourage the infantrymen to attain the perceived level of battle expertise of their German other-ranks counterparts. In a both literal and figurative sense, British superiority in artillery and air power leveled the field.

Manpower shortages meant that the section was understrength in wartime. Colonel T. N. Glazebrook wrote after the Sicilian and Italian campaigns that the

> riflemen of the section rarely have the opportunity to use their rifles effectively either by day or by night. I consider that the nucleus of the section is the Bren group of three men, carrying eleven magazines. These men were the pick of the platoon and every effort was made to train them in weapons and fieldcraft to perfection. Second only in importance in the platoon is the 2-inch mortar team. . . . Sections rarely went into battle stronger than seven or eight, including the commander. A section reduced below six, including the commander, suffers by reason of the undue strain of finding double sentries at night. . . . [T]o maintain a minimum strength of seven . . . the present [War Establishment] of one and nine is adequate."[70]

Each equivalent German *Gruppe,* or squad, that the British section expected to meet in battle formed around a belt-fed machine gun, either MG 34 or MG 42, giving the German squad an extraordinary advantage in the rate of fire to achieve local fire superiority, allowing both an effective automatic fire element and a maneuver element within itself.[71] The psychological effect of the more numerous German automatic weapons, with three to five true machine guns in each platoon, was considerable: "The machine-gun's chilling, unmistakable sound—almost like cloth being ripped—always sent a shiver through me, even when it was not being fired in my general direction."[72]

The British Army relied on the training principle that "there is nothing to equal engagement with the enemy for learning the best way to fight. Everything else is make-believe."[73] British prewar tactical doctrine highlighted mobility and the value and primacy of the infantry arm, and demanded the assignment of the highest-value men to the most difficult task in war: close combat.[74] The loss of the regiment's culture and its ability to instill the social expectations for cohesive behavior, demanding the tactical reliance on massive firepower to move the infantry forward, reduced the importance of tactical cohesion in the infantry small unit. These policies placed into the hands of its most junior leaders, the subalterns, the section, platoon, and company commanders, the necessity to overcome the failure of British Army intent. Their task was to instill and foster cohesive battle performance in the primary groups of the British infantry in the experience of combat itself.

Junior Leaders

"From the beginning of the war it was the consistent policy that the supply of officers should be drawn from the ranks and that the road to a commission should be through an Officer Cadet Training Unit. . . . Commissioning after service in the ranks and training at [this unit] always produced the majority of officers, and it was almost the sole source for the supply of the young regimental officer of the combatant arms."[75]

The army did not adhere rigidly to this formula, finding several additional methods to find officers for noncombat positions. Through the army officer emergency reserve the army commissioned 39,076 (30,712 by 1941) former officers, or direct civilian commissions for technical requirements, such as for civil, electrical, and mechanical engineers, barristers, accountants, and such "white-collar" positions. Some deserving soldiers from the ranks also received direct commissions, such as senior warrant officers and NCOs. Early in the war, university students passing the requirements of a particular level of military skill went directly to an officer candidate training unit upon induction or as a volunteer.[76]

The selection of officer cadets from the ranks advanced through several forms during the war, with the primary method from 1939 through 1942 accession through a command interview board controlled by the local senior command headquarters with a single permanent member, the president, a colonel. To appear at the unit board, a soldier received a recommendation from his commanding officer. The standards noted that "the primary qualities required of a candidate are powers of leadership, personality, and where necessary, technical knowledge. Considerations of birth or social status are not considered."[77]

In the summer of 1941, the War Office removed the command interview boards from local command headquarters and placed them under its direction with the new designation War Office selection boards. These implemented a standard formal interview procedure, and introduced selection-group testing, among other reforms. The failure rate of over 25 percent of officer candidates at the officer candidate training unit was a large source of dissatisfaction with the command interview board process, which rejected 14 percent of potential officer candidates. The central War Office selection boards, however, rejected potential officer candidates at a rate of 40 percent or higher, resulting in a lower failure rate and greater numbers of successful graduations at the training unit. "By April 1942, all Command Interview Boards had been replaced by some 17 War Office Selection Boards."[78]

This opened the door for the subsequent experimentation with, and adoption of, what they believed were the processes used by the German Army in the selection of its officers since 1926. A single selection board experimented with the "German" method during the first three months of 1942. All boards adopted this new procedure during 1942. "On the psychological side, group intelligence tests, a biographical questionnaire, a psychiatric interview and several 'laboratory' tests modeled on the German pattern were given to all candidates. The military side consisted of an interview by the President and a series of outdoor tests."[79]

After the war, British psychologists criticized the laboratory and outdoor psychological tests along the "German" model. The very characteristics upon which so much weight centered—character and will—they believed, after the war, were indefinable through this process.

> The claim that such and such a test revealed will-power, imagination, or other vague faculties was entirely a priori and in the

> absence of empirical substantiation, German psychologists had little defence when they fell into disfavor and the work stopped entirely in [1942]. . . . The reader may conclude that [War Office selection board] methods were essentially similar to those of the German military psychologists which we condemned. . . . But in Britain no use was made of elaborate reaction time and other tests alleged to reveal a whole series of faculties of the mind and character. Moreover, much more regard was paid here to the need for validation and follow up. . . . It is the M.T.O. [military training officer] side of the [selection board] procedure which most closely resembled the German counterpart, and which is probably most open to criticism.[80]

Soldiers received the selection boards well for their perceived fairness and freedom from personal prejudice compared to the command interview boards. Final acceptance as an officer cadet came from the selection-board president after a review of all results and the opinion of the junior officer–board member (a unit commander with experience in leading officers at the battalion and lower level), and the psychological opinion. The role of the psychiatrist came under criticism from a very conservative officer corps as well as parliament, which introduced a special expert committee to review the methodology. The parliamentary committee reported with favor to keep the psychology-based component as a principal piece of the evaluation of potential officers.[81]

There existed the belief that many qualified candidates were not put forward by commanding officers, because proffering one's best soldiers as officer candidates meant their loss to the battle preparation of the unit, as the individuals were then lost to the greater army.[82] "There was in some quarters a feeling that it was not worth while being an officer; there was a lot more responsibility, hard work and trouble, and very few advantages. Officers, it was said, had to pay for amenities which other ranks got free, while in certain sections of Press and Parliament officers were constantly being held up to public reprobation as an idle class who took all the best accommodation and food. If such was the public view of officers, why bother to be one?"[83] The potential officer must first receive a recommendation to the War Office selection board from his organization, through a unit-level interview, and units continued to be reluctant to part with their best soldiers. With the adoption of the general service corps in July 1942, a new policy sought to surmount this difficulty by requiring the interviewers at primary training centers and wings to determine, report, and forward all those regarded as possible officer candidates to a selection board.[84]

Due to the identification of fewer and fewer quality candidates at initial training centers, the acceptance rate at the selection boards declined in the course of the war (from 60 percent in 1942 to 39 percent in September 1943). The officer candidates came from the general induction population.

The length of officer candidate training for infantry officers was four months, and there was also a pre-training preparation period lasting from six to ten weeks depending on the candidate's state of military tactical proficiency.[85] Overall, the system worked. A bare sufficiency of regimental officers was found for the ever-shrinking number of field divisions as the war progressed. In the opinion of the postwar studies, quality not quantity was the key indicator of success, and in this the output appeared less than satisfactory in the judgment of units in the field. "There was throughout the war a continual shortage of the regimental officer fit for active service in a frontline unit . . . but the real trouble was that there was never a really sufficient supply of good material coming from the ranks. Reference has already been made to the responsibility of commanding officers for finding and putting forward suitable candidates, but it is pertinent to refer here to the reluctance of many very suitable men to follow their plain duty and accept commissions. Mistaken loyalties to units or friends held back many who might have been officers, and the wider loyalty to the cause was too often forgotten."[86]

Once commissioned and assigned to his platoon, the development of the officer as troop leader and small-unit tactician was the responsibility of his commanding officer.

> The response of the regimental soldier to the demands made upon him is an indication less, perhaps, of the degree of efficiency achieved by the War Office, than of the quality of the society that produced him. The War Office had to work with the material that came to its hand. Of this, it is enough to say that the shortage of good officer material was perhaps the gravest of enemies to military morale, and "man-management" was the lesson that a large proportion of officers found it hardest to learn. Unprecedented also were the demands made upon the capacities of the individual officer: he had to be an expert in mastering a large and ever-increasing variety of weapons and tactics, and a still larger variety of complicated administrative rules. Moreover, the Army suffered, as it did not in the War of 1914–1918, from the fact that the Air Force diverted

> from it thousands of potential officers, including a large proportion of the best material—men with enterprise, a high degree of intelligence, and the gift of leadership. Both for officers and men, the Army had to be content to take, along with much excellent material, a great deal that represented the leavings of the other Services. To create from this mixed mass a sufficient number of first-class officers—whatever pains were spent on perfecting the methods of selection and training—was an impossible task.[87]

Schools were available after officer candidate training, but for the infantry officer, combat preparation occurred in his unit. In the British Army, "schools existed to train instructors, but [commissioned officers] wanted to retain their autonomous power to train the subordinates they lead in battle."[88] After four months of officer candidate training, large numbers of young officers arrived at their infantry platoons, many directly into combat, where their performance and survival relied upon the quality and experience of their NCOs to help them through a period of intense scrutiny by the men assigned to them. In units where the officer arrived during combat, this gentler entry into combat leadership was often denied as well. Dependent upon the tradition of a commander's responsibility for the education of their own officers, many had "been given no specific training in leadership."[89]

The army commissioned 210,508 officers during World War II. Of the total, 126,467 officers, or 61 percent, received commissions through the officer candidate training unit. Of the junior officers commissioned as subalterns into the army through the training unit, the majority, over 72,000, were the result of the oft-maligned command interview board process prior to April 1942. The remainder received commissions following the introduction of the War Office selection methodology. Direct commissioning from the ranks produced 28,800 officers.[90]

In combat, the NCO was quite often called upon to replace an officer fallen in battle. The lack of common training in a uniform doctrine for minor tactics worked against the success of a good NCO thrust into the officer's role. "The most dangerous job in the army was commanding a rifle platoon or rifle company. In Northwest Europe between June and November 1944, about one-third of rifle platoon and company commanders became casualties each month."[91]

Many commanders believed that too many NCOs were not up to taking over after the loss of their platoon commanders, as noted in an after-action report on the Sicilian campaign in January 1944.

> Battle throws up only a limited number of natural leaders. . . . [T]he supply of suitable [NCO] personnel did not meet the demand. A high proportion of my serjeants had been fusiliers or lance-corporals only a few months previously. Platoon serjeants were often required to command platoons in battle and were rarely equal to the task: those that were, were given direct commissions. It is obvious, therefore, that the standard of leadership amongst NCOs deteriorated as the campaign progressed and it became increasingly difficult to fill the gaps. . . . The root of the matter appear[s] to be the low standard of the infantry intake as compared to other arms.[92]

British Army custom and policy were the primary causes of the leadership shortcomings of the NCO cadre. There was no standard NCO battle school, nor curriculum, for combat leadership. The battle preparation of NCOs was dependent on the commanding officer of his battalion and company, not a common system of training, or one understood and enforced throughout the army. The army rejected standard junior officer tactical preparation (for NCOs as well) in any form of formal school setting as interfering with a commanding officer's prerogatives. The chief of the Imperial General Staff in 1937, Field Marshal Sir Cyril Deverell, disapproved the idea, noting: "I have a strong prejudice against Young Officer courses. If a young officer cannot learn his job in his own unit, there is something wrong with the unit. He should learn it there. Schools are to produce instructors, not to teach the elements."[93] The Tenth Corps of the British Eighth Army in Italy, in its lessons from the Italian campaign, noted in January 1944 that "a platoon commander requires considerable military knowledge, energy, and courage. The platoon commander influences his command in a greater degree than a commander at any other level, and a platoon reflects the character of its commander to a remarkable degree. Serjeants have commanded platoons for limited periods well, but every permanent platoon commander should be an officer. . . . The highest possible standard is required, and only the pick of our young men is good enough."[94]

Discipline and Sanctions

The method of selecting officers and NCOs detracted from the ability of inexperienced leaders to mitigate the worst aspects of a system for compulsion at odds with the expectations of the conscripted citizen of a democratic society. The extraordinary use of formal discipline indicated that insecurity and uncertainty among these junior officers and noncommissioned officers led to an overreliance on courts-martial and commanding officers' administrative penalties, rather than a stronger form of internal leadership. The inducted citizen-soldiers expected what many considered fundamental rights—the dignity of his acceptance of service to the nation, even unto death, and to receive that respect from his seniors. The incidence of military transgressions as in desertion, cowardice, mutiny, and unauthorized absence were primary indicators used to assess the morale of the soldiers. These indicators were a sort of temperature gauge of the willingness of the soldiers to stay in the fight:

> On five occasions during the war, in North Africa in the summer of 1942, in the Italian mountains during the winters of 1943/44 and 1944/45, in the Anzio beachhead in 1944, and in Normandy in the summer of 1944, the rate of desertion and psychiatric collapse was such that the senior commanders began to worry that their pre-war fears that the morale of their troops would be too fragile to sustain combat might be about to come true.[95]

At each of these times, commanders requested the reinstatement of the death penalty as a deterrent.[96] Of over 210,029 British Army court-martial convictions during World War II, 106,064 were for unauthorized absence (75,157), cowardice (167), and desertion (30,740).[97] When considering the normal wastage associated to combat units, the significant desertion rate of the infantry also posed a challenge to manning these formations as the war came to its end. Over 22,000 deserters were at large in May 1945. Of more than 99,000 men struck from the rolls for desertion, 22,000 of these men were still absent at the conclusion of the war. With more than 75,000 courts-martial for desertion, and 90 percent of deserters from the infantry divisions, when placed against 81,000 riflemen in all divisions, the toll on combat effectiveness was profound.[98] The concern over these numbers was the obvious loss of manpower among the infantry, who represented the most critical combat shortage in the army.[99] "Some months spent in daily contact with deserters in Italy taught me that although there were a few very sad cases caused by nervous strain or domestic troubles, the offence was usually a carefully thought-out plan, and the belief in a pardon after the war was widespread."[100]

Commanding officer punishments. For the soldier, his most prevalent contact with the disciplinary system came at the hands of his immediate supervisors, both commissioned officers and noncommissioned officers. A corporal, William C., a section leader transferred from the Seventh Battalion Oxford and Bucks Regiment, became a member of the 2/5 Queens Royal Regiment during 1944 when, due to high casualties, the Seventh Ox and Bucks' survivors dispersed to other infantry units. Held in high regard as a combat leader and recommended twice for awards for valor (though none awarded as his commanders were lost to death or wounding), he committed twenty-four company-level offenses from the beginning of his enlistment in 1940. His transgressions and awards for misbehavior included:[101]

- "Having a rusty charger [and] dirty [small arms ammunition]"; seven days' confinement to barracks.
- "Disobedience of Regtl Orders in wearing a scarf without a greatcoat"; five day's confinement to barracks.
- "Leaving his rifle unattended"; seven days' confinement to barracks.
- "Slackness in training"; seven days' confinement to barracks.
- "Being absent from breakfast parade"; seven days' confinement to barracks.
- "Not having his bed made up before going to breakfast"; reprimand.
- Six offenses for various short periods of AWOL (absent without leave) for which his discipline ranged from loss of ten days' pay, to fourteen days' confinement to barracks, to 168 hours in a detention facility.

Other examples of commanding officer awards for minor disciplinary infractions, in 2/5 Queens, included:[102]

- Private William C. of 2/5 Queens: "Being improperly dressed; i.e. Having no cap badge in his hat"; five days' confinement to barracks.
- Private Joseph C., 213 AA Training Battery: "Improperly dressed, i.e. Wearing civilian shoes and not wearing web gaiters with Battle Dress"; four days' confinement to barracks.
- Private Clifford W., 2/5 Queens: "Conduct to the prejudice of good order and military discipline, i.e. leaving his bed cot in an

untidy condition"; five days' confinement to barracks. "Being at Pew St Birmingham without a pass or permission from his [commanding officer] at 1510 hours"; ten days' confinement to barracks.

- Private Cecil T., 2/5 Queens: "Absent from Padres Hour parade and leaving Coy lines w.o. permission to do so"; seven days' confinement to barracks. "Insubordination, i.e. making an insubordinate remark to his superior officer"; deprived seven days' pay.
- Private Victor M., 1/6 East Surrey Regiment: "Using threatening language to an NCO"; twenty-one days' field punishment (barracks).
- Lance Bombadier Archie G., 471 Btry LAA RA: "Urinating on the Barrack Square at about 2150 hrs"; reprimand.
- Private Bernard D., 2/5 Queens: "Conduct prejudice to good order and military discipline i.e. Falling out of the ranks whilst on line of march without permission"; deprived three days' pay.
- Private Pat C., 2/5 Queens: "1. Improperly dressed i.e. not wearing gaiters with battle dress; 2. Wearing civilian shoes in Elliot St. about 1400 hrs"; seven days' seven days' confinement to barracks. "Improperly dressed at 0615 hrs Drill parade, i.e. no field dressing"; five days' seven days' confinement to barracks. "Being absent from Coy lines after lights out (2215 hrs)"; seven days' seven days' confinement to barracks.
- Private Allen B., 2/5 Queens: "Being late on 0645 hrs parade"; five days' seven days' confinement to barracks. "Being improperly dressed i.e. Having the bottoms of his shorts turned up on a Coy parade"; five days' seven days' confinement to barracks.

Field general courts-martial adjudicated those transgressions considered most egregious in military discipline, emphasizing unauthorized absence of any extended duration or during battle:[103]

- Private George R., 2/5 Queens: Following the Salerno battle, tried for "deserting His Majesty's service in that he, in the Field on 16 Sep 43 [during the Salerno battle] absented himself . . . at about 1600 hrs 16 Sep 43 until 0900 hrs 17 Sep 1943"; sentenced to nine month's field prison.
- Private Leslie C., 2/5 Queens: Following the Salerno battle, tried for "deserting His Majesty's Service, in that he . . . at approximately 1330 hrs 21.9.43 [immediately before the battle

> to capture Naples] absented himself at approximately 1130 hrs 27.9.43"; sentenced to eighteen months' detention.

All of these soldiers who encountered the British Army disciplinary system were members of a detachment of thirty-one reinforcements tried and convicted for conspiracy to mutiny and mutiny on November 15, 1944. Their actions took place in the rear echelon of 2/5 Queens Royal Regiment (West Surrey). In essence, these thirty-one men refused orders as a group to move to the front to "be allocated 'to companies some of which are in the line.'"[104] The battalion was part of the 169th Brigade of the 56th (London) Infantry Division, and saw significant combat and losses from the Salerno landings in September 1943, including devastating losses in the assaults on Gemmano and Croce on the Gothic/Rimini lines during September and early October 1944.[105]

The thirty-one soldiers were a mix of infantrymen with an original assignment to 2/5 Queens upon enlistment, alongside men reallocated from disbanded infantry formations no longer considered combat-capable due to losses. This replacement unit provided a unique window into the methods used by the British Army to replace casualties in the infantry divisions through both redistributions of survivors of disbanded battalions such as the Seventh Ox and Bucks, and the reassignment of recovered wounded, as well as the intake of retrained Royal Artillery and other soldiers to fight as infantrymen.

Reassigned soldiers from anti-aircraft and artillery formations made up for the overall shortage of British infantry, and for the losses sustained by 2/5 Queens in their fighting. Ten of these thirty-one soldiers transferred into the unit during 1944 from disbanded infantry units. Eight transferred from the Royal Artillery (anti-air, anti-tank, and artillery) or ordnance. Twelve spent between one and four months in hospital either to wounds or illness. Five received battle wounds. Two had previous field general courts-martial and twenty had minor disciplinary offenses on their records, seven with multiple instances of absence without leave.[106] Their defense counsel stated to the court:

> For all of them the strain, the strength and strain of the battle for the Gothic Line had been too much. They broke down. They were not a new draft going up to the line. Many of them had been in the Gemmano and Croce battles [4–13 September 1944] which we all know were among the fiercest yet fought in Italy, and the Division from

> which they come was in the thickest of the fighting. As Defending Officer it was my duty to find out if there were any witnesses as to character. I had the greatest difficulty in doing that, not because none were unwilling to give evidence, but because their platoon commanders or company commanders had been either killed or wounded, and were not therefore able to come and give evidence. Practically all their platoon and company commanders were either killed or wounded. Casualties were very heavy in that battalion.[107]

All received convictions and penal servitude sentences ranging from three to twelve years, with a single exception. The soldier twice recommended for valor, Corporal William C., a former member of the Seventh Ox and Bucks, received eighteen months' detention, as special leniency for his battle record. The concerns of senior leaders in the British Army were well stated in the handwritten "Comment of the Court":

> The accused have been found guilty of conspiring to mutiny. The evidence had clearly established an understanding or agreement not to go into the line. That this could be done without leaders or organization shows the feeling towards this type of crime which is becoming dangerously prevalent in the army. Due to the seriousness of the situation the court have the following suggestions to make.
>
> 1. That steps be taken to instruct presidents of courts martial that desertion is both a serious and prevalent crime and as such requires a far heavier punishment than is normally given.
>
> 2. The common opinion among men at the moment is that an amnesty will be declared at the end of the war & that they will be released & go home almost as early as the fighting soldier. They therefore lose nothing by desertions & are in no way alone or conspicuous in their crime. Not only should this comprehension be dispelled but action should be taken at this stage to ensure that men who are guilty of desertion, mutiny, and such crimes are not given an early release after the war.[108]

There were no executions in the British Army of World War II for military offenses. British discipline did not falter. The British soldier endured as ordered. He carried the kernel of his sense of purpose and achievement to the end of the war. Field Marshall Bernard Law Montgomery was known as one who perhaps knew best the British soldier and it was said of him that he "understood this 'Civilian' Army as few before him. The rigid old type of discipline was not enforced. Human weaknesses were fully appreciated and the man's lot was made as easy for him as possible. This is why he was so lenient in the matter of dress and why a certain amount of 'personal commandeering' was winked at. All in all he realized that the Prussian type of discipline was not suited to the civilian soldiers of the Empire."[109]

While the "Prussian type of discipline" was not the way of British Army implementation of discipline, the nature and frequency of the minor military offenses listed for the soldiers of one platoon of 2/5 Queens replacements spoke to what many enlisted, as Paul Fussell noted, met the definition of "chickenshit": "Frequent unnecessary inspections—of personal appearance, barracks, weapons, vehicles, kit—remain the commonest ways of indulging in chickenshit. . . . As Louis Simpson observes, 'The aim of military training is not just to prepare men for battle, but to make them long for it. Inspections are one way to achieve this.'"[110]

The failure of the regimental system, the assumed capability to instill a sense of family and cohesive behavior norms based on tradition, combined with the method of selection for these junior leaders, created the opposite of willing cooperation among the soldiers, and a reliance on coercion through formal discipline. The average number of courts-martial to soldiers in the British Army was one for every ten soldiers. The US Army exceeded the British Army in its litigiousness, with one court-martial for every eight soldiers. If, as seen in the German example, courts-martial represented 13 percent of disciplinary measures carried out against soldiers, the 87 percent remaining being company-level discipline, amounted to well over 1.6 million cases statistically, not counting the types and forms of informal "chickenshit" the soldiers noted filled their days while in the control of the junior leaders of the army.

Recognition and Awards

In 1945 the subject of special recognition of the infantry soldier become a matter of concern and discussion at the highest echelons of the British Army. The adjutant general responded on this subject in July 1945, critical of the recognition granted the US Army infantry with the combat infantryman badge:

> We do not think that individual recognition of Infantry soldiers for their work in the front lines is appropriate, not only because of the difficulties of making such awards retrospective would be too great, but also because awards for good work and gallantry are already provided for by honours and mention in despatches.

> We do feel, however, that the introduction of a badge to be worn by all ranks of the Infantry in such a way as to be recognized by the public, would be a fitting recognition of the gallantry and steadfastness of the infantry.[111]

The British Army at the highest levels knew and regretted the condition of the infantry in terms of its quality, training, and morale. At the same time, no solution appeared to remedy the problems of infantry prestige. Only after the conclusion of the war in Europe did an investigation of a special badge for the combat infantryman entertain serious consideration. In keeping with the recognition of the power of the regiment in offering such acclaim, and as no corps of infantry existed, all infantry belonged to the individual infantry regiments and wore the unique badge thereof. In response to the adjutant general, the chief of the Imperial General Staff concluded in July 1945 that the infantry needed no special recognition, nor any increase in pay, although he did not object to an infantryman receiving the same pay as a tradesman. He preferred the award of decorations for whole units, as in the French practice for any especially meritorious action.[112]

The British military system of individual awards for valorous or otherwise meritorious actions remained throughout the war the primary visible recognition of the prowess of an infantry soldier, the Victoria Cross chief among British awards for valor. These awards were not worn on the battle dress of the soldier. British Army policy remained throughout the war that no badge of distinction outside the military honors system for individual bravery was either necessary or needed.

Replacements and Wounded

British Army design and tradition were to provide replacements for combat infantry units from men receiving basic and advanced infantry training at a regimental training center to its battalions overseas in combat. The strength and cohesive power of the regiment based the idea of esprit de corps and morale of the regimental soldier on this tradition. The inability, however, to maintain the numbers of the army in the face of the overall manpower shortage, and the requirements for combat casualty replacements, caused the failure of this intent. By 1942 the policy of individual replacement from any initial training center was paramount and universal, and entire classes of infantry training center graduates were sent to battalions in the field without regard to region or regimental affiliation.[113]

An individual drafted overseas with other new soldiers and recovered wounded went first into a replacement center in the theater assigned. From there, he received assignment to a combat unit and moved up to the line with other individual replacements. If moving to the Allied Armies Italy, the long transit at sea and the indifferent conditions in the transit camps reduced his physical and mental preparation for combat, while the impersonal nature of his movements increased his anxiety about the combat he faced. In general, he joined a group of men to whom he was a stranger, and for whom he felt no connection.[114] An British Eighth Army psychiatrist made the following observations concerning the problems encountered by the individual replacement in September 1944:

> Men who are of the 18–22-year age group who have not seen any action are a special problem. The officers and NCOs of the unit receiving them need to display considerable skill and care if the reinforcements are to be properly assimilated by the unit; they also need time for the job. . . . If a system could be devised whereby all large drafts had officers already knowing their men, it would be highly advantageous to morale. Psychiatric experience in this theatre suggest that in many cases the morale of the soldier suffers considerably from any but the shortest of stays in a Transit Camp or [replacement transit depot]; sometimes the deterioration is disastrous.[115]

Veteran combat soldiers, serving years in a regiment in a theater of war, were not immune to the impersonal transfer. When a division disbanded, as happened with increased frequency as the war progressed, these men also found themselves transferred into strange units to maintain the strength of the receiving division. Similar to the US practice of keeping divisions in the line once committed and feeding individual replacements to maintain their strength, British policy cannibalized existing divisions for reinforcing infantry to keep up the strength of remaining deployed divisions. Experience for the British infantry was the same as in the US (and German) infantry—long and continuous service in the line until wounded or killed.

There were insufficient men to keep divisions manned at War Establishment strength. Circumstances forced the army to keep fewer, nearly fully manned divisions continuously in the line of battle.[116] The disbanding of the Fiftieth (Northumberland) Division in November 1944, to cadre strength for return to England as a training formation, caused the prime minister to object to further organizational loss to the British Army:

> I am much distressed by the impending cannibalization of the 50th Division. We cannot afford at this stage to reduce our stake in the western line of battle. . . . It is a painful reflection that probably not one in 4–5 men who wear the King's uniform ever hear a bullet whistle, or are likely to hear one. The vast majority run no more risk than the civil population in southern England. It is my unpleasant duty to dwell on these facts. One set of men are sent back again and again to the front, while the great majority are kept out of all fighting to their great regret.[117]

Field Marshal Montgomery reminded the prime minister that the purpose for the selection of this division was the determination that it was low in strength and battle weary. Men of the infantry sections, presumed to be those most "weary," were posted as reinforcements to the remaining divisions, causing one to wonder just what part of the infantry division was "battle weary," if not the infantrymen. The infantry went back into the line while the division staff and selected cadres went home to train other men for other divisions.[118]

Physical wounds. An Army Council instruction during the week of June 28, 1939, provided British Army policy on the handling of wounded and recovered soldiers, requiring that after twenty-one days in hospital the soldier be struck from his unit strength and posted to a training unit of his arm of service for subsequent reassignment. The policy allowed for soldiers to be dispatched to their original units if in the home territory; if not, to a training unit for reassignment.[119]

With the shipping shortage as the principal limitation to returning wounded soldiers to the home island, the army established hospitals in the overseas combat theaters for the treatment and convalescence of wounded soldiers to return them to duty in-theater. Unable to transport the wounded outside the theater of operations, convalescent medical care units were the norm in the line-of-communications area. "Convalescent depots are intended for the reception of officers and men who require no further active medical or surgical treatment, and who, though not yet fit, are likely to become so in a reasonable time."[120]

The regulations were modified in 1942 with the Army Council instruction of January 14, 1942. Clearer and more detailed procedures were created for the convalescent depots in the United Kingdom. The new instruction stated:

> Military convalescent depots (hereinafter referred to as "convalescent depots") have been established throughout the United King-

> dom, where soldiers may receive post-hospital training to enable them to be returned to their own units fit for duty without passing through holding or training units.
>
> No soldier will be sent to a convalescent depot unless he is able take an ordinary diet, able to walk and attend to his personal needs, and is likely to be fit for regimental duty within approximately one month after his admission.
>
> Convalescents will not be returned to duty from convalescent depots until they are fit for duty either in medical category C or higher.[121]

The intent of the policy was to return soldiers to their own unit. Category C was described as suited for "men fit for service at home only."[122] Overseas theaters such as that encompassing the Western Desert of North Africa maintained their own convalescent depots and transit areas for the movement of wounded as well as replacement soldiers, expanding to encompass the wounded from operations in Sicily and Italy, in the course of the war.

In 1943 a new system of casualty reporting went into effect to identify and return eligible soldiers to the United Kingdom wounded three times since September 1940, allowing them to remain on the home island for six months, and defining the term "wounded" as the result of discharge of enemy weapons or friendly fire, including the effects of such weapons including splinters, fragments of masonry, building collapse, and other damage. Wounds requiring admission to general hospitals, casualty clearing, and field-dressing stations were qualifying. "Battle exhaustion" cases did not qualify as a wound for the purposes of return to the United Kingdom.[123] "Return to the UK will not be regarded as automatic as of right but will only be granted if and when passages are available."[124]

The lack of shipping meant that convalescent hospitals in theaters of operations such as North Africa, Sicily, and Italy, made such places a detriment to soldier morale:

> A transient population, leavened in base depots with a number of men who, usually on account of their low medical category were "unpostable," and a staff many of whom were "throw-outs" from front-line units, inevitably created an atmosphere of homelessness and gloom which it would have been difficult to overcome even by a first-rate commanding officer with a sufficient staff working in satisfactory conditions. Until late in 1944 no home leave was permitted on compassionate or other grounds. . . . On account of shortage of manpower, no system of home posting on the conclusion of

> a fixed period of service overseas could be introduced until 1943, by which time many regular soldiers had served six, seven, and in some cases many more years overseas. . . . Apparent lack of sympathy in this matter on the part of the authorities at home was alleged by the troops concerned as the main cause of the "Forgotten Army" feeling which grew to be endemic.[125]

George Richards, a private with the Second Battalion Royal Fusiliers wounded on May 13, 1944, in the attack on Monte Cassino, described being taken back to a bridge crossing and left there due to heavy incoming German fire. While there exposed on the bridge, he was wounded again. He returned to his unit four months later, following his stay in hospital and time spent in a convalescent company, and stated that he "didn't know anybody, all new faces, tremendous casualties, didn't have much to do with them, new recruits tend to stick together . . . frightens them, of course it does, to see the injuries." He believed that he was not ready to rejoin his unit as he was still bleeding from his leg wound, noting that the doctor told him, "I have no power to do anything now, we are desperately short of infantry, I cannot help you." Richards asked, "I couldn't climb a mountain could I?" to which the doctor replied, "No, you couldn't." But he went back to his unit regardless.[126]

Private Harry Wilkinson, a soldier with the Sixth Durham Light Infantry, received light wounds twice, once in the Western Desert and again after Normandy, and contracted malaria during the Sicily campaign. Each time he returned to his unit. His recovery from malaria was three weeks in hospital and two weeks in a convalescent camp. "I kept saying to get me the hell out of there . . . started to fiddle with the thermometer to get out of hospital. . . . I wanted to get back to my unit."[127]

"Another perpetual source of complaint, very relative to morale, was the difficulty of ensuring that sick or wounded men on recovery were returned to their own units, or to units of their own regiment."[128]

Psychological casualties. There were two distinct functions performed by British Army psychiatrists of relevance to the combat soldier: first was to remove at induction those deemed least useful to the army; second was to treat soldiers suffering from trauma-induced neuroses in combat and return them to duty. At induction, the first task was the evaluation of inductees and volunteers and "the early discovery of the dull or psychologically difficult who could be transferred or otherwise disposed of through the psychiatrists."[129] It was noted that "psychiatric referrals were to be made of the dull, the illiterate, men with a history of chronic minor ailments or showing marked preoccupation with health, notably anxious or timid, men with overt symptoms such as stammering, tics and tremors, men with a really poor attitude to service and men with conspicuously unstable, drifting backgrounds."[130]

The purpose of these evaluations was to eliminate men who, due to mental or emotional history of instability, did not have the capacity to withstand the mental stress of combat.

The physical and mental strain of constant danger in the infantry small unit demanded a period of relief from the battlefront as the simplest cure. Some commanders took the initiative to allow individuals to take breaks behind the line to rest and recuperate physically and emotionally. When questioned how he himself withstood the strain of close combat, and his observations of men who appeared to be breaking under the stress, Private Wilkinson of the Sixth Durham Light Infantry replied: "Well the best thing to do is shout at him as loud as you can . . . threaten him and if he doesn't stop you give him a beltin' as you would anybody that is hysterical. . . . Quite a lot of noise in battle, a lot of chaps shout to keep contact in battle . . . the fact of a voice, it sounds a bit reassuring to know there's someone fairly near you. . . . First thing you want after a battle. . . . A bit of peace and quiet, a bit of time to gather your own thoughts."[131]

By April 1943, the army directed that immediate care for "battle exhaustion" cases take place at forward casualty-clearing stations, basing this policy on the results of Eighth Army divisional experiments in the Ninth Australian and Second New Zealand Divisions in the Western Desert.[132] "The creation of facilities for forward psychiatry was timely in view of the near collapse of morale in some units of the British Army. Major John Wishart witnessed at first hand a mass breakdown of troops at the battle of Two Tree Hill on 17 January 1943, in which sixty-six of 450–550 casualties were psychiatric. In the 3rd Battalion of the Grenadier Guards there had been a 'generalised panic,' which was attributed to fatigue, inadequate rations, poor selection of personnel, heavy losses, and the fact that many men were inexperienced."[133]

Short periods to rest were the common methods of care, with estimates of the success rate in the units at 30 to 40 percent. This methodology became the norm for the Twenty-First Army Group in northwestern Europe as well.[134] A resolution for the problem of extended time in combat did not appear, as divisions and their infantrymen stayed in the line of battle until wounded or killed. When discussing the rationale for breaking up the Fiftieth (Northumberland) Division in November 1944 over the prime minister's dissent, the secretary of state for war explained the problem: "We did not visualise every division being in the line all the time. This has in fact been the case . . . had estimated infantry casualties at 220 per day. . . . [I]n fact, casualties have averaged 350 a day."[135]

The effect of leadership in the reduction and amelioration of psychological trauma was also noted. "Good leadership also played its part, and the 8th Army especially, was well-known for its relaxed competence and *esprit de corps.*"[136]

Conclusion

It is scarcely to be disputed that the army never received so large a proportion of the highest-grade men in intakes as did the Royal Navy and Royal Air Force. The only method of correcting this would have been to have carried out a job analysis of each service, to have applied selection testing to intakes before allotment to the different services, and to have distributed the various grades in each intake on a scientific allotment between services, giving to no service any opportunity of selecting its entrants. This would have abrogated all latitude of choice to the individual. The British people are by tradition tenacious of individual choice and yet understand how ruthless conscription must be to achieve economic results. "Perhaps it is well that some liberty of choice should remain, but it is well to recognize the cost of that liberty."[137]

The destruction of enemy ground forces and the occupation of territory were the tasks of the infantry. The infantry was 20 percent of the army and suffered 80 percent of the casualties. The disproportionate number of casualties within the infantry attests to one of the tragedies of a reverence for the scientific selection method. If the final purpose of the British Army was to find, fight, and defeat its enemies on the battlefield, rather than to support itself, then science and psychology as the basis for selection went too far. The science ensured that the army supported itself with high-quality mechanics, tank drivers, gunners, radio operators, armament artificers, and mechanical engineers. The science and psychology identified those least capable of serving in any capacity. The assumption that the regimental system might provide the inducted citizen-soldier with a sense of purpose—the norms and standards of cohesive behavior, the traditions needed to withstand the extraordinary stress of infantry battle—proved inadequate and its failure unforeseen. The intent for the infantry soldier's socialization into the small unit was undermined by the disproportionate assignment of the least-capable mentally and physically to the infantry, then choosing its leaders from that same pool of soldiers. The breakdown of the regimental system in terms of replacements further strained the ability of the infantry small unit to express its norms and standards of behavior in the absorption of new recruits and returning wounded from other units or disbanded units. The recurrent attempts at forming such bonds within the unit were difficult in the extreme. The resulting tactical method of the British Army did not rely on the power of its infantry to advance on its own. The massive fires available to the British soldier provided the edge needed to overcome the German infantry through the destruction of that enemy by these fires. German cohesion was overcome by the weight of metal available to the British Army, and the German inability after 1942 to fight combined arms without an air force. In this, the British Army general officers fought the German army as they needed. With the gravest restrictions on incurring combat losses, these leaders saved the lives of their soldiers, paving the way to victory through fires, incurring 16 percent of the battle-wounded and killed compared to that inflicted during the Great War (see Appendix 5).[138]

Notes

1. Wilmot, *The Struggle for Europe,* 233–280; Gilbert, *The Second World War,* 682–707; Keegan, *The Second World War,* 327–362, 503–585.

2. TNA PRO WO 216/101, personal telegram, Prime Minister to Field Marshal Montgomery, no. T.2348/4, 12.12.44; TNA PRO CAB 123/106, extract from War Cabinet meeting, W.M. (44) 161st Conclusions, December 4, 1944.

3. Ungerson, *Personnel Selection,* 96; Overy, *Why the Allies Won,* 282–313; Hart, *Montgomery and "Colossal Cracks,"* 84–102; Ungerson, *Personnel Selection,* 27.

4. Ungerson, *Personnel Selection,* 27.

5. Ibid., 28.

6. Vernon, *Personnel Selection in the British Forces,* 39; Ungerson, *Personnel Selection,* 27.

7. Pigott, *Manpower Problems,* 22–24.

8. Ibid., 20–21.

9. Ungerson, *Personnel Selection,* 40.

10. TNA WO 32/9960 E.C.A.C./P (41)111, December 5, 1942, paper by Adjutant General, "Use of Manpower in the Army," pt. 1, January 1939–December 1941, appendixes A, C; June 1943, paper no. 3 by Adjutant General, "Use of Manpower in the Army," June 1942–May 1943, appendix A; Ungerson, *Personnel Selection,* 63; Pigott, *Manpower Problems,* 15, 54.

11. Pigott, *Manpower Problems,* 40.

12. TNA PRO WO 32/9960, appendix H, "Selection of Personnel, June 1942–May 1943," to paper no. 3 by Adjutant General, "Use of Manpower in the Army."

13. Ungerson, *Personnel Selection,* 35.

14. Ibid., 7.

15. Ungerson, *Personnel Selection,* 54; TNA PRO WO 165/101, War Diary D.S.P., November 1943.

16. Ungerson, *Personnel Selection,* 52.

17. TNA PRO WO 32/9960, appendix H, "Selection of Personnel, June 1942–May 1943," to paper no. 3 by Adjutant General, "Use of Manpower in the Army"; http://www-3.ibm.com/ibm/history/exhibits/builders/builders_hollerith.html.

18. TNA PRO WO 32/10464, letter from Deputy Director of Organization, Personnel, P. N. Crosse to the Controller of Hollerith Equipment (Mr. Lewis), April 5, 1942.

19. Ungerson, *Personnel Selection,* appendix 5, "Job Analysis," 113.

20. Ungerson, *Personnel Selection,* appendix 3, "Job Analysis," "Reconnaissance Battalion."

21. Ungerson, *Personnel Selection,* 51.

22. Ibid., 47.

23. Ibid., 48.

24. TNA PRO WO 365/99, A.G. Stats (War Office), *Ages of British Officers and Other Ranks on the 1st July 1940.*

25. TNA PRO WO 216/66, annex to note from D.A.G. (B) to V.C.I.G.S., August 17, 1942.

26. Ungerson, *Personnel Selection,* 52.

27. TNA PRO WO 216/66, note from Inspector-General, Training, to C.I.G.S., August 4, 1942.

28. TNA PRO CAB 108/19, *Statistical Digest,* series C, no. 21, "Statistics Relating to Manpower November 1944" (issued by the Central Statistics Office), 14.

29. TNA PRO WO 199/461, "Minutes of a Meeting held at the War Office to Consider the Build Up of Forces for Offensive Operations," January 5, 1943.

30. TNA PRO CAB 120/230, Prime Minister personal minute to S. of S., no. M400, December 9, 1940.

31. TNA PRO WO 32/10400, Infantry Battalion (Higher Establishment), Notified A.C.I. [Army Council Instruction], May 19, 1943.

32. By War Establishment, approximately a further 9,000 British infantry riflemen served in twenty-five infantry battalions in British Indian Army divisions, with eleven infantry battalions in the Chindits alone, in late 1944. Indian Army numbers derived from Guy and Boyden, *Soldiers of the Raj.*

33. TNA PRO WO 277/2, T.B.H. Otway, *Airborne Forces,* appendix O, 438; Forty, *British Army Handbook,* 168–169.

34. TNA PRO WO 216/101, personal telegram, Prime Minister to Field Marshal Montgomery, no. T.2348/4, 12.12.44; TNA PRO CAB 123/106, extract from War Cabinet meeting W.M. (44) 161st Conclusions, December 4, 1944.

35. TNA PRO WO 32/10899, appendixes F and F2 to amendment no. 1 to 20/Gen/6166 (SD 4), May 1944, "Review of Army Manpower," June 7, 1944; French, *Raising Churchill's Army,* 189.

36. TNA PRO WO 216/101, telegram, Tactical H.Q. 21st Army Group to the War Office, December 11, 1941.

37. TNA PRO WO 204/10153, G.H.Q. Middle East Force, "Disposal of Personnel ex Disbanded RA Units," October 31, 1944; TNA PRO WO 32/10899 20/Gen/6166 (SD 4), "Review of Army Manpower: Note on Activity Forecast in Relation to Reinforcement Position," May 4, 1944.

38. Ungerson, *Personnel Selection,* 96.

39. TNA PRO WO 203/1068, unsigned letter to H.Q. 33 Indian Corps, 14th Army, from the Major-General Commanding 7th Indian Division, June 17, 1944.

40. TNA PRO WO 231/14, "Notes on the Sicilian Campaign Nov–Dec 1943," by Col T. N. Grazebrook, January 3, 1944.

41. Otway, *Airborne Forces,* 21.

42. Ibid., 41.

43. Ibid., 42.

44. TNA PRO WO 32/9960, "Paper no. 3 by the A.G. on the Use of Manpower in the Army: Period June 1942–May 1943," pt. 1, para. 3; TNA PRO WO 165/101, War Diary of the Directorate of Selection of Personnel (D.S.P.), May 1943.

45. Otway, *Airborne Forces,* 43–45.

46. TNA PRO WO 165/101, War Diary of the Directorate of Selection of Personnel (D.S.P.), August 1943.

47. TNA PRO WO 216/87A, *Interim Report by the Inspector of Infantry Training Establishments on Primary Training Units in Relation to Infantry Commitments,* August 10, 1942.

48. Gibb, *Training in the Army,* 3; TNA PRO WO 231/10, "Lessons of the Tunisian Campaign, 1942–3"; G.O.C. 18th Army Group, Allied Forces in North Africa, "Lessons by Arms," sec. 5, "Infantry," July 20, 1943.

49. French, *Raising Churchill's Army,* 126.

50. Place, *Military Training in the British Army,* 69–70.

51. Ibid.

52. Ibid., 169.

53. Ibid., 60.

54. Ibid., 62.

55. Place, *Military Training in the British Army,* 48–49; French, *Raising Churchill's Army,* 200.

56. TNA PRO WO 165/94, Directorate of Infantry, Progress Bulletin (Infantry) no. 1, July 11, 1943, 1.

57. TNA PRO WO 165/94, Directorate of Infantry, Progress Bulletin (Infantry) no. 16, November 18, 1944, 2, emphasis in original.

58. Gibb, *Training in the Army,* 107; Place, *Military Training in the British Army,* 32.

59. Gibb, *Training in the Army,* 107.

60. TNA PRO CAB 120/32, Secretary of State for War, Sir P. J. Grigg, to Winston Churchill, April 12, 1943.

61. LHCMA Kirkman MSS, Kirkman to Harding, November 25, 1944. Used by permission.

62. Place, *Military Training in the British Army,* 174.

63. Hastings, *Overlord,* 249.

64. Place, *Military Training in the British Army,* 175.

65. General Staff, War Office, *Infantry Training,* pt. 8, *Fieldcraft, Battle Drill, Section, and Platoon Tactics,* 39.

66. TNA PRO WO 165/94, Directorate of Infantry, Progress Bulletin (Infantry) no. 1, July 11, 1943, appendix B.

67. TNA PRO WO 365/203, "Analysis of W.E. Standard Formations," W.E. ruling, March 18, 1943, Infantry Battalion, dated March 26, 1943.

68. French, *Raising Churchill's Army,* 39.

69. Ibid., 81.

70. TNA PRO WO 231/14, Headquarters 7th Army, US Army, "Notes on the Sicilian Campaign, Nov–Dec 1943," notes by Col. T. N. Glazebrook, lately Commander of 6th Batallion, Royal Inniskilling Fusiliers, 78th Division in North Africa, Sicily, and Italy, January 3, 1944.

71. BAMA RH 11 I/70, *Die Schützengruppe im Gefechte, Ausbildungsvorschrift für die Infanterie (A.V.I.), Heft* no. 2, *1st Kompanie, Infanterie Ersatz Bataillon 19,* April 13, 1942, 1.

72. Whitehouse and Bennett, *Fear Is the Foe,* 153.

73. Gibb, *Training in the Army,* 3.

74. General Staff, War Office, *Field Service Regulations,* vol. 1, *Organization and Administration,* 3–4.

75. TNA PRO WO 277/12 A.J.K.; Pigott, *Manpower Problems,* 30; TNA PRO WO 32/9960, A.G. memorandum to the E.C.A.C., "Use of Manpower in the Army," pt. 2, appendix J, "Selection of Army Officers," para. 1, November 21, 1941.

76. Ibid., para. 7.

77. Ibid.

78. Ungerson, *Personnel Selection,* 60.

79. Ibid., 57.

80. Vernon, *Personnel Selection in the British Forces,* 23, 64.

81. Ungerson, *Personnel Selection,* 58–62.

82. Pigott, *Manpower Problems,* 31.

83. Ibid., 38–39.

84. Ungerson, *Personnel Selection,* 63.

85. TNA PRO WO 32/9960, A.G. memorandum to the E.C.A.C., "Use of Manpower in the Army," pt 2., appendix J, "Selection of Army Officers," para. 3, November 21, 1941.

86. Pigott, *Manpower Problems,* 38–39.

87. Sparrow, *Morale,* 21–22; TNA PRO WO 163/50, memorandum by A.G., "Papers and Minutes of the 1st to 10th Meetings of the Army Council Held During 1941," January 13, 1941, 2.

88. French, *Military Identities,* 155.

89. Ibid., 285.

90. Pigott, *Manpower Problems,* appendix D, 81.

91. French, *Raising Churchill's Army,* 77.

92. TNA PRO WO 231/14, "Notes on the Sicilian Campaign," Col. T. N. Grazebrooke, January 3, 1944.

93. French, *Raising Churchill's Army,* 60.

94. TNA PRO WO 231/8, *Lessons from the Italian Campaign,* 10th Corps, 8th Army, January 1944, chap. 3.

95. French, *Military Identities,* 138.

96. TNA PRO WO 163/88, letter, General C. J. E. Alexander, Commander-in-Chief, Middle East Force, to the Undersecretary of State, War Office, April 7, 1942; TNA PRO WO 163/53, E.C.A.C. letter to Secretary of State for War, discussion of letter from General Maitland Wilson, Supreme Allied Commander, Mediterranean Theatre, "Requesting Re-instatement of the Death Penalty," May 16, 1944.

97. McPherson, *Discipline,* appendix I(a).

98. Glass, *The Deserters,* 307. "In February 1953, shortly after he was returned to office as prime minister, Winston Churchill declared a general amnesty for wartime deserters as part of celebrations for the coronation of Her Majesty Queen Elizabeth II. Deserters hiding in Britain were free to take up legal occupations and those overseas could at last return home. The postwar 'manhunt' that had preoccupied the civil and military police for more than seven years was over."

99. TNA PRO WO 277/7 A.B.; McPherson, *Discipline,*, 49, appendix 5, tab. 4, "Summary," 117. In one random study of 2,000 apprehended deserters, 1,770 were infantrymen; McPherson, *Discipline,* 51.

100. McPherson, *Discipline,* 57.

101. TNA PRO WO 71/950, "Proceedings of Field General Court-Martial on Active Service of 31 Men All of 2/5 Battalion, the Queens Royal Regiment, for the Charge of Conspiring with Other Persons to Cause a Mutiny in His Majesty's Military Forces," Brigadier E. A. Aderne, D.S.O., O.B.E., presiding, 8th Army, November 15, 1944, "Service and Casualty Forms."

102. Ibid.

103. Ibid.

104. Ibid., "Service and Casualty Forms," trial record cover page.

105. Ibid., "Service and Casualty Forms," trial record cover page.

106. Ibid., "Service and Casualty Forms."

107. Ibid., "Service and Casualty Forms," record of trial, first day, 51.

108. Ibid., "Service and Casualty Forms," record of trial, handwritten note appended to front page of trial report.

109. McPherson, *Discipline,* 17.

110. Fussell, *Wartime,* 81.

111. TNA PRO WO 32/11532, E.C.A.C. note to Adjutant General, July 3, 1945.

112. TNA PRO WO 32/11532 C.I.G.S., HF/24/C-in-C, to Adjutant General, July 1, 1945.

113. Sparrow, *Morale,* 7.

114. Ibid., 7–8.

115. TNA PRO WO 204/6724, Allied Force Headquarters (Italy), Office of the Surgeon, "Psychiatric Comments on Some Current Issues Affecting Morale," September 19, 1944, 5.

116. TNA PRO WO 216/101 D.C.I.G.S., note through C.I.G.S. to Secretary of State for War, November 12, 1944.

117. Ibid., including Prime Minister personal minute, serial no. M. 1159/4, December 3, 1944.

118. TNA PRO WO 216/101, D.C.I.G.S. note through C.I.G.S. to Secretary of State for War, November 12, 1944.

119. TNA PRO WO 32/2567, War Office, "Procedure for Dealing in War with Casualties, Serious and Dangerous Illness, etc., by the Home Authorities and at Stations Abroad Other Than Theaters of War," June 28, 1939.

120. General Staff, War Office, *Field Service Regulations,* vol. 1, *Organization and Administration,* sec. 122, "Convalescent Depots," 203.

121. TNA PRO WO 165/8, War Office, Army Council instructions, January 14, 1942, para. 2147, p. 4.

122. TNA PRO WO 32/9960, "Paper by A.G. on the Use of Manpower in the Army," E.C.A.C., A.C./P (41)62, appendix B, "Medical Categories," December 3, 1941.

123. Ibid.

124. TNA PRO WO 32/10435, "Return to UK of Personnel Who Have Been Wounded Three Times," *War Office Circular of Instructions Relative to Documentation in Theatres of War,* serial no. 11, October 7, 1943.

125. Sparrow, *Morale,* 8.

126. Imperial War Museum, London, Oral History Accession no. 26634, George Richards, 2nd Battalion Royal Fusiliers, reel 10.

127. Imperial War Museum, London, Oral History Accession no. 13358, Harry Wilkinson, 6th Durham Light Infantry, reel 14.

128. Sparrow, *Morale,* 12.

129. Ungerson, *Personnel Selection,* 24.

130. Ibid., 30.

131. Imperial War Museum, London, Oral History Accession no. 13358, Harry Wilkinson, 6th Durham Light Infantry, reel 14.

132. French, *Raising Churchill's Army,* 140.

133. Harrison, *Medicine and Victory,* 173.

134. French, *Raising Churchill's Army,* 140.

135. TNA PRO WO 216/101, D.C.I.G.S. note through C.I.G.S. to Secretary of State for War, December 11, 1944.

136. Harrison, *Medicine and Victory,* 179.

137. Pigott, *Manpower Problems,* 20–21.

138. TNA PRO WO 277/12; Pigott, *Manpower Problems,* appendix E, 81 (1914–1918: 2,369,753 wounded and killed; 1939–1945: 383,654 wounded and killed).

9

The United States: Improving Infantry Quality

THE OFFENSIVE PHASE OF THE WAR AGAINST GERMANY and Japan coincided with the end of the growth and mobilization of the US Army and the activation of the Sixty-Fifth Infantry Division in August 1943. The first offensive landing operations of World War II for the United States began on August 7, 1942, with the US Marine Corps assault on Guadalcanal and with US and British Army landings in North Africa on November 8. Operations on Guadalcanal ceased with the Japanese withdrawal by February 9, 1943. German and Italian forces surrendered in North Africa by May 13, 1943. The summer of 1943 ended all hope of victory for Germany, Italy, and Japan. From August 1943, the Allies retained the initiative in the offensive, and with few setbacks were always victorious.

German soldiers at Stalingrad surrendered to Soviet forces in February 1943, and the final German offensive of the Russo-German war halted at Kursk in August 1943. The stalemate and struggle for air and naval supremacy over Guadalcanal and New Guinea ended in the spring of 1943 with US naval and air superiority established to the end of the war in the Pacific. The Allies brought the German U-boat menace under control in summer 1943. The British and US combined bomber offensive was under way, with 1,000 bomber raids in daylight and night over Germany. Deep daylight bombing of Germany with long-range fighter escorts in the first months of 1944 ensured the final destruction of the German Air Force. With the postponement of the cross-channel attack in 1943 and a brief respite for stateside divisions to receive the men needed to train, the ground war in Sicily and Italy placed the severest strain on the replacement system. These strains intensified from D-Day, June 6, 1944, to January 1945 as the Allies confronted the German border defenses at the West Wall, or Siegfried Line. Pacific casualties mounted in the Philippines and Okinawa in 1945.

In the offensive stage of the war, finding replacements, both officer and enlisted, to keep the existing divisions at full strength became the major feature of calls for conscription. The process of stripping and filling divisions in training to provide combat casualty replacements overseas and to man divisions moving overseas affected the training status of infantry divisions in an ever-expanding ripple effect on readiness, compounding the existing quality deficiencies of the infantry force in general.

The nature of the infantry crisis in this final stage of World War II was both qualitative and quantitative. Despite a general belief that US manpower was inexhaustible, the numbers available for conscription continued to fall toward the war's end. With the US Army's strength capped by the troop basis of 1943 and its near identical troop basis of 1944, the army looked within itself to find the resolution for both the disproportionate quality assignment to the infantry and the numbers needed to replace combat casualties. The final effects of the classification system on the US Army as casualties increased, as well as the need for quality in the infantry force, became ever more obvious to those charged with the creation of that infantry.

US manpower was just enough to sustain the armies in Europe and the Pacific. Quality and quantity issues highlighted the nature of the last divisions leaving the United States for combat. From February 1945 there were no more divisions in the United States. With all divisions committed, the war ended just in time, with the Allied ground forces having just sufficient power to sustain the victory.[1]

The US Infantry Crisis

By March 1945, the US Army had grown to an actual strength of 8.2 million. Of this number, nearly 1 million were officers and more than 7 million were enlisted men and women, with the highest officer-to-enlisted ratio of any of the three armies studied here.[2] Fewer than 1.2

million soldiers resided in eighty-nine combat divisions.[3] One-quarter of the army was in the combat arms; infantry officers and those enlisted in combat divisions by the table of organization were just 700,000, less than 10 percent of the mobilized army.[4] From the summer of 1943, the ground combat divisions of the US Army played a heavier role in the fighting that followed Guadalcanal and Tunisia. Over 80 percent of ground casualties were in the infantry.[5] With the landings in Italy, War Department policy required that all divisions in combat be maintained at full strength and remain in the line from the time of commitment to combat. On June 10, 1944, General Marshall met with the Combined Chiefs of Staff in London and discussed with them the "new policy by which divisions at the front were being kept at full strength throughout operations with resultant increase in morale and in the length of the periods possible for units to operate without relief."[6] The infantrymen, however, as army psychiatrists noted, did not notice the enhancement in morale, and "it continued to be reported by qualified observers that the front-line infantryman was in a bitter mood, believing himself an unappreciated and forgotten man, kept in combat until exhausted, wounded, or killed, denied the comforts or advantages abundantly provided to rear-area troops."[7]

Calls for individual replacements for infantrymen in the main exceeded the output of replacement training centers. The 1942 shortfall of over 300,000 occurred again in 1943–1944. The army culled and replenished stateside divisions again and again to meet these calls, pulling men from divisions in training for combat replacements and sending them overseas. Men trained in one division filled the ranks of divisions deploying earlier to an overseas theater of operations. "Before the end of the year [1943], 24,541 men were withdrawn from fourteen infantry divisions for transfer either to overseas replacement depots or to alerted units. In 1944 low priority divisions were subjected to three major drafts for enlisted replacements. In February the 63rd Division lost about 3,200 men and the 70th lost about 3,000. . . . In the late spring and early summer seventeen infantry divisions lost 66,411 men and a final draft in July–August took away 12,057 more. The 106th Division [lost] 6,776."[8]

These losses were for overseas replacements and to fill alerted units only. Each division remained responsible to fill army-wide quotas for officer candidate school, pilots, parachute volunteers, and the Army Specialized Training Program. The aggregate losses for divisions were enormous. "The 94th Division, from activation [September

1942] to departure for port of embarkation [July 1944], lost 8,890 soldiers; the 65th Division, [August 1943 to January 1945] 11,782; the 106th Division [March 1943 to October 1944] 12,442; the 100th Division [November 1942 to September 1944] 14,787. . . . The 69th Division [May 1943 to November 1944] . . . lost 22,235 enlisted men . . . [and before its own embarkation had] trained approximately three divisions [of its own infantrymen]."[9]

Officer losses were just as substantial and disruptive to the training and preparation of these units for battle. The 65th Division lost 1,088 officers between activation and final embarkation.[10] The total number of replacements provided to all theaters of war between September 1943 and August 1945 was 1,073,697, of whom 860,867, or 80 percent, went to the infantry of the combat divisions. Infantry replacements to the European theater in this period were 452,285 of 549,975, or 82 percent of the total, representing the total casualties in terms of killed and wounded.[11] "Psychological casualty rates of 120 to 150 percent annually were not uncommon in the infantry battalions,"[12] suggesting that the difference between the total casualty rates and the number of replacements in transit—661,059 infantry casualties in all theaters against 860,867 replacements—reflected the number of infantrymen in desertion, or AWOL, and also those in hospital and convalescence.

To provide such numbers from a shrinking induction base required augmentation by other means, as the troop basis allowed no further growth and the numbers for replacements came from within the existing units of the army: "The Troop Basis fixed the size of the Army at 7,700,000 in 1943, but in fact it continued to grow through the remainder of the war. In order to stay within this limit, the Army had to reduce the calls for induction through selective service, although unable to meet the current need, and faced the challenge of filling the shortage of infantry from within its own strength. Overall the shortage of inductees against the calls amounted to 443,967 between September 1943 and April 1944."[13]

Measures included dissolving the ASTP program; reassigning aviation cadets from aircrew training to casualty replacements in the infantry; inactivating at a more rapid pace anti-aircraft and tank destroyer battalions and other combat units deemed over strength or unneeded for the fighting in Europe, and giving them six weeks of infantry training in the infantry's advanced replacement training centers; and continuing and accelerating the mass withdrawals of trained personnel from existing infantry divisions until the final decision in October 1944 to send all remaining divisions overseas.[14] In addition, 24,000 surplus pilots in training transferred to the Army Ground Forces (AGF) and received assignment to the infantry, as well as the thousands of soldiers from disbanded anti-aircraft, tank destroyer, and coastal defense units who received six weeks of retraining in infantry skills prior to being sent as replacements to an infantry division.[15]

With the essential destruction of the German Air Force and its ability to support the ground combat units of the German Army in any form of combined arms warfare by 1943, large numbers of anti-aircraft battalions thought essential at the beginning of the war became pools for infantry replacements. The quality of the anti-aircraft soldiers was higher than that of the infantryman due to the math skills required of anti-aircraft weapons specialists. As with the Army Air Forces and ASTP replacements, these men, even with rudimentary infantry training in theater, aided in the transformation of US Army infantry quality during 1945. These measures succeeded in keeping committed divisions at close to their full strength to the end of the war. The problem was "a failure to calculate correctly the required number of replacements in time to provide them by the established process of training. A basic factor in this failure was the unanticipated severity of combat for ground forces, particularly for infantry units."[16]

The Army Specialized Training Program

The Army Specialized Training Program served as a metaphor for all that was wrong and right in the army classification system and its effects on the quality of the infantry force. The ASTP, and similar college programs such as ROTC, drew over 219,000 soldiers between 1942 and its disbandment during the first half of 1944, nearly the total number of infantry riflemen in all the combat divisions of the US Army. The largest percentage of these men came from the AGF.[17] The purposes of the ASTP were many, and not all related to a direct army need. Intended to meet army requirements for college-trained specialists, the program also acted to "prevent the complete cessation in the training cycle of such important professions as medicine, and provide some students for hard-pressed colleges, many of which would have been in serious financial difficulties unless they were afforded an opportunity to participate in training servicemen."[18]

The men selected for the ASTP were the hardest loss to their units, diminishing the pool of possible leaders in the infantry and other combat divisions. The requirement for the ASTP, a score of 115 or better on the army's general classification test, was higher than for candidates for officer training. Although the share of these men coming from the AGF was 47 percent of the total, "in reality it was substantially more than an even share, since men with the required score of 115 were proportionately less numerous in the Ground Forces than in the Air or Service Forces."[19] Fifty thousand selectees came from men at induction centers with a score of 115 or better on the test. These men generally attended initial training at a reception center and went to a division for training prior to attendance at college. "Their subsequent selection for ASTP meant that replacement centers trained men who did not become replacements and that units trained men whom they could not keep."[20]

General Marshall advised the secretary of war on February 10, 1944, that the army required 134,000 trained men as replacements for operations in France, and that "the outstanding deficiency currently noted in our divisions is the number of non-commissioned officers who are below satisfactory standards of intelligence and qualities of leadership."[21] To make up this shortfall, the AGF recommended either disbanding ten divisions, three tank battalions, and twenty-six anti-aircraft battalions or cutting the ASTP program in its entirety. The ASTP ended. The program provided 73,000 men of high quality as ground combat replacements and allocated 55,500 to the infantry divisions.

The fate of the ASTP became intertwined with the overall replacement problem for the infantry following the landings at Sicily and the unanticipated and heavy infantry casualties as the offensive war intensified. The inability of the selective service system to meet the calls for induction, the limit on the size of the army in the troop basis of 1943 and 1944, and the heavy and unplanned-for infantry casualties all contributed to a snowballing replacement problem that demanded the continuous stripping and filling of divisions in training for combat casualty replacements and filling the next divisions going overseas. As noted earlier, when faced with both the qualitative crisis in small-unit infantry leadership and the overall quantitative problem, the decision to disband the ASTP was not a difficult one for General McNair, who opposed the program from the beginning.[22] What then to do with these extraordinary, high-quality young men?

Upon disbandment the ASTP students went to infantry division training in the United States as replacements for trained men stripped from the division for other divisions overseas, or those going earlier to embarkation. They were most valuable in the period when eighteen-year-olds were not permitted to go overseas as casualty replacements without prior time in a stateside division in training. ASTP men were by policy not sent to officer candidate training, although their qualifications were higher than the standard for officer candidates. They were not promoted to NCO or specialist rank, even if that was the rank they attained prior to acceptance into the ASTP, as they gave up their rank to attend the program. These extraordinary, intelligent soldiers were "a select group numbering only 2 percent of the Army, for conversion to privates."[23] The majority became riflemen. AGF policy noted that to promote them or assign them to positions of responsibility upon arrival at their units "would force commanders to ignore need, experience, and demonstrated leadership in making assignments, and that ASTP graduates, irrespective of their educational advantages, should demonstrate their ability in the unit to which they were assigned before receiving a promotion."[24] Factors outside of their control meant they could not be promoted for their time in college, "nor could they be allowed to qualify for commissions; with the reduction of the mobilization program in June [1943] the need for additional officers in the ground arms almost disappeared, and the scanty quotas of AGF officer candidate schools were filled with college men of another type, the ROTC students whom the Army was legally obliged to allow to try for commissions."[25]

The sad fact was that if not for volunteering into the ASTP, their quality meant assignment to officer candidate school. The reduction in officer candidate quotas by June 1944 meant that the AGF officer candidate schools produced their smallest number of officers at the very moment when junior officer casualties were to become the heaviest of the war to date. On June 15, 1944, only 577 officer candidates were in training in AGF schools. More than 73,000 of the highest-quality men found in the army became available for officer training, but without quotas there was no opportunity.[26] In contrast, the US Marine Corps college programs existed for the primary purpose of procuring college-educated officers outside the pool of enlisted personnel, and from July 1943 those programs constituted the primary source of its junior officers.[27]

Illiteracy and Selective Service

The US Army inducted over 320,000 men during the war designated as illiterate, providing training for up to thirteen weeks to bring them up to a fourth-grade education level as the minimum required for induction. Over 107,000 illiterates were inducted between August 1942 and May 1943, and a further 217,000 from June 1943 to October 1945, when induction of illiterate men ceased. All of these men spent time in special training units along with certain Class V inductees. "The AWOL rate among [special training unit] personnel was greater than for personnel of other units. They had more trouble understanding their presence in the Army. They didn't show as much initiative or assume as many responsibilities."[28] Due to these concerns, and continuing complaints from the infantry divisions, on July 15, 1942, "the boundary between AGCT classes IV and V was lowered from 70 to 60. In 1943 measures were initiated by which after special training many class V men were upgraded to class IV."[29] After 1943, Classes IV and V were combined when breaking out the classes of soldiers assigned to the Army Ground Forces in a further attempt to reduce complaints of quality.[30]

Improving Infantry Quality

Replacing combat casualties. In March 1944, in addition to the ending of the ASTP, the War Department policy of Army Air Forces recruiting for pilot training within the existing units of the AGF and ASF ended, allowing some higher-quality men to remain in their units. Some of this benefit was offset by the loss of 20,000 infantrymen as volunteers for parachute training during 1944. Men in AGCT intelligence Classes IV and V were not accepted to airborne training. Soldiers converted from disbanded units, such as surplus anti-aircraft battalions, were supposedly men with higher test scores compared with the men they replaced, but many were not. The commanding general of the Ninety-Fourth Division reported that "the quality of this personnel we are getting is awful. Busted down parachutists, guard house addicts from McClellan and Bragg and various replacement centers. Less than 50 percent are physically qualified."[31]

Infantry volunteers were also solicited from the AAF and ASF. These men disappointed. Many were troublemakers or forced out of their units with the threat of losing their ratings. A substantial portion of the men transferred into divisions as replacements for privates were noncommissioned officer specialists, with one division, the Seventy-Sixth, reporting an excess of 1,228 noncommissioned ranks, with others ranging from 250 to 800 in excess.[32] Attempts were made to give new incoming NCOs the opportunity to lead in positions commensurate with their ranks; many of those NCOs failed, and in losing their rank they became morale problems.[33] A process of culling the rear—the AGF zone of the interior—for combat-qualified enlisted men for the field forces found 42,000 soldiers for transfer by October 1944.[34]

The policy for eighteen-year-old soldiers forbade their assignment as combat replacements overseas even after seventeen weeks of initial training, except as part of a division deploying overseas. About half of the inductees in the army were eighteen-years-olds in 1944. The effect of this policy was to lower the average age of the infantry divisions in training in the United States. Older trained men deployed as overseas replacements, or were stripped for units going overseas sooner, and the eighteen-year-olds took their place. In November 1944, the War Department rescinded this rule, and eighteen-year-olds were again shipped overseas as replacements after their seventeen weeks of training, further reduced to fifteen weeks in January 1945.[35]

In the months of March to July 1945, the War Department received all inductees from selective service between the ages of eighteen and twenty. This became policy based on an agreement with the Navy Department over the strongest objections of the US Marine Corps. The age of marine inductees rose by an average of seven years by June 1945.[36] The Navy Department reduced the effect of this agreement by excepting inductees who passed the "Eddy" radio aptitude test, men of the higher intelligence ratings. This allowed the Navy Department, to the chagrin of the War Department, "to be still procuring about half the eligible young men becoming eighteen years of age—the better half in terms of intelligence and physical stamina."[37]

Overall, the addition of high-quality replacements from the ASTP, the AAF, and disbanded units improved the combat infantry in terms of intelligence and leadership potential. "But by February 1944, when large transfusions of high-quality men from other commands began, nearly half the divisions had been shipped overseas; the men who had been received by the Ground Forces during the period of mobilization had been built into the structure, and this had conditioned the training of all divisions activated since the outbreak of the war. The retraining of the divisions that did receive new men in 1944 had to be hasty and fell below the standards of the training program, so that the qualitative improvement did not result in a proportionate increase of combat efficiency."[38]

The physical profile system. Given the increasing concern over infantry force quality as evidenced by the disproportionate assignment of lower-intelligence men to the AGF and infantry, General McNair attempted to increase the importance of the physical classification relative to civilian occupational skills and AGCT score for infantry. He believed that this policy change was of the highest importance to the army and the nation—that the infantry, accounting for less than 10 percent of the actual strength of the army, "was nevertheless the decisive arm, that it must be scrupulously conserved, and that the strongest and healthiest men should be assigned to it."[39]

This suggestion for the physical profiling of infantry inductees was dismissed in March 1944 comments by the War Department G-1 Division: "[Physical profiling] would destroy 'the time-tested' policies whereby civilian occupation and military training were considered in making assignments. Yet the chief of the G-1 Division observed on 14 February [1944] that in consequence of emphasis on civilian skills 'the combat arms got what was left after the Air Forces and Service Forces had selected the pick of the lot.'"[40]

The ASF was perhaps even more indisposed to allow the AGF to achieve an enhancement in the most desirable persons combining higher intelligence and physique. The commander of the ASF dismissed General McNair's proposal in December 1943, noting that the "Ground Forces must unavoidably receive a less intelligent group of men, partly because of War Department priorities for aviation and ASTP students, which militated against the Ground and Service Forces alike, and partly because 'the primary factor' in assignment was to utilize civilian skills, which militated against the Ground Forces in particular. It was natural, Army Service Forces declared, for the Ground Forces to receive a lower percentage of men with high AGCT scores than the Service Forces, 'because of the greater requirements for such personnel in Army Service Forces.'"[41]

General McNair cited the physical profiling systems of the British, Canadian, and German Armies as examples, and "after summarizing the causes of the relatively low quality of personnel in the ground arms, he restated his belief that the sub-average percentage of ground soldiers in the upper intelligence levels contributed to the high casualty rate of infantrymen. He ascribed the sub-average physical quality of the Infantry in part "to the fact that professional men or skilled workers come from the more privileged classes, which are better fed and housed, and, as a result, have better physiques, generally."[42]

General McNair won permission to establish a trial evaluation of the physical profile system in February 1944. In June 1944 the system became policy for assignment to the AGF, AAF, and ASF. As he feared in November 1943, it was too late to materially change the fully mobilized army. The new system was applied with reluctance by the test and medical examiners. The induction stations ignored the quotas—80 percent of the highest-profile for assignment to AGF—if individuals "had a civilian specialty desired by one of these [AAF or ASF] commands."[43] The administrative rigidity and indifference of the classification system through induction to assignment meant that "only 4,726 out of 176,974 trainees processed through RTCs [replacement training centers] between September 1944 and January 1945, or 2½ percent, were transferred under the Physical Profile Plan. In contrast, 17,400 men not qualified for infantry duty had to be sent to infantry centers during the same period."[44]

Along with this administrative recalcitrance there were other factors that militated against its successful use in the US Army in practice. "This classification system was based on a system previously rejected in Canada. One of the difficulties was that even doctors, independently, would not give the same man the same physical profile. The differences were sometimes enough to change a man from Grade I to a Grade IV rating. Men who rated Grade I in one assignment sometimes would exhibit psychoneurotic tendencies and drop to Grade IV when they were shifted to another assignment, as happened to many thousands withdrawn from the Service Forces to be retrained as infantry."[45]

The classification "limited service" ended in 1943. In August 1944 the War Department liberalized the rules, allowing the divisions to clear themselves of men physically unfit for combat. These men resided in their divisions through 1943 and 1944, unusable, not capable of training, and not transferable due to the overall shortage of manpower. More than 3,000 positions within combat divisions were set aside for assignment to physically substandard individuals.[46]

On December 7, 1944, with battle casualties in the European theater running at 3,000 a day, or 90,000 a month, the AGF received just 53,000 soldiers a month from reception centers. Not all of the current inductees were physically qualified for combat. Raising the induction rate appeared to broaden the proportion of physically unfit. "Many men received at the front as infantry or armored replacements in the later months of 1944 were therefore inadequate physically."[47] The problem of replacements in the face of the falling numbers coming through induction meant that the shortages continued in the European theater and led to the ever-increasing use of expedients to fill the fighting divisions. "By sending men below desired physical standards, men with 15 weeks training (or with 6 weeks retraining in infantry) and men in the higher enlisted grades beyond normal proportion, and by cutting allocation of replacements to the Southwest Pacific, the War Department was able, on 8 January 1945, to assure the European theater that about 56,000 replacements a month (85 percent infantry) would arrive from the Zone of the Interior from February to June."[48]

By the end of 1944, all divisions were either overseas or moving to a port of embarkation. The final nine infantry divisions in the United States embarked overseas in January and February 1945. The end of the war came none too soon.

The Rifleman

An August 1943 survey by the Special Services Division, Army Service Forces, showed that when asked what branch of the army they would select for themselves if free to choose, 76 percent chose the AAF; just 11 percent of the infantry named their own branch. This was the lowest total, and the dissatisfaction of the infantry disturbed General Marshall, prompting him to direct General McNair to offer suggestions for improvement. The policy of the War Department was to keep the infantryman and his division in the line of battle, feeding replacements into the division to maintain its fighting strength. There were few remedies or paths to elevate the infantryman's view of his service under conditions that required he stay in combat until wounds or death ended his tour of duty.

The nature of the rifleman's task during the offensive phase of the war was to go forward across no-man's land to close with and destroy or capture an enemy or to cause him to abandon his positions. For the infantry division, Table 9.1 shows the small numbers of riflemen, 2,916 out of 14,253.

The United States brought an enormous weight of metal to the preparation of a rifleman's tactical mission and objective in the form of mortars from company and battalion, artillery from regiment, division, corps and field army, and AAF tactical aircraft bombing and strafing support, as well as, on occasion, strategic bombers. Massive and accurate supporting arms firepower was a hallmark of US combat arms.

No matter how powerful or effective this support, however it was the rifleman who exposed his life to an enemy's fire to move forward. The numbers of these riflemen demonstrated the scarcity of infantry in an army of over 8 million soldiers. The rarity of these men engendered a powerful concern in the commanding general of the AGF for their quality, as he stated in a public address in November 1943: "There is nothing in front of him but the enemy. . . . The only force that can break the hostile infantry is our own infantry. Victories are won in the forward areas—by men with brains and fighting hearts, not by machines."[49]

In the end, it fell to fewer than 3,000 infantry riflemen (see Table 9.1) out of over 14,000 men in an infantry division to move across and into the enemy's positions to achieve that infantry division's mission. If more than 40 percent of the infantry division consisted of Category IV and V individuals by AGCT score, then 5,600 men were of this classification. When extended statistically to the sixty-seven infantry divisions, there were 375,000 Category IV and V soldiers, more than the total of all US Army riflemen in rifle squads. Where were these men serving in the infantry divisions and regiments? The meta-narrative that informed the assumptions about

Table 9.1 Riflemen in Rifle Units of a Single Infantry Division, June 1944 (by 1943 Table of Organization)

Rifle Squad	Rifle Platoon	Rifle Company (193 men)	Infantry Battalion (871 men)	Infantry Regiment (3,118 men)	Infantry Division (14,253 men)
12 riflemen including one Browning automatic rifle per squad	36 riflemen in three rifle squads	108 riflemen in three rifle platoons	324 riflemen in three rifle companies	972 riflemen in three rifle battalions	2,916 riflemen in three infantry regiments

Sources: Army Ground Forces, Historical Section, *Study no. 8,* annex I, 71.

the US infantry failed in the classification method. It denied the infantry leaders and "men of the sort to display the qualities of exceptional physical vigor and great mental alertness in individual combat soldiers, and required considerable powers of leadership in commanders of small units down to the squad. The wits, skill, and stamina of semi-isolated riflemen and small-unit commanders determined not only individual survival on the battlefield but also the outcome of the battle itself."[50] By December 1944, with all divisions committed and the last nine moving to embarkation, the hard reality of the nature and importance of the quality of the infantry soldier became apparent. Of more than 8 million mobilized soldiers, fewer than 223,000 riflemen of the combat divisions of the US Army carried the offensive forward into no-man's land, most from Categories IV and V.

The Question of Elites

The airborne infantry divisions were elite assault units. These divisions drew quality infantry away from the line infantry divisions of the army. The airborne infantry sought the most physically robust men. The airborne command took volunteers and tested their physicality through a strenuous physical fitness qualification test. The first airborne divisions appeared in the AGF in 1942. Airborne infantry self-selected as volunteers. The airborne divisions received an additional advantage: they discarded their AGCT Class IV and V soldiers and limited themselves to recruiting volunteers among the top three classes by order of the War Department and AGF.[51]

The airborne focused on infantry assault from the air by parachute and glider, landing behind enemy lines to seize key localities and hold until relieved by following infantry forces. Airborne infantry withdrew from combat upon completion of the assault mission in general. Replenished in terms of emotional and physical rest and recovery, the divisions added new men, new equipment, and training time before recommitment to battle. These two elements defined the nature of the elite combat forces of the US Army.

The airborne soldier was an example of the connection between physical robustness, intelligence, aggressiveness, and the prevalence of small-unit leaders from among the soldiers themselves. These were the Americans most often depicted in books, documentaries, and films, with the strongest connection to the popular conception of the term "Greatest Generation."[52] These soldiers most exemplified the meta-narrative expectation and assumptions for the US infantry. In combat against the Germans, the quality of the airborne soldiers gave them a staying power unequaled against German infantry. They demonstrated in battle the provenance of the connection between physicality and intelligence in creating a quality soldier.

This was unlike any other infantry or armored division in US service. The principal morale problem for the infantryman of a line infantry division was the unlikelihood of leaving combat unless wounded or killed. The experience of the soldiers in the five airborne divisions was not representative of the sixty-seven infantry divisions.

Cohesion and Battle

Interwar doctrine emphasized, as in the British Army, the key role of the infantry-artillery team of the French battle method of attack that won World War I. Also, as with the British Army, the US Army never adopted the French approach in total. Works such as George Marshall's *Infantry in Battle* maintained the master plot that, when combined with a vigorous and innovative use of artillery, the exceptional initiative of the US infantry was unbeatable. The anecdotes and lessons of Marshall's work stood as a testament to the ingenuity, wit, and stamina of the ideal US infantryman and his small-unit leaders as an example and an assumption of future capability and capacity as well.[53]

With the German example from the defeat of France, and the expulsion of the British Army from the continent in 1940, new doctrine appeared in 1941. It sought to abandon the methods and expectations for the infantry-artillery team, moving away from a fire-driven battle method and toward one more in line with the mechanization and maneuver requirements of modern war. "No longer was the army a primarily infantry-based force with auxiliary arms. The infantry was now an arm of close combat, with very limited offensive capability, dependent on the combined arms for success on the battlefield."[54]

With this new conception, the tactical references for infantry small units created a more dogmatic system for tactics in the infantry battalion. A rifle battalion infantry training manual introduced in October 1940 reduced the expectations for the offensive power of the standard infantry formations, in favor of more fire-based assault. In a manner similar to British Army battle drill, the manual discouraged rehearsing set drills in favor of junior leader initiative, yet provided detailed and definite formations and tactics for the execution of the fire-based offensive.[55] The armored division carried the decisive capability in battle. Infantry was no longer paramount, but rather another member of a combined arms team, using a definitive fire-based drill. This doctrine guided the tactical expectations for the infantry into the offensive phase of the war.

Once infantrymen were selected and placed into small units, the institutional responsibility for the cohesive behavior of the infantry in battle—maintaining the norm that the infantryman was ready to move forward at the risk of his life—lay with the army. Part 2 of this book has described primary institutional intents and effects in the offensive phase of the war through final victory, including how the infantry trained for battle, who led the infantry small units, the effect of the application of the formal and informal system of discipline, the policy for replacements, and the treatment for wounds and the return of wounded soldiers to combat. The institution understood and wrote into doctrine the requirement for cohesion and how to achieve it. The manner and method of soldier classification, the disproportionate assignment of men with lesser capability mentally and physically to the most demanding mental and physical form of combat—face-to-face infantry battle—denied the ability to build and sustain that cohesion.

Basic Training

Upon the introduction of conscription in September 1940, the inducted infantry soldiers were sent to their individually assigned divisions for training in a rifle company as done since 1922. By March 1941, the construction of facilities for combined replacement training centers allowed all inductees to receive basic training at one of these locations. Four infantry, three field artillery, three coast artillery, one armored, and one cavalry replacement training center were in operation on this date, with three additional coming on line for the anti-aircraft command. In practice, none of these branches were consistent in the basic and specialist training periods. Infantry, cavalry, and armored force training periods were each thirteen weeks; field artillery was twelve weeks; and coastal artillery split between eight weeks and twelve weeks. Basic training varied from as little as two weeks of its twelve-week replacement training center for field artillery, to five weeks of its thirteen-week replacement training center for infantry.[56]

To add further complexity, the chiefs of each of the traditional branches, or "arms"—infantry, cavalry, field artillery, and coastal artillery—ran their own replacement centers under the joint control of the commander of the corps area in which situated. Established in 1940 and 1941, the armored force, the tank destroyer center, and the anti-aircraft command constituted additional ground "arms" or "quasi-arms." These operated their own replacement centers (three in the anti-aircraft command and one each in the armored force and tank destroyer center), as well as instruction for their specialists and their own officer candidate schools. By 1943, the army had created an airborne command with this responsibility for specialists and officers as well. While the traditional arms owned command of no tactical units and no authority for the preparation of the combat divisions and independent combat units, the "quasi-arms" did. In effect, the army was not indoctrinating an individual with an army ethos, but an individual infantryman, gunner, cavalryman, or tanker.[57]

Policy changed in December 1941 when the War Department decided not to expand the replacement training centers to keep pace with the growth of the army. It sent inducted men through two avenues into the divisions: (1) for new divisions, men were sent as "fillers" directly from the induction reception centers for basic training in the division; while (2) older divisions, the regulars and the National Guard, exempted from the basic training requirement, received only trained graduates from one of the replacement training centers. In practice, the replacement training centers never achieved the numbers needed for the older divisions. Nor did sufficient fillers for the new divisions arrive from induction centers. All divisions took both types of new men, complicating attempts to develop training for combat and building cohesive small units.[58]

There was no guarantee that either type of recruit remained in his initial unit after assignment. Thousands of men undergoing initial training in their first combat unit were stripped from divisions to fill the ranks of divisions with a higher priority for overseas deployment. Others might move from a replacement training center to a unit overseas as a combat casualty replacement.[59]

The reorganization of the War Department and the disestablishment of the chiefs of the arms in infantry, cavalry, field artillery, and coastal artillery in March 1942 changed the oversight of training again. The new headquarters, Army Ground Forces, none had this responsibility for all ground tactical units and their preparation from induction to deployment into combat. The replacement and school command, within the AGF, received responsibility for the four traditional arms for individual replacements, follow-on schools, and officer candidate schools. The tank destroyer school was under indirect control, but not the armored force and the anti-aircraft command schools. The armored force, when replaced by the armored center in February 1944, came under the control of the replacement and school command; the anti-aircraft command retained its independence through the course of the war.[60]

As with the British Army, the US Army organized itself for entry instruction around the various "arms" or corps branches of the service. There the similarity ended. The British Army maintained the importance in spirit, if not in fact, of the value of the regiment for the esprit de corps of the unit and the bonding of the individual to his peers. The US Army recognized no regional or regimental connection among recruits and made no attempt to establish such relationships. The process of induction and assignment to initial military units was impersonal and industrial in both principle and method.

Combat Training and Organization

Lieutenant General Lesley J. McNair believed his primary role was the development of combat readiness in the deploying divisions, and he gave virtual autonomy to the replacement and school command in regard to all basic and specialist training matters in the replacement training centers.[61] The War Department's mobilization training program from October 30, 1941, provided individual and small-unit combat training guidance up to and including the regimental combat team.[62] The training plan prescribed two live-fire tests to heighten realism for the progression of the infantry within the regiment: the platoon combat proficiency tests and an infantry battalion field exercise test.[63]

Small-unit tactical operations manuals provided the guide for the infantry division battle training process. These included company manuals from 1942 and 1944, and a battalion manual.[64] The larger AGF plan for the combat preparation of divisions envisioned a standard progression: the first phase was seventeen weeks in length covering individual and small unit up to battalion level; the second period was thirteen weeks in length for progressive unit instruction from squad through regiment. The objective was "to develop each unit into a fighting team capable of taking its place in the division team and fulfilling its own particular role in battle."[65] The unit phase was to be followed by fourteen weeks of combined arms training, the objective of which was "to weld the several units of the division into a division team capable of acting as a concerted whole and maintaining itself under any and all battle conditions."[66]

It was unfortunate for the small unit, the individual soldier, and during the divisions that this forty-four-week progression had little basis in reality for most of the war. Postponements of the cross-channel invasion, first from 1942 (Operation Sledgehammer) to 1943 (Operation Roundup, later Operation Torch), and then to 1944 (Operation Overlord), worked against the achievement of these very appropriate aims. In 1942 and again in 1943, invasion planning necessitated the rapid expansion of Army Service Forces and Army Air Forces units at the expense of the AGF in both numbers of inductees and the quality of those recruits.[67]

From the summer of 1943 through to the end of the war, combat casualty replacements for committed divisions reduced time and again the opportunity for planned, progressive training.[68] "Regular Army units were robbed repeatedly for cadres and replenished by untrained inductees or partially trained men from replacement training centers; National Guard units contributed whole battalions, regiments, even combat teams to satisfy urgent requirements for task forces and defense missions; all types of organizations were subjected to repeated strippings of personnel and equipment to meet the needs of alerted units."[69]

> But old organizations suffered more than new ones. . . . It was not unusual in 1942 [1943 or 1944 or 1945] for an organization, as a result of repeated losses for cadre, officer candidate school, and transfers, to have to go through basic training several times. Then when finally alerted, it was often so far below authorized strength as to require it in turn to rob some unit of lower priority before leaving port. This circle was a vicious one inimical alike to orderly training and morale.[70]

Policy established that all infantry divisions were centers of basic instruction, unit combat training centers, and replacement pools for the higher priority or urgent replacement needs of the greater army. "A vicious circle of turnover and retraining was established,"[71] and "the result of this unfortunate situation was to make a mockery of the teamwork which Army Ground Forces had from the beginning viewed as the main objective of its effort."[72]

Army Ground Forces took the extraordinary step on April 27, 1944, of creating a special program for the twenty-two divisions (seventeen infantry divisions) activated after November 1942 and expected to bear the brunt of the final stripping and replenishment prior to embarkation for overseas. "This directive, based on a careful computation by General McNair of the maximum stripping which divisions could stand, provided for an adjustment of training on the basis of the division's readiness date and the sources of its filler replacements."[73] The special training program called "for six weeks of individual training time and tests for replacements received from other units or replacement training centers of an arm or service other than that to which assigned; thirteen weeks of individual training and tests for replacements received from [induction] reception centers; five weeks of unit training; four weeks of combined training; seven weeks of maneuvers; and, six weeks of post-maneuver training."[74] None of the twenty-two divisions fulfilled this curtailed twenty-two-week training program. All but one of the divisions received up to 4,000 fillers after alert for movement to embarkation, and none had "progressed far beyond the level of basic training."[75]

This was not the case with all of the new divisions, those not from either active division cadres or activated National Guard units. The Eighty-Fifth Infantry Division and the Eighty-Eighth Infantry Division went through the entire forty-four weeks of the AGF-prescribed training regimen with no stripping prior to embarkation and subsequent initiation into combat on the Italian peninsula in 1944. "From the beginning they 'fought as veteran units,' and for General Marshall gave 'the first confirmation from the battlefield of the soundness of our division activation and training program.'"[76]

Each division received an inspection conducted as part of preparation for overseas movement prior to its embarkation. With this report, the receiving theater of operations—the commanders of the large organizations from the Supreme Headquarters Allied Expeditionary Force (SHAEF) to the army groups, field armies, and the corps due to receive the divisions—knew the training and personnel status and the expected combat capability of each. The nature of the infantryman that the army placed in each of the divisions was a known quantity.[77]

Under the 1944 AGF table of organization for the infantry division, the basic unit of the infantry was the rifle squad of twelve men, armed with a single .30-caliber Browning automatic rifle and ten .30-caliber M1 Garand semiautomatic rifles. In a previous reorganization during 1943, the number of Brownings in a regiment dropped from 189 to 81 (the number of squads in a regiment) for the course of the war.[78] The squad structure allowed for a degree of fire and maneuver as well as the ability to take casualties and still function. Like the British Bren gun, the Browning was a magazine-fed automatic weapon with the same disadvantage of requiring a magazine change after each twenty-round magazine, a key limitation when pitted against the belt-fed machine gun in each squad (*Gruppe*) of German infantry counterparts. Unique to the US soldier and marine was the fact that infantrymen carried a semiautomatic rifle. No other nation equipped its forces with such large numbers of these weapons, and together the Browning and the M1 Garand offset somewhat the extraordinary rate-of-fire advantage of the German squad's machine gun.[79] The US Army infantry squad was complex when compared to either the British or German small unit or the marine rifle squad. This created a more difficult challenge in learning the harmonization of actions required in combat for each of four distinct groups within the rifle squad: two leaders, two scouts, one three-man Browning team, and five riflemen. British and German squads allowed a straightforward single fire support and single movement element, in contrast.

There were three squads in a rifle platoon, each led by a sergeant and a corporal, with three rifle platoons in a rifle company organized with a weapons platoon of two .30-caliber light machine guns, three 60-millimeter mortars, three anti-tank rocket launchers, and one .50-caliber machine gun. The rifle company stood at 192 men[80] and had seven officers: two at the company headquarters, a captain in command, and a first lieutenant as executive officer, with each of the three platoons commanded by a second lieutenant. Key NCOs in the rifle company included senior leaders at the company level and a platoon sergeant in each platoon.[81] Parachute infantry companies had three platoons with two squads of twelve men each until December 1944, when they went to three rifle squads per platoon;[82] the glider infantry companies had two platoons with three squads of twelve men each.[83]

Three rifle companies, a headquarters company, and a heavy weapons company composed the battalion, with all ranks at 871 men. Three battalions along with a headquarters company, a service company, a cannon company of towed 75-millimeter and 105-millimeter howitzers, and an anti-tank company filled out the regiment, with all ranks at 3,118 men. Three regiments, along with divisional artillery—twelve 155-millimeter howitzers and thirty-six 105-millimeters howitzers—as well as service, ordnance, and medical units, completed the infantry division, with an all-ranks strength of 14,253 men.[84] This organization, when supported by the massive artillery available from corps, army, and army groups, as well as the tactical air forces of the Royal Air Force and the Army Air Corps, meant that the infantry small unit expected extraordinary fire support in both offensive and defensive combat. It was the responsibility of the junior leaders of the infantry—the NCOs, lieutenants, and captains as the squad, platoon, and company leaders—to create the conditions for cohesive behavior in the test of combat itself.

Junior Leaders

The rifle battalion organization and training manual of October 1940 established expectations for leading the US infantryman. "Leaders must develop the physical vigor, self-denial, will-power, and knowledge that enable them to master difficult situations. The salient American characteristics of courage, self-reliance, initiative, and vigor provide excellent foundations for leadership."[85] Leaders were expected to demonstrate the soldierly qualities that earned the respect of their men and also to share in their hardships. They did this through the care of their troops, force of will in execution of discipline and orders, and firmness in the application of formal discipline and justice. They had to possess a firm character and acceptance of responsibility, exhibiting the highest initiative and encouraging the same among subordinates.[86] With these ideals in mind, the junior infantry officer of World War II found that meeting the expectations for his characteristics and behavior in battle was one of the most important factors supporting the development of cohesion in the small infantry unit. The key weakness in junior officer selection was that they came from the same ranks of the enlisted soldiers provided to the infantry through the army's classification system. Exceptions were the graduates of the US Military Academy at West Point and ROTC programs, the smallest source of accessions during the war.

Finding leaders. General Marshall advised his army commanders in February 1942 of the immediate need for large numbers of junior officers. General Ben Lear, the commanding general of the Second Army, directed his subordinate corps and division commanders to establish interview boards to canvas their units for candidates. "Army expansion requires tremendous increase in number of officers. Our greatest source of officer material of suitable character is now in the ranks of organizations. You will personally and immediately explore your source of supply and recommend to this headquarters every noncommissioned officer and private, particularly college and high school graduates, who is qualified to take the officer candidate course no matter what duty he may be fulfilling now."[87]

Unlike the British Army, there were no centralized, formal War Department boards such as the British War Office selection boards to govern the evaluation of officer material from the ranks. Each division owned responsibility for its own selection board, and their decisions were final for their personnel.[88]

General Lesley J. McNair, commanding general of Army Ground Forces, believed the soundest principle for the selection of army officers was to take them from the ranks of enlisted men. This policy remained in force throughout the war (see Table 9.2). He opposed college commissioning programs, including the expansion of ROTC and the ASTP. McNair believed the immediate needs of the army for

Table 9.2 Output of AGF Officer Candidate Schools, 1941–1945

			1943		1944			
Branch	1941	1942	Enlisted	ROTC	Enlisted	ROTC	1945	Total
Antiaircraft	341	9,637	13,385	936	303	507	—	25,109
Armored	177	4,998	4,590	216	164	471	733	11,349
Cavalry	181	1,536	1,468	179	60	40	—	3,464
Coast artillery	—	913	899	51	29	72	—	1,964
Field artillery	233	10,905	9,898	1,840	417	1,255	595	25,143
Infantry	457	24,190	19,452	1,874	4,143	2,786	8,300	61,202
Tank destroyer	—	1,565	3,261	161	77	194	—	5,258
Total	1,389	53,744	58,210		10,518		9,628	133,489

Source: Keast, *Study no. 6,* 47, tab. III.

platoon commanders outweighed the requirement for delay in their arrival through ROTC or ASTP. He also felt that the ASTP diverted men needed in the combat arms from officer candidate school, and feared their acceptance into college denied to the battle arms the benefit of their leadership potential. Unlike the US Navy and the Marine Corps, the size of the US Army prevented it from making a college education a prerequisite for officer candidates.[89] "The candidate schools were by far the largest source of commissioned officer personnel. By the end of 1943, with mobilization nearly complete, 108,000 officers had graduated from officer candidate schools of the Army Ground Forces. They constituted about two-thirds of all officers serving in ground arms. . . . By the end of the war in the Pacific 133,000 officers had graduated from the AGF officer candidate schools."[90] Only 25,000 additional AGF officers graduated from officer candidate school from the end of 1943 to the conclusion of the war, with 15,229 commissioned into the infantry.

The officer procurement process in the US proceeded in three major phases. An acute shortage of officers characterized the first phase, from December 1941 through mid-1943, emphasizing the rapid acquisition of large numbers of junior officers to keep up with the mobilization of all divisions. This phase ended with activation of the last infantry division during August 1943. The decision in the 1943 troop basis reduced the planned combat army by twelve divisions and 400,000 soldiers. A large surplus of officers distinguished the second phase, from August 1943 to June 1944. The officers for these planned, but not mobilized, divisions were already in training at officer candidate schools when the decision occurred for the reduction in forces. The effect of the officer surplus was a reduction in quotas for officer candidate school, creating the critical shortage of infantry officers that coincided with the full commitment of the US Army ground combat division from June 1944 to September 1945. This final period witnessed the twin problems of surplus in some arms (e.g., anti-aircraft and tank destroyer) and the climactic failure of infantry officer replacements. Meeting the battle-casualty-driven need for infantry junior officers demanded a general redistribution of officers from other branches.[91]

In the first phase, the army faced an immediate scarcity of officers in its dramatic expansion after December 1941. The source of these officer candidates was from mobilized divisions already experiencing great difficulty filling their own small units with junior leaders. The divisions were often reluctant to provide the officer candidates at a level and quality necessary to fulfill the army's requirements for the new officers demanded by the high tempo of mobilization and training for new divisions coming on line. Competition with the navy and marines, as well as the AAF, for high-quality men left the AGF at the bottom of the quality ladder.[92] The extraordinary preference granted the AAF in 1942, with the 75 percent rule and its follow-on policies through November 1943, denied the AGF tens of thousands of high-caliber men.[93]

In order to fill its officer quotas under these conditions, the AGF departed from the fundamental basis of the officer procurement policy almost as soon as it was stated. "The prevailing Army policy was to secure new officers from the enlisted ranks of the Army, except for certain specialists appointed directly from civilian life, and to award commissions for general duty, including combat leadership, within a branch. Both these fundamentals were compromised in 1942."[94] This was the genesis of the volunteer officer candidate program, which went outside the normal army requirement that officers come from the enlisted ranks. This method reached out to men deferred from the draft due to dependent children, with the expectation of greater maturity and capability in these men. The army accepted over 38,000 for attendance at officer candidate school after basic training, with 27,000 attending school by December 1942, without lowering quality.[95]

Another measure taken to fill needed officer positions was the direct assignment to officer candidate school of older men for positions in billets designated as administrative officer positions. These graduates were identified as "commissioned for administrative duty only," anticipating noncombat duty. This conflicted with the officer candidate school mission "to produce platoon commanders for units of the field forces," and in practice units ignored the restriction, assigning officers where needed. The army disbanded the program in 1943.[96]

Providing the bulk of officer candidates from the ranks of activated divisions stripped those divisions of their best NCO material. The Second Army illustrated this difficulty for the combat preparation of its Thirtieth Infantry Division, noting that "men are now being examined for officer candidate schools who have had, in some cases, one year of high school, and in some instances none. . . . [T]he units of the Second Army are just about down to rock bottom in regard to candidates for the Field Artillery, Infantry and Engineers."[97]

It was noted that "we are scratching the bottom of the barrel now for officer candidates. We are short of the right material for noncommissioned officer leaders. We will pay dearly for this in battle."[98]

General McNair reported in September 1942 that officer candidate school output, in terms of the number of officers graduating, was one of the bright spots in the AGF program to prepare divisions for combat.[99] The inspector general, in contrast, reported to General Marshall in January 1943 that the quality of recent officer candidate school graduates fell to such a degree that solving the problem required sweeping changes in the selection and development of officers. The inspector general recommended lengthening the course of instruction and changing the minimum AGCT score to 115 from 110, as well as establishing preparatory schools for each arm to better gauge an ill-defined requirement for leadership qualities among the selectees.[100] General McNair rejected all of the inspector general's recommendations, believing the quality problem was not a result of the length or type of training, but of the quality of the officer candidates themselves. General McNair judged that "the best way to make a leader was to keep the candidate in school for the shortest practicable period and then to throw him on his own resources in a unit where the habits of leadership and command could be developed in actual practice."[101] The War Department directed that all officer candidate school courses be lengthened to seventeen weeks, despite General McNair's recommendation.[102]

The War Department decided in 1942 not to expand the recruit training centers for enlisted soldiers as the army grew. At the same time, it called for the output of officer candidate school courses to grow at the same rate as the expansion of army divisions. This mismatch created the primary problem for the second phase of officer procurement, from August 1943 to June 1944. With the reductions of the 1943 troop basis, many more officers received commissions than there were positions for them.[103] This added to the problem of quality the compounding problem of surplus officers. This overabundance manifested itself in several difficult ways.

The recognition of the surplus caused a drastic reduction through 1943 and the first half of 1944 in the output of officer candidate schools. This decision became the major factor in the replacement crisis following the invasion of the European continent at Normandy in June 1944. Further, the surplus created an over-strength of officers in the divisions, requiring the creation of "pools" to hold these men. By

AGF policy, officers were not sent overseas as replacements unless completing at least three months in an actual troop unit. These overstrength officer pools created an overabundance of supervisors, further diminishing in turn the responsibilities of the NCOs and increasing confusion over the scope and range of their authority in tactical units preparing for combat. The wealth of officers was in the specialties least needing casualty replacements—anti-aircraft, coastal artillery, and tank destroyer units—all of which were facing large cutbacks in order to balance the infantry, field artillery, and armored forces.[104]

In the third and final phase of officer procurement, from June 1944 to war's end, the overproduction of officers resulted in the shutdown of officer candidate school locations and the total reduction of output to match the administrative demands of the troop basis, reaching its nadir in June 1944, just as the landings in northwestern Europe augured the highest junior officer casualties yet experienced in the war. As final preparations for the assault on the European continent were being made, the AGF officer candidate system fell to its lowest point since 1941. By 15 June 1944, only 577 candidates were attending the AGF schools, in comparison to 21,000 the year before and 9,000 four months later."[105]

The lag time between determining the need for officers and getting them to officer candidate school and then to their troop units was twelve months. With output down to less than 600 at the time of the invasion, few solutions appeared viable to relieve the coming need for officer casualty replacements through officer candidate schools alone.[106] The decision for redistribution of officers, taken in February 1944, led to the conversion of 9,000 anti-aircraft, coast artillery, and tank destroyer officers to infantry in the United States and another 3,000 overseas by the end of 1944.[107]

The Army Specialized Training Program, ended by order of General Marshall on February 18, 1944, held the potential to relieve the shortfalls. "In the ASTP group, none had scores (on the AGCT) lower than 110, 62 percent had scores of 125 and higher, 28 percent had scores of 135 and higher, and 7 percent had scores of 145 and higher."[108] Marshall's order stated that "ASTP trainees may apply for officer candidate school and will be selected upon graduation on the same basis and standards as applicable to all enlisted men of the Army," and that "ASTP graduates, assigned to units, will not be promoted to a technical or non-commissioned grade until they have been assigned and then only upon the recommendation of the unit commander."[109]

The arrival of these men in the rifle squads of the infantry divisions increased the quality of the combat small units.

Based on War Department planning figures from the 1944 troop basis, the AGF estimated a shortage of over 13,000 officers, most infantry, by mid-1945. Once again, the officer candidate school quotas ramped up to provide more infantry officers. But with the majority of combat units already overseas, AGF selection could only draw on the dwindling divisions that remained in training in the United States and on the output of the recruit training centers. Quantity again overcame the need for quality. Further exacerbating the problem was the fact that the officer surplus in 1943 denied qualified NCOs during that year the opportunity for a commission, because no quotas were forthcoming.[110]

By 1944, with the stateside divisions available to the AGF shrinking, the War Department sought candidates from units deployed overseas. How better to lead small units in combat than with combat veterans of those same units? Hard-pressed to maintain their numbers in combat, however, divisions overseas were reluctant to lose a combat-tested NCO or private to officer candidate school just when their units needed them more than ever, and battlefield promotions were never a significant factor for finding officers for the wartime army.

> Because of the poor response from unit commanders, approximately 40 percent of recent [1944] OCS quotas had to be filled from the replacement training centers. The high rate of failures in the candidate schools indicated that much of the material uncovered was below standard. An analysis of five recent classes showed that 45 percent of the men enrolled had been relieved. . . . Though the causes were different, the effects were those observed in late 1942: last-minute urgent calls for officers exceeded the available supply of qualified candidates; to fill quotas poor candidates had to be accepted; these failed in large numbers, the original program was only partly fulfilled, units were bereft of good noncommissioned officers, and other units received mediocre officers or disgruntled rejects.[111]

The final plan called for units overseas to receive limited numbers of officer replacements from the United States, furnishing most of their needs from within. This plan was overcome by the events of December 1944, which left most fighting units too short of men to consider this solution, and with the collapse of Germany and the end of fighting in May 1945 the usefulness of the plan never received a test.[112]

The quality of the junior officer force suffered in the crises of both 1942 and 1944–1945. The key difficulty through the war was the reliance on the pool of enlisted soldiers in the infantry divisions, given the circumstances of their downward selection, which limited the opportunity to find the quality candidates for officer candidate school from that same source. The focus on officers was not unusual for an army that during the interwar years paid little attention to the quality of the enlisted force or its NCOs, with the expectation that small-unit leaders were to come from those entering the army during the war, not from the NCOs of the old army. If officers were the key to the leadership of small units in combat, then the importance of the NCOs was as helper to the officer in his wartime duty, not to take his place if he became a casualty.

The War Department recognized that the importance of training the NCO in wartime for leadership duties aligned with the requirements of combat as learned in World War I. General George C. Marshall commented later in the war that, unlike their German counterparts, who with few men fought on as if a lieutenant general were in command, the US sergeants "when new at the game would think somebody ought to come right away and reinforce them or take over."[113]

General McNair rejected in March 1943 a suggestion to create a platoon sergeants course outside of the divisions "on the general principle that the best place to develop noncommissioned leadership was in the unit, and that responsibility for developing it reposed with the unit commander. This principle was regarded as fundamental by General McNair, and when a year later (May 1944) the War Department expressed interest in a school for NCOs, modeled on that for officer candidates, Army Ground Forces reaffirmed its belief in the principle's soundness."[114]

Wartime policies also created a distinction between line and specialists ranks. The line NCO was a leader in a troop unit, carrying the rank of corporal or sergeant and occupying a position within an authorized table of organization for that rank. Specialist ranks were intended to differentiate between persons whose expertise rated pay commensurate with line NCOs but did not have NCO responsibilities. In 1942, certain specialist grades became equal to line ranks and were allowed to assume the privileges of the line NCO in matters of reduced fatigue or guard duty. This increased the resentment of the private soldiers, whose duties increased, further diminishing their respect for their junior leaders. These measures were symptomatic of the confusion and uncertainty that existed with regard to the roles and responsibilities of the NCO.[115]

By 1945, 50 percent of all enlisted men in the US Army were NCOs.[116] With the officer surplus of 1943 through June 1944, extra officers stacked up in unit over-strength pools to earn three months of required time with the troops before transfer overseas, to the detriment of the role of the NCOs in the unit. The AGF G-3 operations officer, Brigadier General John M. Lentz,[117] stated in remarks to the staffs of two new divisions in May 1942 that "develop[ing] good NCOs and junior officers . . . is at present one of our major problems"; and again that "the No. 1 problem is leadership."[118] General Robert L. Eichelberger, commanding the US Eighth Army in the Southwest Pacific area in December, noted that "the sins of our military system rise up to haunt us. Where are trained corporals, sergeants, and lieutenants who can lead men?"[119] General McNair did not waiver in his policy, refusing to create a standard AGF curriculum for NCO battle leadership.

As in the British Army, the development of NCO leadership was the responsibility of the unit, not the army. The prerogative for promotion and training was up to the unit commander.[120] Where were the officers taught, and who taught them? What was essential so that an NCO could lead in battle? Who taught junior officers what was necessary to lead in battle? The junior officers leading platoons and companies had little more time in the army than did their soldiers. Reaction to the recognized quality shortfalls created a greater reliance on officers, to the detriment of NCO leadership development. The US Army infantry NCO found himself often in the position of leading his unit in combat when his officer became a casualty. The NCOs who succeeded in making the transition in combat learned the lesson in battle, not from any attempt to prepare them for the role in either training or expectations.

Selection and preparation of NCOs in the elite formations, as in the airborne infantry, differed in no appreciable manner from the process in the line infantry divisions, with the exception that their raw material received a higher-quality selection. Airborne units allowed no intelligence Category IV or V soldiers in their ranks. As in other matters concerning tactical precepts and doctrine, the elites benefited from their unique form of entry and narrow focus on infantry combat missions. These organizations relied on the initiative and leadership of junior and noncommissioned officers to carry their units to victory without the benefit of the massive fire power available to the line infantry divisions. The quality of the airborne infantry NCO highlighted success in every engagement against the German infantry as well. This was the type of fighting on which their battle practice focused. Its success depended on aggressive, well-trained junior officers and noncommissioned officers who possessed to an unusual degree the qualities of alertness, drive, and fitness demanded by their employment in combat.

Discipline and Sanctions

"American soldiers shared the common supposition that for the war to be won, many would have to do as they were told. They accepted the necessity of a basic martial discipline and of their subordination as individuals, to the cause of success in combat. But they came to denounce scathingly the ethos of command; the orders it gendered often seemed less intent on enhancing their effectiveness as fighting men than on humiliating them as enlisted men."[121]

These young Americans, as their British counterparts, came to resent the relentless, restricting nature of the military culture which they found themselves in. They appeared most to resent what they considered the "chickenshit," or "Mickey Mouse," nature of the structures and individuals imposed on their lives and thoughts and actions. Paul Fussell defined "chickenshit" as the "behavior that makes military life worse than it need be; petty harassment of the weak by the strong; open scrimmage for power and authority and prestige; sadism thinly disguised as necessary discipline; a constant 'paying off of old scores'; and insistence [more] on the letter than the spirit of the ordinances."[122]

The postwar report on officer-enlisted relations noted that "soldiers who are conscripted into service, often against their will, are not inclined to look kindly on the life that resulted from this conscription. From the beginning, there was a manifest dislike for regimentation and for the feeling of loss of individual identity in the military machine. By reason of their historical dislike of the military system, Americans have a deep-seated feeling against and strongly resist any growth of an old-world type of military caste because such would be out of keeping with our democratic government."[123] This day-to-day experience was reinforced by reliance on the formal disciplinary system of the court-martial and on unit commanding officer authority for the administrative or nonjudicial punishment of transgressors of minor military infractions.[124]

The postwar report noted "that the feeling of superiority on the part of commissioned personnel most rankled the enlisted personnel."[125] There was the belief among the enlisted ranks that these social distinctions were the basis for the unnecessary indignities experienced by the soldiers without any positive effect on discipline or military efficiency. "It was in this category that abuses were most rampant, violations occurred most frequently, irregularities were most apparent. . . . The largest differential, which brought the most criticism in every instance, was in the field of military justice and courts-martial procedure which permitted inequities and injustices to enlisted personnel."[126] Here the enlisted men felt the abuses were most evident in what many believed "faulty selection, insufficient orientation and indoctrination, and inadequate training in responsibility, [which] increased the number of abuses and arrogance displayed by officers."[127]

In terms of military offences, the US Army was the most penalized of the three armies studied. The US soldier knew of the punitive measures available through repeated readings from the Articles of War, reminding him that certain offenses, particularly those regarding misbehavior in battle, were "punishable in time of war by death or such other penalty as a court-martial may direct."[128]

Similar to the British soldier in this regard, however, "generally speaking, then, the death penalty was not used [for military offenses]. . . . In regard to heavy prison terms, many men may have thought it likely that the more drastic sentences would be revised in response to public pressure at the end of the war. So the combat man did not in fact face a choice of possible death from the enemy versus certain death if he refused combat."[129]

US military discipline appeared as a major obstacle toward cohesive behavior. The common revulsion for military life, and the anger created by incessant referral of soldiers to disciplinary procedures, indicated at least a common cause in disgust with the system. Overreliance by inexperienced officers and NCOs on formal discipline hindered the development of the small combat unit and diminished the authority of its leaders.[130]

There were over 2 million court-martial proceedings of all three types for military offenses (ranging from offenses such as absense without leave to capital crimes such as rape or murder) between

January 1942 and September 1945, or one court-martial for every eight of the more than 10 million men and women serving to September 1945. The period 1942 to 1945 produced 66,993 US Army general courts-martial. At the end of the war, 45,000 military personnel were confined as a result of general courts-martial, not counting the thousands confined as a result of "the far more numerous special and summary courts-martial."[131] The conviction rate for soldiers was 97 percent.[132]

The former governor of Vermont, serving as a lawyer in the US Army during World War II, noted that he and his peers "were advised, not once but many times on the courts that I sat on, that if we adjudged a person guilty we should inflict the maximum sentence and leave it to the Commanding General to make any reduction. . . . I was dismissed as a Law Officer and Member of a General Court-Martial because our General Court acquitted a colored man on a morals charge when the Commanding General wanted him convicted—yet the evidence didn't warrant it. I was called down and told that if I didn't convict in a greater number of cases I would be marked down in my Efficiency Rating; and I squared off and said that wasn't my conception of justice and that they had better remove me, which was done forthwith."[133]

Convictions from general courts-martial tended toward the most serious offenses, and a sample of general courts-martial in the European theater of operations is indicative of the relative rarity of such major transgressions. As with the British Army and the German Army, desertion and AWOL constituted the greatest number of offenses. For the US Army during World War II, 21,000 were apprehended and convicted of AWOL and desertion by general courts-martial,[134] and a further 60,000 by special and summary courts-martial.[135] There were 183 death sentences awarded for military offenses in the European theater of operations, of which 110 were awarded for desertion. In the European theater, the total number of AWOL cases tried by general courts-martial was 3,857, and for desertion 1,963.[136] By July 1945, the US Army had executed 142 soldiers for crimes of rape and murder, 63 in the European theater.[137]

The military penalties for desertion were reported by an American psychiatrist to be a "positive incentive to desertion."[138] As in the British Army, desertion was a major problem for the infantry on the line and at the ports of embarkation. A February 1943 memo from the adjutant general noted that "absences without leave and desertions, especially from units which have been alerted for overseas movement, have reached serious proportions."[139] The numbers were so large as to deny the ability to put them on trial. Many deserters preferred prison to duty overseas, and there were so many in stockades that staging-area barracks had to be used to house offenders. The memo provided a list of deserter's "tricks," including maiming themselves, feigning physical and mental illness, going AWOL in order to stand trial and be confined, throwing away their clothing and identification tags, and "when an officer approaches the area, the word is passed along and they dash for the woods through windows and doors, even jumping from upstairs screened windows, taking the screens with them."[140]

In the theater of operations for northwestern Europe and Italy, desertion sometimes led to criminal behavior via the black market in purloined military supplies. The problem became twofold. Many infantry soldiers receiving leave for Paris or Rome did not return to their units, and in order to stay away from the front, they connected with criminal soldiers running the black markets. Supplies needed for the fighting were lost, as were the infantrymen, requiring greater numbers of replacements and exacerbating the problems of both the supply and manning of the combat units.[141] The commanding general of the Thirty-Sixth Division reported that his soldiers were losing their efficiency and aggressiveness in September 1944, with the number of desertions the key indicator of the decline in morale. He attributed desertions to "the heavy loss of officers and noncoms and their replacement by those lacking both field experience and acquaintance with the men they commanded. Desertion was an indicator of poor leadership."[142] The United States announced no general amnesty for military offenses after World War II.

Overall, for the US armed forces, the internalization of discipline as a means to instill and enhance the power of informal sanctions within the small combat unit, in order to increase the likelihood of the bonding necessary for cohesive battle behavior, was unsuccessful. As Gerald Linderman concluded on discipline and American young men of World War II: "The worth of discipline in and of itself; obedience as the indispensable standard; trust in one's superiors, and a high confidence that their commands welded intelligence, courage, and true concern for the lives of soldiers—these were not the values of American society. In World War II, its young men resisted a system of military discipline that had been designed as if they were."[143]

Recognition and Awards

During 1943, a survey was conducted of soldiers in all branches of the US Army to determine satisfaction with their job assignments. The infantry had the lowest satisfaction level by far of all branches, and in an August 4, 1943, letter, the chief of staff, General George C. Marshall, sought recommendations to improve the infantry's morale. The Army Ground Forces commanding general, Lesley J. McNair, responded with three principal measures. The first was to increase the pay of the infantry to reflect their proximity to the enemy; the second was to create a combat badge; and the third was to change the infantryman's appellation from "infantry soldier" to "fighter." Marshall rejected the increase in pay on the grounds that other branches would want the same. He also rejected the term "fighter."[144]

On October 27, 1943, a combat infantry badge received approval for wear. The reasons stated for the introduction of the badge were:

- The need for large numbers of well-trained infantry to bring about a successful conclusion to the war and the already critical shortage of infantrymen;
- Of all soldiers, it was recognized that the infantryman continuously operated under the worst conditions and performed a mission which was not assigned to any other soldier or unit;
- The infantry, a small portion of the total Armed Forces, was suffering the most casualties while receiving the least public recognition;
- [It would] provide special recognition of the unique role of the Army infantryman, the only soldier whose daily mission is to close with and destroy the enemy and to seize and hold terrain. The badge was intended as an inducement for individuals to join the infantry while serving as a morale booster for infantrymen serving in every theater.[145]

The requirements indicated that the soldier "must be an infantryman satisfactorily performing infantry duties, must be assigned to an infantry unit during such time as the unit is engaged in active ground combat, and must actively participate in such ground combat. Campaign or battle credit alone is not sufficient for award of the [badge]."[146] And further clarification in 1944 "specified that 'action against the enemy' for purposes of award of the [badge] was to be interpreted as 'ground combat against enemy ground forces.'"[147]

After the war, all soldiers awarded the badge also received the Bronze Star Medal.[148] As with the British Army, the fundamental method of awarding recognition for valor remained the system of medal awards for individual acts of bravery, with the Congressional Medal of Honor the highest among these. US soldiers did not wear decorations on their battle uniforms.

Replacements and Wounded

Individual replacement was the policy of the US Army throughout World War II.[149] By late 1944, the apparent shortcomings of the replacement system were a heavy concern:

> Long periods of unrelieved combat produced large casualties; these in turn laid heavy demands on the replacement system. Even when replacements were provided promptly, they usually had to be fed into units while actively engaged with the enemy. New Men, under these circumstances, suffered heavy casualties because they were not adjusted to battlefield conditions or acquainted with members of their units. The dwindling stock of veterans was more rapidly reduced because these men had to expose themselves more often in providing needed leadership for the large drafts of replacements.[150]

The poor morale of the replacement soldiers came to be understood as a result of the failure to instill a sense of unit and cohesive behavior before the replacements entered combat:[151] "Our whole system of the employment of divisions for long periods and continuous replenishment of these divisions by replacements while they are in action has created a vicious cycle with respect to battle fatigue which no system of individual relief can overcome."[152]

The bonding within units was often as problematic for the veteran as it was for the replacement. The fealty of the veterans for one another was often a hindrance to the acceptance and survival of the replacements. In the words of one such new man: "You go up as a replacement and your chances are awful. The men who are there are all friends, they feel responsible for each other, they'll do anything to save each other. That means every dirty, dangerous job they hand right over to the replacements. The Sergeants don't even bother to learn your name. They don't want to know anything about you. They just trade you in for their friends and wait for the next batch of replacements."[153]

A US Army Ranger described the reception of replacements to the author James Jones after the war, as quoted by Gerald Linderman: "One day at Anzio we got eight new replacements into my platoon. We were supposed to make a little feeling attack that same day. Well, by the next day, all eight of them replacements were dead, buddy. But none of us old guys were. We weren't going to send our own guys out on point in a damned fool situation like that. . . . We were sewed up tight. And we'd been together through Africa, and Sicily, and Salerno. We sent the replacements out ahead."[154]

It was the same situation for all US soldiers going into combat as replacements. The elite units, such as the US Army Rangers, all utilized a system of individual replacements, with results like this one, related by an airborne replacement soldier:

> R. did his initial training with F Company of the 501st Infantry, an outfit that eventually became part of the 101st Airborne Division. He learned early on that the Army was a collection of teams that worked together. . . . The team was both family and workplace. That reassuring identity had been taken away from R. when, instead of returning to the 501st after airborne school, he was sent as a replacement to North Africa. R. was stuck in the replacement detachment, shortened in GI slang to the "repple-depple," while the 505th trained in the North African heat. The replacements did not receive further training, but endured the heat and boredom in a state of suspended animation.[155]

Physical wounds. The US Army created no policy to return wounded men to their original units. US soldiers, if wounded and sent beyond the division-level medical station, upon recovery went to replacement depots in the theaters of operations overseas. Reassigned where needed, the recovered soldier—termed a "casual"—returned to duty in any unit as determined by immediate need.

Psychological wounds. As in the British Army, the US Army performed two fundamental psychological tasks: first, at the induction stage, to identify and either discharge or treat men with psycho-neuroses or other mental and personality disorders deemed inconsistent with military service; and second, to treat soldiers suffering mental neuroses as a result of battle trauma.

> In all, an estimated 1,992,950 men, or more than 30 percent of all rejections, were found by the Selective Service to be unfit for general duty because of mental and educational deficiency and neuropsychiatric conditions.

> Of those who served in the Army during the war years, 1942–45, 379,486 were separated from the service for neuropsychiatric reasons. These accounted for 45 percent of all discharges for disability. An additional 356,000 were separated for inaptness, lack of required degree of adaptability, or enuresis.[156]

For the US Army, as in the British Army, the greatest attention was paid to the problem of neuropsychiatric casualties, "battle exhaustion," and the means to return men suffering from this condition to battle: "Psychiatric casualties winnowed the ranks at an alarming rate. The armies in Europe evacuated 151,920 cases of neuropsychiatric disorders to hospitals in 1944 and 1945, and combat units discovered that on average, for every three men killed or wounded, one other soldier became a psychiatric casualty."[157]

The greatest toll of psychiatric disorders took place among the infantry. In fifty-nine days of combat in Italy, the Third Infantry Division sustained 3,144 battle casualties and 5,446 nonbattle casualties.[158] Major General A. W. Kenner, chief medical office at SHAEF, estimated that 15 to 20 percent of all nonfatal combat casualties were neuropsychiatric, with losses in the North African theater estimated at 1,200 to 1,500 per 1,000-man strength per year in the rifle battalions. He concluded that "practically all men in the rifle battalions who were not otherwise disabled became psychiatric casualties. . . . The average point at which this occurred appears to have been in the region of 200 to 240 regimental combat days."[159]

British Army policy called for pulling individual riflemen out of the line of combat every twelve days or so for a few days' rest. In contrast, the US soldier in Italy stayed in the line twenty, thirty, forty, or up to eighty days. US Army divisions stayed in the line from the time they entered combat, with few exceptions.

> Under present policy [1944] no man is removed from combat until he has become worthless. The infantrymen consider this a bitter injustice. He feels that he is expected to do ten, or even 100 times as much to win the war as anyone else, but he can look forward only to death, mutilation, or psychiatric breakdown. . . . In itself, winning the war is unimportant to the American soldier. . . . The doughboy fights because he has to. . . . He fights for his buddies and because his self-respect will not let him quit. . . . After several months of combat he looks around to find that most of his buddies

are gone. He is one of the old men. For whom can he fight now? What incentive has he to go on?[160]

Majoe General Kenner's report recommended several steps to enhance prevention of these types of casualties, including the need for a break from the line. Army policy did not permit the rotation of units, so as with replacements, the rest period was based on an individual need. A British medical officer cited the results of US policy in the perception of combat units: "It is currently stated in Italy that there is no longer any such thing as cowardice in the U.S. Army, for any man who runs away from the enemy falls into the hands of a psychiatrist before he can be court-martialed and is thereupon declared not responsible for his acts on the grounds of psycho-neurosis."[161] Kenner's strongest recommendation was group rather than individual replacements. This report cited "the strongest force which keeps a man going in combat is his self-respect and pride. Its strength depends upon the bond between himself and his fellow soldiers."[162]

The War Department and the AGF agreed on the deficiencies of the replacement system: "To keep a minimum number of divisions almost constantly in contact, refilling them from a stream of individual replacements, resulted in excessive losses, weakened units, and hopeless veterans."[163] In 1945, General Joseph Stilwell, new commander of the AGF after the death of General McNair in July 1944, recommended that group replacements become the norm. Theater commanders' reactions were so unfavorable that he abandoned the idea, and replacements continued to be trained and assigned as individuals, grouped for disciplinary and movement purposes.[164]

Intended treatment for neuropsychiatric cases, defined as mental trauma due to exposure to the stress of battle and the concussive effects of high-explosive weapons, was as far forward—as combat-oriented—as possible. Results achieved indicated the same limited success the British experienced in returning these men to the fighting line.

The two primary factors identified by Major General Kenner for the prevention of "combat fatigue" were the quality of the officers at the company level and the bond, the esprit de corps, of the unit. Caring, competent, veteran leadership enhanced unit comradeship. The tightness of the link between men and their expectations of each other kept men together and fighting. Bonded units provided a buffer for those within the group whose will weakened.

In all armies the loss of combat power to neuroses found expression in many forms. The most common were removal with diagnosis as mentally defective for purposes of combat, or courts-martial for desertion or cowardice. Following the war, the American psychiatrist Matthew D. Parrish, who served in combat aircrews during World War II, found unit cohesion was of primary importance in both preventing and treating psychoneurosis. He suggested the term "membership" for describing this preventative and curative principle, and he recommended adding it to the other principles of forward treatment. Parrish considered the importance of group cohesion to be the final and perhaps most important lesson of World War II in preventing psychiatric breakdowns and creating combat effectiveness.

American psychiatrists concluded after World War II that psychiatric casualties were an inevitable consequence of life-threatening hostilities; that they defied efficient screening out ahead of time, that their numbers depended on individual, unit cohesion, and combat environmental factors; and that appropriate interventions can return the majority to combat duty.[165]

Conclusion

In the offensive phase of the war, the full impact of the failure to focus on measures to build and sustain cohesion in the small infantry unit became apparent. The critical assumption that the American man would perform in the infantry units of World War II as he had in World War I, the Civil War, and the Revolutionary War was unfounded in fact. The characteristics of warfare prior to World War II were the paramount importance of the infantry arm and its unique requirement in history and doctrine to achieve victory. The importance of the infantry in prior wars led to the placing of the bulk of American men into that arm. This ensured a proportionate availability of men possessing those characteristics deemed essential in the infantry battle.[166] The surprise was that the classification method made these expectations invalid. The disproportionate assignment of the men of the highest physical and mental quality to the AAF and ASF denied the infantry the presence of the leaders expected to come from these ranks. By the time the offensive phase of the war arrived, the character of the US infantry was set and allowed no substantial modification in the course of the final months of the war.

Notes

1. Weigley, *History of the United States Army,* 729–730; Overy, *Why the Allies Won,* 321–325.
2. Palmer, *Study no. 5,* 1.
3. Palmer, *Study no. 4,* 38–39.
4. Ibid., 25.
5. Statistical and Accounting Branch, Office of the Adjutant General, War Department, *Army Battle Casualties and Non-Battle Deaths in World War II,* final report (December 7, 1941–December 31, 1946), June 1, 1953, 5.
6. Bland and Stevens, *The Papers of George Catlett Marshall,* vol. 4, 476–478; Glass, *Deserters,* 69–70. Glass quotes Marshall's "Biennial Report of the Chief of Staff of the U.S. Army, July 1, 1943, to June 30, 1945, to the Secretary of War," reported in *Yank* magazine, October 19, 1945, 7.
7. Palmer, *Study no. 4,* 28.
8. US Army Ground Forces, *Study no. 12,* 31.
9. Ibid., 34.
10. Ibid., 35.
11. Palmer, Wiley, and Keast, *The Procurement and Training of Ground Combat Troops,* 222–223, tab. 9, and 218, tab. 8.
12. Ibid., 228.
13. Keefer, *Scholars in Foxholes,* 266, "Table: War Department, Army Service Forces Monthly Report," sec. 5, "Personnel," July 31, 1944.
14. Palmer, Wiley, and Keast, *The Procurement and Training of Ground Combat Troops,* 599.
15. Palmer, *Study no. 4,* 20–21.
16. Palmer, Wiley, and Keast, *The Procurement and Training of Ground Combat Troops,* 238.
17. Ibid., 32.
18. Ginzburg, *The Lost Divisions,* 48–49; Keefer, *Scholars in Foxholes,* 266; Palmer, Wiley, and Keast, *The Procurement and Training of Ground Combat Troops,* 212.
19. Palmer, Wiley, and Keast, *The Procurement and Training of Ground Combat Troops,* 34.
20. Ibid., 35.
21. Palmer, *Study no. 5,* 18.
22. Palmer, Wiley, and Keast, *The Procurement and Training of Ground Combat Troops,* 31.
23. Ibid., 39.
24. Ibid., 37.
25. Ibid., 35.
26. Keast, *Study no. 6,* 24.
27. Furer, *Administration of the Navy Department in World War II,* 573.
28. Houle et al., *The Armed Services and Adult Education,* 169–190; Hershey, *Special Monograph no. 16,* 159–163.
29. Palmer, *Study no. 5,* 51, n. 17.
30. Ibid.
31. US Army Ground Forces, *Study no. 12,* 37.
32. Ibid., 20.

33. Palmer, Wiley, and Keast, *The Procurement and Training of Ground Combat Troops,* 79.
34. Ibid., 81.
35. Palmer, *Study no. 4,* 27; Palmer, *Study no. 5,* 38.
36. Memorandum for Director of Personnel, Marine Corps, subject: Effect upon the Average Age of White Inductees of U.S. Marine Corps Caused by the Loss of Men in the Eighteen- to Twenty-Year Age Group, June 11, 1945; US Marine Corps, Office of the Commandant, General Correspondence (1939–1950), Record Group 127, NACP.
37. Palmer, Wiley, and Keast, *The Procurement and Training of Ground Combat Troops,* 75.
38. Ibid., 85.
39. Palmer, *Study no. 5,* 30; Headquarters Army Ground Forces, memorandum for Chief of Staff (attn: G-1 Division), subject: Utilization of Available Manpower Based upon Physical Capability, December 21, 1943, WD, AGF (1916–1954), Ref. Files L. J. McNair (1940–1944), RG 337.3, NACP, 1.
40. Ibid., 30.
41. Palmer, Wiley, and Keast, *The Procurement and Training of Ground Combat Troops,* 65.
42. Ibid.
43. Palmer, *Study no. 5,* 33–34.
44. Keast, *Study no. 32,* 26.
45. Lerwill, *The Personnel Replacement System in the United States Army,* 316.
46. Palmer, *Study no.* 5, 36; Wiley and Govan, *Study no. 16,* 103.
47. Palmer, *Study no. 4,* 27.
48. Ibid., 24.
49. US Army Ground Forces, *Study no. 11,* 68.
50. Palmer, *Study no. 5,* 1.
51. Palmer, Wiley, and Keast, *The Procurement and Training of Ground Combat Troops,* 20.
52. Brokaw, *The Greatest Generation,* 11–12.
53. Marshall, *Infantry in Battle,* xi.
54. Bonura, *Under the Shadow of Napoleon,* 254.
55. Ibid., 251.
56. Keast, *Study no. 32,* 2–3.
57. Keast, *Study no. 7*; US Army Ground Forces, *Study no. 9,* 16.
58. Keast, *Study no. 7,* 2.
59. Ibid., 5.
60. Keast, *Study no. 7,* 5.
61. Ibid., 3.
62. Ibid., 1.
63. Palmer, Wiley, and Keast, *The Procurement and Training of Ground Combat Troops,* 446.
64. US Army, War Department Field Manual 7-20, *Infantry Battalion,* 1; US Army, Infantry Field Manual 7-10, *Rifle Company, Rifle Regiment,* 214–65; US Army, Infantry Field Manual 7-10, *Rifle Company, Infantry Regiment,* 260–305.
65. US Army Ground Forces, *Study no. 12,* 7.
66. Ibid.; US Army Ground Forces, *Study no. 11,* 26.
67. US Army Ground Forces, *Study no. 11,* 21; Palmer, Wiley, and Keast, *The Procurement and Training of Ground Combat Troops,* 568.

68. Keast, *Study no. 7,* 2.
69. US Army Ground Forces, *Study no. 12,* 1.
70. Ibid., 23.
71. Keast, *Study no. 7,* 2.
72. US Army Ground Forces, *Study no. 11,* 72.
73. Ibid., 64.
74. Ibid.
75. Palmer, Wiley, and Keast, *The Procurement and Training of Ground Combat Troops,* 481, "Table: Data on Seventeen Infantry Divisions Heavily Stripped in 1944."
76. US Army Ground Forces, *Study no. 11,* 68.
77. Palmer, Wiley, and Keast, *The Procurement and Training of Ground Combat Troops,* 481–482.
78. US Army Ground Forces, *Study no. 8,* 22, and annex 1, 71; Mahon and Danysh, *Infantry Part I,* 57.
79. US Army Ground Forces, *Study no. 8,* 52.
80. Ibid., 22.
81. Mahon and Danysh, *Infantry Part I,* 57.
82. US Army, Table of Organization 7-37, *Infantry, Rifle Company, Parachute;* US Army, Table of Organization and Equipment 7-37T, *Infantry, Rifle Company, Parachute.*
83. Rottman, *U.S. Airborne Units in the Mediterranean Theater,* 34.
84. US Army Ground Forces, *Study no. 8,* 25, and annex 1, 71.
85. US Army, FM 7-5, *The Organization and Training of the Rifle Battalion,* 4.
86. Ibid., 5.
87. Wiley and Govan, *Study no. 16,* 94.
88. Ibid., 93.
89. Keast, *Study no. 6,* 3.
90. Ibid.
91. Ibid., 2–4.
92. Ibid., 8.
93. Ibid.; Palmer, *Study no. 5,* 13; War Department, memorandum for the Adjutant General, subject: Recruits for the Air Corps, November 29, 1942, WD, AGF (1916–1954), Ref. Files L. J. McNair (1940–1944), RG 337.3, NACP; Headquarters Army Ground Forces, staff routing sheet (memo slip), subject: Reception Center Procedures in Assignment of Personnel to Army Air Forces, November 10, 1943, WD, AGF (1916–1954), Ref. Files L. J. McNair (1940–1944), RG 337.3, NACP.
94. Keast, *Study no. 6,* 8.
95. Ibid., 9.
96. Ibid.
97. Wiley and Govan, *Study no. 16,* 94.
98. Keast, *Study no. 6,* 9.
99. US Army Ground Forces, *Study no. 11,* 31.
100. Palmer, Wiley, and Keast, *The Procurement and Training of Ground Combat Troops,* 356–357.
101. Keast, *Study no. 6,* 10; US Army Ground Forces, *Study no. 11,* 42.
102. Keast, *Study no. 6,* 10.
103. Keast, *Study no. 7,* 2.
104. Keast, *Study no. 6,* 25.

105. Ibid., 24.
106. Ibid., 27.
107. Palmer, Wiley, and Keast, *The Procurement and Training of Ground Combat Troops,* 135–136.
108. Keefer, *Scholars in Foxholes,* 165.
109. Ibid., 167.
110. Keast, *Study no. 6,* 26–29.
111. Ibid., 31–32.
112. Ibid., 42.
113. Ibid., 258.
114. US Army Ground Forces, *Study no. 11,* 43.
115. Fisher, *Guardians of the Republic,* 260.
116. Ibid., 251.
117. Lentz went on to command the Eighty-Seventh US Infantry Division during December 1944 through May 1945, relieving during the siege of Bastogne.
118. US Army Ground Forces, *Study no. 11,* 42.
119. Ibid.
120. Fisher, *Guardians of the Republic,* 258.
121. Linderman, *The World Within War,* 185.
122. Fussell, *Wartime,* 80.
123. Doolittle Board, *Report of the Secretary of War's Board,* 8–9.
124. US Department of War, *The Articles of War,* 4–6.
125. Doolittle Board, *Report of the Secretary of War's Board,* 18.
126. Ibid.
127. Ibid.
128. Stouffer, *The American Soldier,* 112.
129. Ibid., 113.
130. Linderman, *The World Within War,* 194.
131. Willis, "The United States Court of Military Appeals," 40.
132. Ibid., 41.
133. Ibid.
134. Rose, "The Social Psychology of Desertion from Combat," 614; US Department of Army, *Army Almanac,* 625, 634–643.
135. Glass, *The Deserters,* 203.
136. General Board, US Forces, European Theater, *Military Justice Administration in Theater of Operations,* appendix 4, November 20, 1945.
137. Willis, "The United States Court of Military Appeals," 43.
138. TNA PRO WO 204/6724, H. R. Alexander, General Commander-in-Chief, Allied Army in Italy, "Status of Infantry Soldiers," November 21, 1944.
139. Glass, *The Deserters,* 59.
140. Ibid., 60.
141. Ibid., 195–196.
142. Ibid., 203.
143. Linderman, *The World Within War,* 234.
144. Memorandum for the Chief of Staff from Lt. Gen. L. J. McNair, Commanding Officer, AGF, subject: Lack of Popularity of Infantry, August 9, 1943, WD, AGF, Ground Hist. Sec. 1942–1947, RG 337.6.1, NACP; Memorandum for the Chief of Staff from Lt. Gen. L. J. McNair, Commanding Officer, AGF, subject: Reflection of Pride in Organization and Satisfaction with Job Assignment, August 28, 1943, WD, AGF, Ground Hist. Sec. 1942–1947, RG 337.6.1, NACP,

tabs. B and C to enclosure 1; Memorandum for the Chief of Staff from the Assistant Chief of Staff (G-1), subject: Reflection of Pride in Organization and Satisfaction with Job Assignment, September 14, 1943, WD, AGF, Ground Hist. Sec. 1942–1947, RG 337.6.1, NACP.

145. US Army, Army Regulation 600-8-22 (Military Awards), 1995, paras. 2–6.

146. Ibid.

147. Ibid.

148. US Army, Army Regulation 600-8-22 (Military Awards), 2006, paras. 3–14, 41.

149. Palmer, Wiley, and Keast, *The Procurement and Training of Ground Combat Troops,* 171.

150. Keast, *Study no. 7,* 32.

151. TNA PRO WO 219/942, A.G. Records SHAEF, Neuro-Psychiatric Casualties, Technical Memorandum 13, "Prevention of Loss of Manpower from Psychiatric Disorders," April 4–November 16, 1944, extract from report, Office of the Surgeon General, 5.

152. Keast, *Study no. 7,* 32–33.

153. Linderman, *The World Within War,* 288.

154. Ibid., 329.

155. Ruggero, *Combat Jump,* 104–105.

156. Forster et al., *Physical Standards in World War II,* 37.

157. Doubler, *Closing with the Enemy,* 242.

158. Mansoor, *The GI Offensive in Europe,* 116.

159. TNA PRO WO 219/942, report of Neuro-Psychiatric Casualties, Supreme Allied Commander, SHAEF, memorandum over attached report to all commanders, October 1944, 2–3.

160. Ibid., 2.

161. Ibid., 5.

162. Ibid.

163. Keast, *Study no. 7,* 34.

164. Ibid.

165. Jones et al., *War Psychiatry,* 16.

166. Palmer, *Study no. 5,* 1.

10

Germany: Fighting the Anvil

THE ATTACK AGAINST THE SOVIET UNION CONSTITUTED the main effort of the German Army for the remainder of the war and was the defining experience of the German infantry soldier. All other fronts from this time became secondary and strategic economy of force operations.

Operation Barbarossa began the invasion of the Soviet Union with 3,050,000 men, 625,000 horses, 600,000 motor vehicles, 3,350 tanks, and 11,906 pieces of light and heavy artillery in 149 divisions (of 208) on the morning of June 22, 1941. The main effort of the army groups centered on the *Schnellen Truppen,* with 19 Panzer divisions and 11 infantry divisions (motorized) out of 149 divisions, or 20 percent of the attacking force and composed 14 percent of the entire army.[1]

The tenacity of the Soviet soldier surprised the army. Engaged with Red Army soldiers trained to competent standards in the peacetime Soviet army, German infantry casualties were beyond any expectations in the initial weeks of battle as the Russian armies retreated into their hinterland. Significant casualties and illness returned as the brutal winter and the December 5, 1941, Soviet counteroffensive unfolded. Overall for the first three months of the war in Russia, the German Army transported 368,000 wounded soldiers home to hospital.[2]

On November 30, 1941, prior to the Soviet counteroffensive, Germany's Eastern Army, or *Ostheer,* reported a shortage of 340,000 men with company battle strengths in the infantry down to fifty to sixty, with a nominal replacement strength remaining in the homeland of 33,000 men. Normal strength of the infantry company was 191 men of all ranks during 1941. Within a few days following the opening of the Red Army attack on December 5, the deficiency reached 500,000 men.[3] In December 1941, each *Wehrkreis* established for the first time military district boards to manage the transfer of deferred men into the army, and for the first time the full meaning of a "total war" cast its shadow over the German population.[4]

The *Ostheer* suffered grievous losses from June 22, 1941, through March 20, 1942. With 228,000 sick in hospital, including 100,000 frostbite cases, by February 1942,[5] casualties included 225,559 killed, 796,516 wounded, and 50,991 missing, for a total of over a million casualties.[6]

Infantry in their rifle units declined to an average of less than 50 percent by April 1942, with just 450,000 soldiers gained, including returning wounded and the arrival of replacements from year-group 1922.[7]

The 253rd Infantry Division (fourth wave, *Wehrkreis* VI, Munich) exemplified this decline in strength. Entering the Russian campaign with 16,194 men, with an average infantry battalion strength of 817 men, by April 1942 this division, not slated for the 1942 summer offensive, was down to 9,970 men and an average infantry battalion to 254 men, or a combined regimental battle strength of 2,286 riflemen.[8] A Fourth Army (Army Group Center) report of April 1942 described the situation:

> Nowhere are sufficient reserves available, the last man is permanently deployed. From an existing hole to plug from here and there fighting groups and smaller detachments must literally be scraped together.
>
> As a rule of thumb, by the middle of February [1942], the trench strength [*Grabenstärke*] of a division was a lie at 1000–1500 men. The trench strength of the 52.Division [second wave, *Wehrkreis* XI, Kassel] bottomed at 500 men, the 17.Division [first wave, *Wehrkreis* VII, Munich], at the time the most battle-tested in the Army area, had not nearly so many. As a further rule of thumb, the defense of the main line of resistance (*Hauptkampflinie* or H.K.L.) was approximately 80–90 men per kilometer.[9]

The Sixth Infantry Division (first wave, *Wehrkreis* VI, Munich) recorded actual, or *Iststärke,* strength of 4,755 officers, NCOs, and men on April 10, 1942, down from its full strength of 17,734. Its reports for this period reflect daily contact with the enemy including tanks and artillery. Although no longer in major attacks, casualties, as wounded and killed, occurred each week. The Eighteenth Infantry Regiment of the division reported its trench strength as 350 men, from a normal strength of over 2,400 infantrymen in its three battalions. The division recounted the loss of seventeen battalion commanders and thirty-nine company commanders from December 29, 1941, to April 12, 1942. Of note in each narrative was the continuing training within each battalion for junior leaders behind the front line during this same time period.[10]

The Shock of Stalingrad, April 1942–July 1943

By July 1942, the army in 149 divisions of June 1941 increased to 179 divisions with 2.7 million men, a portion of whom took part in the summer offensive of 1942.[11] Of the 179 divisions in the *Ostheer,* Army Group South became two army groups, reinforced with 20 divisions to make up 68 divisions earmarked for the summer offensive. The organization of the eighteenth and nineteenth waves produced nine new infantry divisions, alongside two new Panzer divisions. Full refreshment occurred for six infantry divisions in the west, and two Panzer and two motorized divisions received the same improvement behind the front in the east. "At the front" (meaning they were not brought out from the line of battle) in the east, an additional twelve infantry, three light, two mountain, five Panzer, and two motorized infantry divisions prepared for offensive operations in the same complete manner. Sixteen infantry divisions and one light division secured partial strengthening at the front, and three infantry divisions received no strengthening prior to taking part in the summer offensive. One army each from Italy and Hungary, and two from Rumania, supported the two army groups in the attack.[12]

Hitler forbade the divisions of Army Groups North and Center from being reinforced, and demanded a reduction of one infantry battalion in each division of these army groups. Reductions affected sixty-nine of the remaining seventy-five infantry divisions. Ten reduced Panzer and seven motorized infantry divisions remained with these army groups.[13]

The army continued to grade its divisions on their readiness for offensive combat. In June 1941, the army considered 136 divisions of 149 as fully manned, equipped, and trained for any tactical mission; by April 1942, a mere eight divisions fit this description prior to the reinvigoration of units for Operation Blau, the summer offensive of

1942.[14] By May 19, 1942, Soviet divisions totaled 546, numbering 4.7 million men, with all but twelve facing the line of German contact.[15] The German summer offensive of 1942 moved the front over a thousand kilometers east to Stalingrad on the Volga and toward Baku on the Caspian Sea and the Caucasus Mountains. Spectacular in distance, the Red Army learned to avoid major encirclements, and traded space for time to bring reserves from the Moscow theater of operations to the Stalingrad theater.

The German summer offensive of 1942 culminated at Stalingrad and in the Caucasus Mountains. Launched in a vain hope to occupy the Russian Caspian Sea oilfields, the offensive died upon the Russian counteroffensive, Operation Uranus, on November 19, 1942. Operation Uranus resulted in the ultimate destruction of the Sixth Army at Stalingrad by February 1943, and the retreat of German forces to near their starting point in April 1942. The defeat at Stalingrad initiated what many historians believed the psychological turning point of World War II.[16] Hitler abandoned 200,000 men when he ordered the Sixth Army to stand fast at Stalingrad;[17] 91,000 of them went into captivity in February 1943. Fewer than 5,000 of these men returned to Germany by 1955.[18] Another 275,000 Germans and Italians surrendered in Tunisia on May 13, 1943, of whom some 130,000 were German soldiers in ten divisions.[19]

Launched to coincide with the counteroffensive at Stalingrad, Soviet Marshal Zhukov's winter offensive, Operation Mars, begun on November 25, 1942, was a complete failure. Marshal Zhukov planned and directed Operation Mars south of Moscow to eliminate the Rzhev salient. The defeat of Operation Mars entailed enormous losses in tanks and men, demonstrating the power of even weakened German infantry divisions at this stage of the war to repel a Soviet offensive.[20]

Hitler dismissed General Halder in September 1942 for his disagreements over the direction of the armies, replacing him with General Zeitzler, who did not question the *Führer*'s military genius,[21] admonishing the staff, "I require the following from every staff officer: he must believe in the *Führer* and in his method of command. He must on every occasion radiate this confidence to his subordinates and those around him. I have no use for anybody on the General Staff who cannot meet these requirements."[22] Hitler had declared the Soviet enemy defeated in October 1941, and despite the enormous casualties of December 1941 and defeat before Moscow, he again assured the nation of victory in 1942.[23]

At the end of November 1942, Fromm also fell from grace. He no longer believed in the war or the possibility of victory with the resources available to Germany. Hitler left him in his position, but no longer allowed him the personal access gained through political calculation the year prior.[24] As a result of the strategic and tactical disasters of early 1943, Hitler took this view one step further and in March 1943 separated the Panzer troops of the *Ersatzheer* away from Fromm, and gave them to the new general inspector of Panzer troops, General Heinz Guderian, also relieved of command during the retreat before Moscow in December 1941.[25]

General Friedrich Fromm and the Continuity of the Ersatzheer

Friedrich Fromm was a central figure in the uninterrupted functioning of the system and characteristics of the induction and enlistment process for the German Army (and *Wehrmacht*) from his assignment to the General Army Office, *Allgemeines Heeresamt,* in 1934, through the original expansion of the army past the Versailles Treaty limitations, and all the variations of the *Ersatzheer* from the beginning of conscription, until his removal and arrest as a conspirator in the *Attentat,* the attempt on Hitler's life on July 20, 1944. Born in 1888 into a military family, his father was a major general in the Prussian Army. Fromm was an artillery officer, serving in various field and staff positions from 1906, including as a member of the Great General Staff, attending an abbreviated course of instruction at Sedan in July and August 1918. The waxing and waning of his influence with Adolf Hitler during the war lessened the control he maintained on certain offices, as in the establishment of a separate office for matters of motorization in the Supreme Command of the German Armed Forces and the *Panzertuppenabteilung* of Guderian. He retained the core of induction, training, and equipping, for the army, in particular, and the *Wehrmacht* in total, until his arrest. He remained the key and uninterrupted guide in all things regarding the quality of the manpower of the German Army until July 20, 1944. A "People's Court" disgraced him and ordered his execution, done on March 12, 1945.[26]

General Fromm's endurance in oversight of the training, entry, armament, materiel, and finance of the replacements for the army provided extraordinary stability during the period 1933 to 1944. General Fromm positioned himself to become indispensable to Adolf Hitler by taking advantage of the reduction in influence and command oversight of the Army General Staff in the midst of the winter crisis of 1941–1942. Fromm believed he was "the strong man in the homeland" that Hitler told his adjutant was needed. In the course of the first weeks of the disaster before Moscow, by December 22, 1941, Fromm succeeded in aggrandizing to his office an independence from the staffs of both the Supreme Command of the German Armed Forces and army headquarters. This further lessened the power, prestige, and scope of authority of the Army General Staff. Fromm did this by projecting a sense of optimism and confidence in the midst of the pessimism surrounding Hitler, and promising to procure needed additional divisions from the replacement army when no one else in the high command believed this possible. In particular, he offered a way for Hitler to lessen the impact on the German people by lowering the needs of the army to 442,000 for replacements from deferred workers in 1942. This alleviated Hitler's fears of a repeat of the morale collapse of World War I and offered a way to avoid total mobilization. Fromm's approach did not restore the losses the army suffered prior to December 1941 and the continued hemorrhaging between December 1941 and April 1942.[27]

Fromm benefited from the peculiar Nazi form of management, pitting one organization against another, or in this case parts of one organization, the army, against itself. This "social Darwinism" played itself out in a continual quest for Hitler's ear. Fromm succeeded in cutting his organization from the normal chain of command through attaining personal access to Hitler. He gained the freedom to pursue his own direction in fulfilling the army's needs for manpower and materiel in a struggle between the army, industry, and the other services. His achievement ensured that through almost the entirety of the war, the manpower process remained intact. The effect on the army was overall to maintain rather than diminish the quality of the recruit and to protect the process of his entry and delivery to his unit.[28]

Wartime General Induction in the Ersatzheer

High casualties in 1941 initiated the first comb-out of all headquarters and unit staffs within the replacement army. The army stripped ten divisions in France of one battalion from each regiment during October 1941. These measures succeeded in moving 250,000 men and 25,000 NCOs who were physically combat-capable from army rear establishments and occupation duty to the front.[29] About 450,000 men of a year-group of 600,000 qualified as physically combat-capable. As 50 percent received assignment to the infantry units, the army estimated 225,000 new soldiers for the combat units.[30] In October 1942, a refinement (see Table 10.1) of the instructions for physical examination gave the physician the latitude to take additional factors into consideration in determining fitness for any form of service and to err on the side of usefulness for service. If inferiority or missing teeth were the only physical faults preventing a combat-capable designation, this no longer constituted a reason to deny the combat-capable finding, pending correction before assignment to a field unit.[31]

A typical example of a wartime mustering event occurred in the city of Munich, Westfalia, in *Wehrkreiskommando* VI, in December 1941, for year-group 1924 (eighteen-year-olds during 1942). It encompassed one portion of the city and surrounding rural areas over a period of one month and completed the muster of 3,445 men. The notice for enlistment went out six weeks before recall of the individual.[32] This year-group mustered for entry into service either after Easter 1942, if the recruit completed school and any apprenticeship, or after Easter 1943, if such were not. The requirement for six months' service with the Reich Labor Service (*Reichsarbeitsdienst*) prior to enlistment into active service became the norm with the 1915 year-group in 1936. Once mustered, the 1924 year-group did not return for entry into the army until after completion of their six-month labor service,[33] followed by initial recruit training of up to three more months prior to movement in a march battalion to their frontline unit—a process taking up to one year after mustering. The general flow of recruits from all year-groups gave the army an indication of their approximate arrival dates with the field army. Before entry into the unit's rolls, the new soldiers received up to four more weeks of training, behind the front lines, in the infantry division's field replacement training battalion.[34]

The army sought new soldiers from deferred men to make up the losses where the youngest year-groups were too few, beginning in 1942 with the first 442,000. In general, the year-groups from 1908 and younger, if combat-capable, received assignment to the field army. If not combat-capable, these men went into rear/support and supply functions of the active divisions. Year-groups of 1900–1907 (and older) received postings for stationary units and homeland *Ersatzheer* duties, not entailing field combat unit support.[35] The struggle between the army and industry was impossible to resolve. It

Table 10.1 Physical Qualification Standards, 1942

Physician's Findings	Possible Employment and Assignment	Assignment After Mustering and Before Enlistment
k.v. (definition expanded)	*Kriegsverwendungsfähig:* War and combat service–capable. The inductee must overall and for each suitable assignment be able to cope with all demands of military service in the battle area including perseverance, resilience against stress, privation, weather, and the like, in relation to physical and mental capacity. In many cases also, mental factors can offset a less developed body, and especially with volunteers offer the possibility of designation as k.v.	*Ersatz* Reserve I
g.v. *Feld* (no change)	*Garnisonverwendungsfähig Feld:* Field garrison service–capable. Only for limited assignment with front units except combat medical units, and namely for office, cooks, and the like. Fully suitable for rearward organizations, construction units, homeland defense units (*Landesschützen*), homeland flak units, and units in the rear of operational areas.	*Ersatz* Reserve I
g.v. *Heimat* (no change)	*Garnisonverwendungsfähig Heimat:* Homeland garrison service–capable. Only in the homeland and not in an operations area. Suitable as trainers, in watch units, cooks, workshops/garages, in an orderly room/office, in an administrative office, and the like. Not, however, suitable as fully fit for service.	*Ersatz* Reserve II
a.v. (modified)	*Arbeitsverwendungsfähig:* Labor service–capable. Suitable same as for g.v. *Heimat*. Not, however, suitable as a trainer or in watch units. The decision to award duty in the homeland as g.v. *Heimat* or as a.v. is the responsibility of the mustering leader on the basis of the doctor's ruling. For a.v. the most important consideration is the work history and suitability for usefulness in the labor force as a skilled worker versus the military.	*Ersatz* Reserve II
Zeitlich untauglich (added)	Temporarily not suitable for service. Physician must annotate how many months before reevaluation.	*Ersatz* Reserve II
w.u. (replaces a.v.u.)	*Wehruntauglich:* Not war service–capable. Suitable for no beneficial, regular labor tasks.	Mustered out as unfit for war service.

Sources: BAMA ZA 1/1778, Personnel and Administration (Project 2b), P-006, Teil I: *Das Rekrutierungssystem der deutschen Wehrmacht in Frieden und im Kriege*; Anlage I: *Durchführung einer Musterung und Aushebung,* 1948, 152–153.

remained for the youngest year-groups to fill the ranks of the army. As the war increased in intensity, deferred workers were as critical to the production of the weapons needed by the army as were the workers needed at the front as soldiers.[36]

Finding Replacements, 1942–1943

The next major step beyond the enlistment of 650,000 deferred workers (442,000 for the army) arriving during the fall of 1942 occurred on the same day the Red Army counteroffensive Operation Uranus encircled the Sixth Army at Stalingrad, December 19, 1942. Hitler ordered the Exchange-Actions (*Tausch-Aktionen*) to gain an additional 200,000 soldiers for the field army. Known as "Exchange 1943," it directed the army to send 150,000 soldiers who were fit for duty in garrison or labor duty in the homeland, to industry. In exchange, industry gave up 100,000 deferred workers and 30,000 homeland civilians (all combat-capable) of year-groups 1906–1922. The order required the army to comb-out another 20,000 combat-capable soldiers from rear-area units for the front.[37]

On March 13, 1943, Hitler demanded the identification and enlistment of 800,000 of the remaining members of the year-groups 1919–1922 no later than August 1943. He ordered Göring, the leader of the air force, to exchange 100,000 of his combat-capable soldiers of year-groups 1914 and younger with the army on March 20, 1943. In return, the air force received 120,000 older soldiers who were less than combat-capable.[38]

Up to this point in the war, manpower for the entire *Wehrmacht* came from laborers in the noncritical work force, such as consumer goods, power, and construction, and the youngest age groups. About 5.5 million deferred men working in critical war industries remained unavailable for enlistment.[39] As late as May 1943, a 10 percent cutback in consumer goods enraged women, whose outcry in 1942 caused Hitler pause; he noted that "one should better proceed with quiet reduction of production [of consumer goods] rather than prohibitions."[40] Nazi philosophy did not value women working outside the home, and although nearly 8 million women worked in some form in skilled crafts and trade, transportation, the energy sector, and domestic work from home, their strength in these work force categories fell between 1939 and 1940 by 270,000, and work rules forbade women from night work. In the armaments industry, women in the work place declined by over 82,000 between July 1939 and May 1941.[41]

Luftwaffe *field divisions.* On 9 September 1942, Hitler ordered the *Luftwaffe* to provide 200,000 of its combat-capable base personnel for ground combat on the Eastern Front as replacements for army field formations. Göring demurred from executing Hitler's order and stated his fears for the air force soldier, a new type of National Socialist man, being exposed to the reactionary leadership of the old army. Hitler agreed and authorized the formation of the short-lived air force field divisions (*Luftwaffe Felddivisionen*). Göring sought his own army in the same manner Himmler was building a private *Schutzstaffel* army, and agreed to furnish twenty divisions of volunteers, which grew to twenty-two divisions. With little time for training, little combat equipment, and less supply, the first three divisions deployed into frontline combat following the encirclement of the Sixth Army at Stalingrad at the beginning of January 1943. Destroyed in the retreat to the Don River,[42] by September 1943 the surviving divisions moved to the army by Hitler's order, with few retaining their *Luftwaffe* designations.[43]

Reorganization and training in the replacement army. Within the *Ersatzheer,* Fromm continued to consolidate the higher training organizations of the replacement army in response to the ever-increasing sum of divisions and their requirements for regional bases for recruitment and training organizations. The first step in this direction was the formation of the reserve divisions in 1940–1941 within each *Wehrkreis* to consolidate the division training regiments. The reserve divisions moved into the occupied territories in France and the Low Countries, Norway, Poland, and the Balkans during 1942. Reserve divisions formed as training units and not as combat-capable.[44]

Fromm began a further consolidation of the division training and replacement battalions in the homeland with the organization of five training and replacement divisions (*Feld-Ausbildungs-Divisionen*) during August 1942. Bound for duty in the east behind the three army groups of the *Ostheer,* the *Ersatzheer* planned for the staggered arrival of these units commencing during the spring and summer of 1943. The intent of the placement of army group–level training divisions in the field was to shorten the travel time for replacements, beginning with the 1924 year-group (nineteen-year-olds), between basic training and arrival at their tactical units in the field army. The replacement regiments of the 6th, 26th, 86th, and the 253rd Infantry Divisions, from *Wehrkreis* VI, Münster, received assignment to the

637th *FeldAusbildungs* regiment, of the 390 *FeldAusbildungs* division for basic training.[45]

Over the course of 1943, in the general retreat following the Kursk offensive, these divisions never fulfilled their purpose. The commanders of the army groups found the experiment a complete failure. The *Ersatzheer* stopped its recruits and support personnel, with the exception of their command elements—all either disbanded or were kept in the field to command the remnants of infantry or security divisions from those disrupted or destroyed during the post-battle retreat.[46]

The focus of the manpower process remained finding the best man for combat duty. Quality received additional emphasis as directed to the induction office (*Wehrmeldeamt*) in Bielefeld, *Wehrkreis* VI, for April 19, 1943:

> It is crucial in the selection of replacements for the infantry that each man must be a singular fighter with the Rifle Company of the Inf. Regt., the Inf. Regt. (mot.), and the [Armored] Rifle Regiment. These units must therefore obtain physical and mentally fully qualified replacements, in doing so the increased stress of those not in motorized Rifle Companies is especially to be borne in mind. It is the duty of the Commander of the Region [*Wehrbezirkskommando*] hereupon to turn their particular attention. . . . Recruits with leadership qualifications are first and foremost for the fast troops [*Schnellen Truppen*].[47]

In the struggle to replace field army losses, the *Ersatzheer* continued to grow. This phenomenon was the result of two primary factors. First, conscription continued throughout this period and accelerated in 1942 and 1943 for the 1923 and 1924 year-groups. This included an increased emphasis on the deferred classes after November 1943. The manner and timing of replacements was a factor of the priority given a particular division. The 253rd Infantry Division did not receive full refreshment during 1942 or 1943, or indeed at any time for the remainder of the war after June 1941. Its replacements, a combination of new inductees and returning recovered wounded soldiers, maintained at a low, but steady rate, during the entire war.[48] The 111th Infantry Division (seventh wave, *Wehrkreis* VIII, Breslau), in contrast, did receive full refreshment during 1942. It received from its home region a sufficient number of recruits to bring the division back to full strength for the 1942 campaign.[49]

The replacement army was the responsible agency for all soldiers in hospital and recovering from wounds or illness until fully combat-capable again. This process went from the stay in hospital, to convalescence at home, to return to active duty and continuing recovery in a training unit as an instructor until receiving assignment to a marching unit. In this group of men from his home region, he returned to his original field army unit upon full recovery (see Table 10.2).

Ever more strident demands for combat capable soldiers continued for the duration of 1943:

- October 25, 1943: Personnel savings in the homeland to provide the exchange of office soldiers fit for duty in garrison or labor duty with the reserve divisions in the homeland.
- November 3, 1943: *Führer* Directive 51 mandated the building of divisions in the west in anticipation of Allied landings, and further reducing the reserves available to the *Ostheer.*

On November 27, 1943, Hitler acknowledge the requirement for full war mobilization of the economy for the first time. He called for a million men for immediate enlistment: "The war and the existence of the German people and the future of Europe has reached its culmination. . . . I order that steps be taken immediately to collect not less than one million men from appropriate sources for the armed forces and the *Waffen-SS* and placed into frontline service."[50]

On December 16, 1943, front help (*Fronthilfe*) required the reporting of all combat-capable and fit-for-duty soldiers in any *Ersatzheer* headquarters for replacement by labor-duty soldiers or civilians.[51]

Table 10.2 Total Strength of the *Wehrmacht* by Organization, 1939–1945 (approximate monthly average in thousands)

	Feldheer	*Ersatzheer*	Total Army	*Luftwaffe*	*Kriegs-marine*	*Waffen-SS*	Total
1939	2,741	996	3,737	400	50	35	4,222
1940	3,650	900	4,550	1,200	250	50	6,050
1941	3,800	1,200	5,000	1,680	404	150	7,234
1942	4,000	1,800	5,800	1,700	580	230	8,310
1943	4,250	2,300	6,550	1,700	780	450	9,480
1944	4,000	2,510	6,510	1,500	810	600	9,420
1945	3,800	1,500	5,300	1,000	700	830	7,830

Source: Adapted from Müller-Hillebrand, *Der Zweifrontenkrieg,* annex A, 254.

Initiative Lost, July 1943–June 1944

By July 1943 the *Ostheer* increased to 3.1 million men in 151 infantry divisions and 22 Panzer divisions. The final German strategic offensive to cut off the Kursk salient, July 5–16, 1943, resulted in complete defeat and the loss of the initiative on the Eastern Front for the remainder of the war. From July 1943, German retreat was never-ending, pausing as the Soviet Army regrouped and replenished for its next leap forward. German ability to cut off Soviet spearheads at their vulnerable flanks ended in 1943 when the availability of thousands of US Lend-Lease jeeps and trucks allowed the Russians to follow their leading breakthrough forces with immediate flank reinforcement of strong anti-tank units to blunt the inevitable German counterattack.

The Red Army moved to the offensive in the north of the Kursk salient in Operation Kutusov, from July 12 to August 18, and in the south of Kursk in Operation Rumjancev, from August 3 to 23;[52] their drastic mobility differential spelled defeat for the foot-mobile German infantry—one result of the massive aid provided by US Lend-Lease, and noted by General Zhukov:

> The Americans provided us with so many materials, without which we could not have formed our reserves and would not have been able to continue the war. . . . We received 350,000 motor vehicles, and what vehicles they were! . . . We had no explosives or powder. There was nothing with which to fill rifle cartridges. The Americans really rescued us with powder and explosives. And how much steel plate they sent us! Could we have really managed our tank production if it weren't for American steel? And now the matter is being presented as if we had had all this in abundance on our own.[53]

By December 1943, the Red Army moved the front line in Russia back to the original Polish border of 1939 and crossed the Dniepr River in the south, with the Crimea abandoned and Soviet forces nearing the Balkan states. A pause for winter in Russia allowed the German Army Group Center to regroup, and in January 1944 the Red Army lifted the siege of Leningrad. By the middle of 1944, the *Ostheer* fell to just over 2 million men. The *Ostheer* rose again in June 1944 to 148 divisions, with the reconstitution of some divisions lost during the winter. Hitler's design for Army Group Center envisioned reinforced strong points all along the front to stop Soviet penetrations. Fewer than 400,000 soldiers remained in Army Group Center in June 1944, and faced more than 3 million Russians preparing to launch their summer offensives, moving from north to south all along the German front sequentially until the end of 1944. German division strength in France was sixty-five, with another twenty-five in Italy. The German Army in Italy held back repeated Allied attacks at the Cassino line during the winter of 1943–1944, losing that line in May 1944. By June the German Army in Italy was once again in retreat and the divisions in France awaited the coming invasion of the Western Allies.[54]

Major Military Operations, June 1944–December 1944

On June 6, 1944, Operation Overlord brought the Western Allies ashore at Normandy. June 22 brought forth the Soviet summer offensive, Operation Bagration, and the destruction of Army Group Center. August witnessed the Allied landings on the southern coast of France. On all fronts the German Army was in retreat. Between June and December,[55] despite the constant loss and replenishment of the infantry divisions, the army managed to reconstitute its divisions along the line of contact each time the Soviets or the Allies halted. Outrunning of their supply lines was the principal reason halting the Red Army in 1944. Willing to sacrifice extraordinary numbers of soldiers, Stalin drove the army forward, steamrolling all substantial German resistance.[56] At the end of the Soviet era the official tally of military dead during World War II was 11.3 million. Upon the opening of the Soviet military archives by Russian historians, some raised the unofficial total to between 23 million to more than 26 million military losses alone in the Red Army.[57] The Soviet method of war appeared as the driving of masses of "random, untrained" infantrymen, supported by vast multitudes of tanks and artillery, against the German divisions.

> Thus, Marshal A. I. Eremenko characterizes the features of the "military art" of "Marshal of Victory" Zhukov as follows: "It must be said that Zhukov's operational art was a five-six-fold superiority in forces; otherwise he would not fight, he was unable to fight without numbers, and he built his career on blood."
>
> . . . One need only recall Zhukov's account to Eisenhower about how Soviet troops overcame minefields: first came the infantry, who at the cost of their own lives exploded anti-personnel mines; then came the combat engineers who removed anti-tank mines so that tanks could overcome the minefield without losses.[58]

As historians noted, Red Army tactics under Stalin led

> from victory to victory, at the price of unimaginable casualties, [and] lay today as a gigantic burden on the shoulders of the people.

> . . . Evidently, Stalin believed that our human resources were inexhaustible. That was completely not the case. In 1942 and 1943 we were forced to call up to the front seventeen years [*sic*] olds and those who had recently turned fifteen. . . . We withdrew from the reserves hundreds of thousands of peoples from Ural and Siberian enterprises, among whom were many unique specialists, and we dressed them in soldiers' greatcoats.[59]

Massive Soviet encirclements required ever greater expedients on the part of the German Army to escape. In October 1944, the Red Army cut off the thirty-three divisions of Army Group North in the Courland pocket. These divisions held their position with the sea to their backs for the duration of the war. By October the Russian Army was poised on the border of East Prussia, and outside Warsaw.[60]

On the Western Front, the US breakout from the Normandy hedgerows allowed the bulk of the German armies to retreat to the frontiers of the *Reich.* The winter produced a stalemate along the old Siegfried line. British and US airborne assaults in Holland failed to secure a crossing of the Rhine at Arnhem in September, Aachen fell to the Americans, and the bloody battles of the Hürtgen Forest began in November. Taking advantage of the Soviet delay before Warsaw, in conjunction with the Warsaw rising, Hitler gambled on an attack in the Ardennes to split the British and US Armies. With the defeat of those last German reserves, and the resumption of the Soviet offensives across the Vistula and Oder, the fate of Berlin and the German state was sealed.[61]

Military Operations at the End, January–May 1945

By October 1944, the enemies of Hitler arrived at the borders of the *Reich,* both east and west. Following the disastrous conclusion of the Ardennes offensive of December, the remnants of the attacking divisions returned to the German border. After a pause of one month the Western Allies attacked toward the Rhine River on February 2, 1945, reaching it along a broad front north of Koblenz by March 10, a distance of 60 kilometers. Attacking again on March 29, by April 4 the Allies crossed the Rhine and surrounded German forces under Army Group B in the Ruhr pocket, and by April 18 advanced to the Elbe River and occupied most of Germany west of the Elbe no later than May 5.[62]

The enormous losses endured by the Red Army, and the out running of its supply bases, caused a pause in the advance from its first arrival on July 31, 1944, at Warsaw and by August 29 at the East Prussia border. The army advanced over 700 kilometers—600 kilometers since June 22, 1944. The Soviet main effort, with the mission to seize Berlin, halted from the end of August until January 12, 1945, nearly four and a half months. The attack shattered German defenses. They arrived at the Oder River only 30 kilometers from Berlin on February 2, only to pause again to recoup losses until the final assault on April 16, a further period of two and a half months. Berlin was surrounded by April 22 and US and Soviet soldiers met on the Elbe River at Torgau no later than April 25. Germany surrendered on May 8. In the end, there was no measure available to overcome Allied superiority. Every soldier found went into the line as more and more draconian measures evolved to ensure loyalty and service in the death throes of the *Reich.*[63]

The Plot to Kill Hitler

The final attempt by armed forces conspirators to kill Hitler failed on July 20, 1944. Hitler survived the bomb planted by Claus von Stauffenberg and exacted a terrible revenge on anyone connected to those involved in the attempt on his life. The German officers planned to put Ludwig Beck into the leadership of Germany with the intent to end the war by negotiation. Beck was allowed to commit suicide, and von Stauffenberg was shot on the orders of General Friedrich Fromm in order to hide Fromm's own complicity in the plot. Fromm was arrested and died in March 1945 in a concentration camp.[64]

As a vacillating and tepid conspirator in the July plot, Fromm lost his nerve. He sent a message to the army indicating he was in charge in Berlin during the emergency and that he had had the chief conspirators shot. This was corrected within hours by Heinrich Himmler, who arrested Fromm shortly thereafter. Hitler placed Himmler in charge of the replacement army. Himmler created the *Volksgrenadier* divisions and increased the size and strength of the *Waffen-SS.* Himmler brought a leadership style to the *Ersatzheer* devoid of administrative ability or patience with any process, substituting a maudlin and mawkish appeal to ancient, Aryan and Germanic patriotism.[65]

Joseph Goebbels (1897–1945) became the *Reich*'s de facto director for the "total war" effort due to his immediate support of Hitler in the aftermath of the events of July 20, 1944. Together with Himmler, the new commander of the *Ersatzheer,* Goebbels demanded that the

Ersatzheer "empty the barracks," calling on July 25 for the enlistment of 1 million men from industry at the rate of 300,000 for duty by September 1, 450,000 in October, and 250,000 in November. Albert Speer, directing the war economy, delayed yielding these men each month. "The total war [effort] could not become operational. It was all too late and completely insufficient."[66]

To take one man from industry demanded one additional foreign worker, prisoner of war, or slave laborer be added. Already there were millions of foreign and slave laborers in German industry. In addition to the demand for 1 million more men from industry, already impossible to implement, on August 7 the Supreme Command of the German Armed Forces directed the *Wehrkreis* headquarters to examine again their staffs and find any combat-capable soldiers and send them to the front. With the continual scouring of these units, very few remained of any combat-capable soldiers, as evidenced by the response from the *Wehrersatzinspektion* and *Wehrmeldeamt* recruiting offices within *Wehrkreis* VII, Munich, on August 20. Tasked to provide 5,000 men through the following actions, they found fewer than 300 men after years of scrubbing various headquarters and support units.

- August 7, 1944: Action to remove combat-capable and conditional combat-capable men from the staffs of training units for assignment to the field.
- August 20, 1944: "Total war" direction of the army to reexamine physically deficient personnel for upgrading for the field. Hitler demanded the "mobilization of the last reserves."
- November 20, 1944: Amplification of the preceding original "total war" to reexamine soldiers fit for labor employment to move into conditional combat-capable, and to move conditional combat-capable to full combat-capable for duty, including clearing those with stomach and hearing problems for duty with special units.[67]

Distrustful of the traditional replacement process, Himmler created six *Volksgrenadier* divisions of the thirtieth wave (July 30–September 3, 1944), and twenty-five in the thirty-second wave (August 26–September 16, 1944). Regardless of Himmler's desires to separate these formations from the regular process of induction, all of these formations found their base organizations for soldiers in the replacement army's regimental formations already in place and continuing to function when their divisions were disbanded or lost to combat in all theaters. New recruits filled many of these units, and as employed in the defense of Aachen, and during the Ardennes offensive of December 16, 1944, received a bare two weeks of training and, with no heavy weapons, suffered severe casualties as a result.[68]

The Ersatzheer *and the Youngest Age Groups*

In the final months of the war, the *Ersatzheer* lost control of the induction process. The advance of enemy formations into Germany caused the piecemeal activation for battle of the various border *Wehrkreis* institutions for recruit training and all schools for the various arms, including those for noncommissioned officers and officers. This happened first in *Wehrkreis* VI as the US Army advanced to take Aachen during October 1944. In September the Infantry Replacement Regiment no. 253, the home for the replacement personnel of the 253rd Infantry Division, was activated for combat purposes under the Division no. 526. This division was one of a number of divisions with the "number" designation and intended to provide a headquarters for replacement and training units in a *Wehrkreis* territory. These divisions were not considered combat-capable. When formed during late 1942, they served the same security and training purpose within the homeland as the reserve divisions did in occupied nations on the German border. The final march battalion of replacements and recovered wounded soldiers sent from Division no. 526 arrived at the 253rd Infantry Division on the Eastern Front in January 1945. Red Army advances into East Prussia eliminated the recruiting base and all facilities for *Wehrkreis* I during January 1945.[69]

The Supreme Command of the German Armed Forces called for the activation of all police and militia (*Landsturm*). On January 27, 1943, the first call went out for volunteers from year-groups 1926 and 1927 as *Luftwaffe* flak helpers for manning flak batteries in the homeland. This call continued for year-group 1928 on July 1, 1944, for this duty. This duty was not allowed to interfere with labor duty following mustering.

For the 1928 (seventeen-year-olds) and 1929 (sixteen-year-olds) year-groups, although many were already employed as flak helpers with the *Luftwaffe,*[70] the notification process failed in 1945. On February 15, 1945, Hitler ordered the army not to bring year-group 1928, nor to interrupt their labor duty for combat, except in the event of an enemy parachute landing.[71]

A variety of measures such as *Goebbels-Aktion* and *Aktion Leuthen* in 1945 sought to enlist and deploy the youngest year-groups into the final battles. These attempts were generally failures, as the numbers gained represented only 10–20 percent of the demand. The simultaneous requirement for activation of the entire replacement army for combat, and for the continued enlistment through the same organization for year-groups 1927 through 1929, failed.[72] These plans included the enlistment (*Einberufung*) of year-group 1928 for the first quarter of 1945 for 200,000 men, but identified for mail notification only 58,000 by March 20, 1945. By then the Red Army occupied all of Germany east of the Oder River, and the US Army was across the Rhine River as far east as Frankfurt. The numbers demanded, and results reported, in these plans demonstrated the inability to muster either the 1928 or 1929 year-groups in any significant numbers as a result of the ongoing failure of the entire recruiting apparatus of the *Ersatzheer*. The chaotic end of the regime included the activation of the units of the *Ersatzheer* for combat in stages beginning in 1944 and 1945, ended the process of induction and mustering for year-groups 1928 and 1929, preventing their entry into combat.[73]

During 1944 the Supreme Command of the German Armed Forces activated the reserve divisions in the occupied countries, returning them and their training functions to the home *Wehrkreis* as replacement and training divisions (*Ersatz- und Ausbildungsdivisionen*).[74] With the field army reeling backward into Germany, the *Wehrkreise* received ever more strident, draconian orders to get men to the front. On March 1, 1945, the training units of the *Ersatzheer* came under command of the army within those *Wehrkreis* expecting the advance of the Allied armies or the Soviet Army. On April 27, 1945, the Supreme Command of the German Armed Forces ordered all remaining formations and offices of the *Ersatzheer* to the command of the field army, and directed them to form into tactical units for commitment to the front.[75]

Planning continued through the end of April, with recruitment plans extending into November 1945. Illustrating the difficulty, Passau, *Wehrkreis* XIII, Nurnberg, notified his headquarters on April 24, 1945, of his inability to follow orders to collect and induct members of year-group 1928, as well as the roundup, organization, and reequipping of retreating soldiers (received on April 16, 1945) "due to the cessation of rail connections and the nearness of the enemy."[76] On April 23 he had ordered the following of his seven *Wehrmeldeamten,*[77] or recruiting offices:

> In light of the closeness of the enemy, all [recruiting offices] are to destroy their classified material immediately, except as these are essential to the necessary functioning of the office. Therefore, these [materials] can only be important orders principally from the year 1945. Personnel records are under the last order in the stack to protect against moisture [damage]. On the basis of new orders . . . the [unit] diary for 44/45 is to be destroyed on the approach of the enemy.[78]

The US Army occupied Passau on April 26, 1945.

At this time the policy of sending men back to their own units failed through much of Germany. Lieutenant Gottlob Bidermann was a member of Infantry Regiment no. 437 of the 132nd Infantry Division (*Wehrkreis* XII, Wiesbaden), fighting in the Courland pocket, cut off by land from Germany by Soviet Army forces. Receiving an award for bravery, he was allowed to return by sea to Germany for a home leave to Leipzig.

On his return trip on February 8, 1945, and intending to return to his unit in the Courland pocket, he was held up by officials at the train station in Berlin and ordered to board a bus for duty on the Oder River front. Refusing to obey this order, he slipped away to the home of his cousin, but when he reached the train station to return to Courland, he saw patrols inspecting the documents of all military personnel. Along with his cousin, and trying to avoid such control, he noticed a general arriving. "I approached him saluted, and introduced myself. I then explained that I was attempting to return to my unit in Courland, and asked if I might be permitted to remain with him for a short period. 'Of course, my son!' he exclaimed. 'I understand how important it must be for you to return to your comrades.' With my irreproachable position next to the general I was no longer delayed by police patrols."[79]

In this final act of the grim catastrophe of World War II, the extinction of the *Reich,* Hitler demanded the ultimate sacrifice of the entire German nation. He had predicted in *Mein Kampf:* "Germany will either be a world power or there will be no Germany."[80] German soldiers found themselves collected from all over Germany and thrown together without mutual training, outside of the bounds of their former regional affiliations and thrust in front of the US,

British, or Russian army masses pouring across Germany's frontiers in 1945. As the Red Army demonstrated throughout the war, random, unconnected soldiers without a cohesive bond marched to their deaths.[81]

A Reaction to Defeat

Fears of the *Führer,* Adolf Hitler, concerning the failing war, possible influence of propaganda from the captured officers of the League of German Officers from the Soviet Union, and continued fear of a morale collapse in reaction to defeat as in 1918, led Hitler to initiate a closer partnership between the Nazi Party and the *Wehrmacht.* He established the National Socialist Leadership Staff of the Supreme Command of the German Armed Forces on December 22, 1943, including the creation of the National Socialist Leadership Officers (NFSO).[82] On January 7, 1944, the *Führer* held a personal meeting to direct the coordination and purpose of the effort.[83] Thus the Nazi Party attempted to inject its ideology to strengthen the staying power of the soldiers, to answer the question, "Why are we fighting?" The concept of the introduction of a Soviet-style Kommisar to guide the political reliability of the soldiers demonstrated this fear in defeat,[84] and fits with the concept of a messianic explanation for the ideologically sustained cohesion of the German soldier.[85]

The primary duty for an army NFSO was given to a single officer at the division staff; all other ideological stiffening of the troops was to be directed by a single officer down to battalion level as an additional duty. There were between 1,074 and 1,251 officers assigned the full-time duty as NFSO for the entire *Wehrmacht,* including the Supreme Command and the headquarters of the *Heer, Luftwaffe,* and *Kriegsmarine.*[86] Is there evidence of the success of such ideological reinforcement in the performance of the troops or the words of after-action reports that indicate the power of such a structure?

Questions concerning the useful effects of the NFSO program, created with so much expectation, were more than justified according to Manfred Messerschmidt. The plan was hampered by the usual chaotic Nazi methods of management, the intrigue and distrust between Martin Bormann and Heinrich Himmler in implementation, the distrust of the traditional officer corps of political interference in the roles and functions of the officer, and the lack of the NFSO's acceptance of the military staff, all amid the continuing military disasters and retreats between the time of the inception of the program in early 1944 until the capitulation in May 1945.[87]

From its very beginning, the NSFO program was slow to find traction within an environment of infighting and jealousy at the higher echelons and resistance at the lower ones. In the resulting disunity and confusion, NSFOs were denied the ability to produce fanatical Nazis out of the German soldiers. The best judges of the program, of course, were the soldiers themselves. Even when they were aware of the lack of privacy in their mail because of censorship, their letters home showed confusion in their minds as to the purpose of the NSFO program.[88]

> But there was nothing about the impact of the introduction of the NFSO and the intrusion of the party on the leadership of the *Wehrmacht* from the summer of 1944 that indicated this effect on the soldiers. There are scores of letters and reports of soldiers that reveal the slight degree of any actual indoctrination. Certainly the military situation (by the summer of 1944) hardly allowed for the strong implementation of such directives, and the so-called "Fuhrer"-Myth, as in their prior experience (in society) had the most important effect (*"Hitler ja—die Partei nein"*), while training in the tenets of national socialism and its doctrine found little interest.[89]

Conscription in "Total War" and Reduction in Physical Standards

For the *Heer,* the key event of this period was the first lessening of the physical standards for the German Army in January 1944, and made official in April 1944. This was the first significant change to German soldier standards since their introduction in 1935 and consonant with the 1922 standards. This was the first recognition of the inability to replace the casualties sustained to this date without lowering the criteria. The announcement for the justification of the change on December 18, 1943, indicated that the lowered standards appeared justified because the return of so many wounded soldiers to full duty required the adjustment of more rigid standards.[90] From the first publication of the physical requirements of the army for the combat arms, the *Tauglichkeitsgrad,* or grade of physical capability for combat service, in March 1935, based on those from the 1922 standards for the *Reichsheer,* until January 1944 (and then a single modification), the emphasis on finding the best, most physically and mentally robust men for combat duty remained virtually unchanged. That standard held over twenty-two years of peace and war.

The Supreme Command announced the change on December 18, 1943, for application on January 14, 1944, to the mustering of year-group 1926. The change to the examining physician's handbook[91] became effective from April 1, 1944.[92] From this date, if a recruit had one fault from the list, it became permissible to designate him as conditional combat-capable, and, depending upon the nature of the fault, as combat-capable for assignment to one of the eight combat specialties—infantry, artillery, pioneers, and so forth. The new designations were:

- k.v. (*kreigsverwendungsfähig*): Recruits with certain faults only and without physical restrictions.
- *bedingt* k.v. (combined fit-for-duty and g.v.H. [*garnisonverwendungsfähig Heimat*]) Physical condition allowing service only with homeland-stationed units.
- Recruits with faults were conditional combat-capable, in general. By Supreme Command order, recruits with one such fault were eligible for the combat-capable designation. Otherwise they were limited to assignment with front units except combat medical units, and namely for office, cooks, and the like. They were fully suitable for rearward organizations, construction units, homeland defense units (*Landesschützen*), homeland flak units, and units in the rear of operational areas.
- a.v. (*arbeitsverwendungsfähig*): Recruits with significant faults only in the homeland, not in an operations area, and suitable as a trainer, in watch units, as cooks, in workshops/garages, in an orderly room/office, or in an administrative office.
- z.u. (*zeitlich untauglich*): Recruits conditionally incapable of war service. These recruits were found physically unqualified upon mustering with the probability of qualification upon return for reexamination.
- w.u. (*wehruntauglich*): Recruits not capable for war service.[93]

The general officers of the German Army understood and accepted the operational effect of these new standards as necessary, if regrettable.[94] Examining physicians received their guidance in the form of *Fehlertabelle,* or "faults" tables, as the basis for decisions on fitness for combat duty. An example from these *Fehlertabelle* provided for a single measure from a total of seventy-eight on the full examination list.

Although this change appeared at first glance to alter the physical nature of the German soldier, it was not clear that there was a measurable impact in the final sixteen months of the war. By January 1944, the total number of men enlisted into the *Wehrmacht* was more than 15.7 million, more than 90 percent of the over 17.2 million men in total who served during the war, and 80 percent of those served in the army (*Heer*). During the next sixteen months to the end of April 1945, 1.5 million more men entered service, 8.9 percent of the total *Wehrmacht* entry, with the possibility that some of them were combat-capable under the less restrictive physical guidelines allowed to the examining physicians.[95]

Cohesion and Battle

The replacement regiments discussed earlier had by 1942 moved into the occupied territories for the combat preparation of recruits and recovered wounded, as well as to free combat units for duty at the front. These units trained the recruits in infantry skills using the experiences gained in the occupied territories and turned them into lesson plans for the soldiers. A staff tasked with this responsibility and located at Marie ter Heide in Belgium in 1943 demonstrated this process of individual instruction in close combat skills and squad-level drills. The intent was to place each man in a leader position in attack and defense. As was noted: "These duties and responsibilities present again and again the opportunity for the individual to demonstrate independent action in the setting of the squad, and for him to display his skill and capability in each situation."[96]

The Sixth Infantry Division, first wave, *Wehrkreis* VI, Münster, demonstrated the types of combat training conducted in the field by the field replacement battalion (*Feldersatzbataillon*) during 1942 and 1943. On April 1, 1942, the division published a plan for the arrival of *Marschbataillon* II/6 of 13 officers and 971 NCOs and men—infantry, artillery, pioneer, and anti-tank (*Panzerjäger*), designating the training periods for each group:

- Infantry: Twenty-one days.
- Artillery: Twenty-one days of infantry training and then to the artillery regiment for gun training.

- Pioneer: Four weeks of training, one and a half infantry and two and a half special pioneer.
- *Panzerjäger:* Four weeks of training, one infantry and three special anti-tank.

Infantry Training:
- First week: Individual fighter training, weapons, terrain utilization, camouflage, post and security duties.
- Second week: Squad and patrolling training.
- Third week: Platoon and company training.

As was noted: "The point of main effort for [infantry] training constitutes the 'attack' and Russian battle tactics with special emphasis on [the use of] their tanks as infantry support [*Sturmgeschütz*]. Each week use live ammunition at least once (including hand grenades!)."[97]

During January 1943, the division conducted a twenty-three-day junior leader course in the rear of the division for forty-six NCOs and men from each grenadier regiment, as well as six each from the pioneer battalion, the artillery regiment, and the supply column.[98] The division conducted thirty-one leader courses from October 25, 1942, to March 10, 1943,[99] while engaged almost daily in combat operations, including the defensive during Marshal Zhukov's failed Operation Mars of November 25 to December 13, 1942.[100]

Otto G., previously introduced as an infantryman of year-group 1926, spent eleven days with Grenadier Replacement Battalion no. 116 beginning on January 25, 1944. He transferred to another replacement battalion and was immediately transferred to the newly established 272nd Infantry Division near Lyon, France, where he received six weeks of infantry instruction within the division. The elements of the 272nd Infantry Division came from the disbanded 216th Infantry Division, third wave, *Wehrkreis* XI, Kassel, lost in December 1943 in southern Russia, with its regiments 348, 396, and 398 supported by the *Infanterieersatzregiment* 216 from Lower Saxony. The new 272nd Infantry Division, forming in France, built the strength of its new regiments 980, 981, and 982 from this same source in *Wehrkreis* XI, maintaining its regional identity from Lower Saxony.[101] Otto G. described the infantry he found in 1944 as an eighteen-year-old recruit and the preparation he received from the veteran leaders of his division:

> The soldier, the individual fighter, the farthest in front, always and everywhere, must bear the brunt of war, as I later learned personally,

> and therefore could also judge this. We infantrymen were in all things disadvantaged: whether it was supplies, equipment, quality and fit of the uniforms, or, or . . . ! Always to the other arms of the service we were the suckers, and must be derided as "dirt-bunnies" (*Sandhasen*) or "stubble-jumpers" (*Stoppelhopser*).
>
> From the beginning our [field] training was molded by two fundamental criteria: it [training] occurred with the troops with whom one would later serve; it occurred through front-experienced NCOs and soldiers, who also led us in our coming unit assignment and became our comrades. These trainers instilled in us in six weeks what was important to hold one's own and to survive as an infantryman. They did this thoroughly and intensively and also with the required toughness, but without harassment. And this brought about a trusting and comradely atmosphere that withstood the test of the deployment into Normandy.[102]

Battle preparation in the field was a strength of the German infantry division. The provision of combat lessons learned, and the modification of battle methodologies, were continuous throughout the war, ensuring the continued relevance of tactical concepts in the different theaters of war. Guidance from battle experience went from army headquarters to tactical units and units of the replacement army to guide the formulation of drills and exercises, and was most effective when combined with the intent to refresh units through periods of rest, replenishment, and training, behind the front. The former commanding general of the Third Panzer Grenadier Division described the "normal" process for a unit withdrawn from the line for refreshment and reconstitution prior to an attack in 1942: "A well-prepared reconstitution furthermore includes the readiness of supplying personnel and material in the reconstitution area on the arrival of the unit, so that the most important task, i.e., the assignment of the replacement—also—weapons and equipment—can begin immediately, and the goal of creating homogenous units (solidarity, comradeship of arms, cooperation, coordination) can be reached."[103] Full reconstitution became less and less the norm during the course of the war. Many experienced combat divisions, with long service on the front line, such as the 253rd Infantry Division, never received full refreshment.[104]

As the war drew to its conclusion in late 1944 and 1945, replacements were often directed away from their parent units when moving to the front. The Third Panzer Grenadier Division illustrated the determination of these soldiers to return to their "home":

> Almost always individual officers, noncommissioned officers and enlisted personnel would suddenly arrive at their "home" unit—they had "deserted" from the alien unit in order to get to *their* unit. Since this involved an especially valuable increase, this breach of discipline always remained unpunished and the unit then proceeded to "safeguard" itself against the intercepting of its valuable personnel replacement transfer battalions [*Marschbataillonen*] by sending an experienced officer to meet the battalion.[105]

Changes to Pre-Induction Training

Otto G. described his labor-duty experience from mid-September to December 10, 1943: "Basic training in [labor duty] was hard, very hard; much harder than in the barely controlled toughness training camp [*Hitler Jugend*], and also harder than later in the infantry."[106] Labor-duty training changed during the war to include more specific training with weapons and hand grenades, reflecting changes to the pre-military indoctrination laws during 1944.[107] Otto G. received his first introduction to the handling of personal weapons and hand grenades in labor duty. This military-specific education expanded in September 1944 with instructions to begin military basic instruction in labor duty. The final step in this process occurred on March 30, 1945, when Hitler ordered the creation of two labor-duty infantry divisions from the youngest year-groups then in labor duty for the purpose of countering parachute landings in the interior of the country. As with the youngest year-groups, the ability to organize these units at the end of the war was essentially nonexistent. Planned for 7,500 members, these were never created.[108]

Junior Leaders

An example of the wartime officer accession was Gottlob Biderman, an infantryman inducted into the 132nd Infantry Division, *Wehrkreis* XII, Wiesbaden, on its formation in 1940. He entered the Soviet Union with his division in late summer 1941. He experienced combat on the Crimean Peninsula, participating in the taking of Sevastopol, before his division transferred in summer 1942 to the Leningrad front at Lake Ladoga. Wounded in action several times, he received his notification of promotion to officer-aspirant while in hospital and transferred in September 1942 to infantry officer school near Prague. His time in officer preparation was three months before promotion and return to combat as an infantry officer for duty with the same division and regiment he served with as an NCO, the 437th Regiment, in January 1943.[109]

The key characteristic was the requirement for service as a private soldier for all officer-aspirants prior to consideration and recommendation by their regimental commander for officer rank, both before and during the war. With the expansion of the army, the army reduced the four-year program of the *Reichsheer* to less than two years. For NCOs whose service overlapped the *Reichsheer* and the *Heer,* the time was spent in officer training lessened due to their experience. For recommendation toward advancement to officer rank during the war, combat service while leading soldiers was the first prerequisite.[110]

High officer casualties on the Eastern Front (see Table 10.3), and the resultant widespread use and success of NCOs as platoon commanders in combat, provided a key rationale for the revised table of operation for the 1943- and 1944-type infantry companies. Each of these reorganizations resulted in the removal of an officer platoon commander from two of three platoons in the rifle company. Quality in the officers mattered more than quantity. For NCO promotion to officer-aspirant during the war the key requirement was the observed leadership of a squad or similar small unit in combat, *vor dem Feind* ("in front of the enemy"). In March 1942 the army demanded a minimum of one year in service, and no less than two months observed as a squad or other small-unit leader in battle, before nomination to officer candidate status (*Offizieranwärter*).[111] The presence of an officer who shared and excelled in the same battle experience as his soldiers enhanced the cohesion of the small infantry unit.[112]

Table 10.3 German Officers Killed, Missing, and Imprisoned, June 22, 1941–March 3, 1944

	Eastern Front	All Fronts
Killed	33,403	37,869
Missing	7,619	7,987
Prisoners of war	—	2,650
Total	—	48,506

Sources: Absolon, *Die Wehrmacht im Dritten Reich: Band VI*, 394; Overmans, *Deutsche Militärische Verluste im Zweiten Weltkrieg,* 56; Müller-Hillebrand, *Der Zweifrontenkrieg,* 255; Creveld, *Fighting Power,* 152.

Note: The numbers for this ending date are based on data created during the war, before the major losses of the final year. Overmans presented a total of four different officer loss figures for a later period, June 6, 1941–August 22, 1944, ranging from 51,845 to 61,937. Overmans's higher figures are based on the postwar reconciliation efforts.

There were 81,314 officers and 374,544 NCOs leading 2.3 million soldiers in September 1939 in the field army. Officers made up just 2.98 percent of the force. By May 1944 the army was short 5,157 infantry officers of a total shortage of 12,887 officers. The infantry shortfall was for the reduced *Kriegsstärkenachweisungen* of a rifle company from five officers in 1939 to two officers in 1943 and 1944. The percentage of officers serving in the field army fell below the 1939 figure, even as the army grew in number of infantry divisions. In comparison, exclusive of the US Army Air Forces, the percentage of the US Army Ground Forces and US Army Service Forces officers serving overseas was 8.5.[113]

There are no figures on the German NCOs who died in the war, nor are there figures on those who took over platoons and companies after the loss of their officers. It was a testament to necessity that rifle companies retained only two officers, one platoon commander and one company commander from 1943, but it was also an affirmation of the role of the NCO in the German Army. The table of organization reflected the reality and necessity to trust the leadership of infantry platoons to NCO platoon commanders.

How the German soldier felt about his officers and NCOs bears out this intent. Officer expectations were high. When discussing whether his company commander was a good officer, one soldier described that commander's actions in always eating last and ensuring his men were fed. Another noted his company commander was a good comrade and brave when the battle let loose.[114] In the transcripts from Fort Hunt "only the non-commissioned officers and sergeants, who received approval ratings of over 75 percent scored more highly [than the officers who scored at 75 percent.]"[115] "The men spoke often of their officers and not infrequently with unmitigated pride": "The lieutenant was always at the front"; "the lieutenant always came back to us"; "he gave me a pat on the back"; "the lieutenant sent an NCO to us"; "our lieutenant was a great guy."[116] German soldiers expected their officers and NCOs to lead them well.[117]

Discipline and Sanctions

The German system of military discipline under the National Socialist regime appeared as the most intensive, intrusive, comprehensive, and impactful of all of the armies studied. Omer Bartov, for example, judged this a primary reason for his conclusion that German infantry cohesion was a creation of fear of punishment, not the development of comradeship among soldiers.[118] Support for this contention was lost in a comparison with the much higher incidence of such interaction with the formal disciplinary systems of the British Army and the US Army. Given the disproportionate assignment of high-quality men to the combat arms and the infantry, the need for such an extraordinary regimen was dichotomous. Such a pervasive oversight of the soldier, and the myriad legal restrictions that were assumed to have surrounded him, might have acted as an institutional deterrent to willing cohesive behavior. The entire German tactical construct rested upon the foundation of self-motivation and willingness to act, with the soldier expected to demonstrate the greatest possible individual initiative in battle. The German system of selection, manning, training, and fighting was one built on the basis of creating and sustaining the morale and cohesive behavior of the infantry and combat arms.[119] The bonding of German infantry soldiers appeared neither forced nor contrived.[120]

For the 18,536 individual German soldiers comprising the *Wehrkreis* VI data, there are 4,643 records with an entry for formal and informal disciplinary action. These records represent 2,456 individuals, the majority (1,511) with a single offense. These individuals represented 13 percent of the total sample. Of 4,643 disciplinary actions listed, 87 percent of the offenses were handled by informal unit measures with a maximum limit of restriction or confinement of four weeks. There are 605 actions recorded that indicated a court-martial as the sentencing authority and represented 13 percent of the total of 4,643 records of disciplinary action. Statistically only 13 percent of the total sample of records indicated punishment at any level, the vast majority remaining at the unit level equivalent to the US or British commanding officer below the regiment. There are 528 individuals recorded in the database who received referral to court-martial.[121] Whereas the British frequency of courts-martial at all levels amounted to one for every ten men, and the US frequency was one for every eight soldiers, the corresponding German figure is closer to one for every thirty-four soldiers.

There were two primary reasons likely for this disparity. First was the emphasis on unit cohesion and the primary role of the commanding officer in keeping punishments in the unit to support morale and unit strength; second was that for the German soldier only a single type of court-martial generally applied to soldiers, and this was at the division commander's level. About 2.8 percent of German soldiers were subjected to court-martial in comparison to 13 percent of US soldiers, and 6.4 percent of British soldiers. Following Omer Bartov's analysis, one might conclude that US and British soldiers were the more likely candidates for being coerced into staying in the line of battle. Christoph Rass noted for the 253rd Division that the majority of the soldiers were in the end not subjected to any lasting negative determinative effect from the threat potential of institutional sanctions on their behavior. Actually, only a fraction of those soldiers who transgressed the norms were permanently lost to the war effort.[122]

Of course, the more severe the infraction, the greater certainty the offender was at risk of the most extreme measures, especially if referred to a court-martial. Extrapolating the *Wehrkreis* VI data for the *Wehrmacht* as a whole, the approximate number of courts-martial was 500,000 for the war, of 17.5 million serving. This falls very near the number recorded through December 1944, with the total of 415,742 for the army (*Heer*) alone.

It was the award and execution of death penalties for military offenses that most distinguished the wartime Nazi military justice system from that of the Western Allies. The closest to, and exceeding, German severity was the Soviet system, sentences of which included sending 442,000 soldiers to clear mines with their bodies during battle. The Red Army executed an estimated 158,000 Soviet soldiers during World War II.[123] If a soldier were subjected to a court-martial, the result was likely severe. Of the nine death sentences of the 528 soldiers who received one or more courts-martial listed in the Rass/Rohrkamp database, or 1.7 percent of those receiving a court-martial, eight were for desertion, and this offense was the most prevalent of those awarded death.[124]

Of 78,883 cases of desertion or AWOL, there were 7,737 death sentences awarded, or approximately 10 percent of all cases. For the offense of cowardice, 769 of 2,245 were given death sentences, or 34 percent. In 13,339 instances of plundering, there were 446 death awards, at 3 percent. For the catchall offense *Zersetzung der Wehrkraft* (undermining fighting power), including self-inflicted wounds, there were 1,166 death sentences out of 12,586 cases, or 9 percent.

In the end, it is likely that it will never be known how many soldiers finally fell victim to execution as the war came to its conclusion. The falling percentage of executions as against death sentences indicated by the available archival data sources appeared to support the conclusion that an overwhelming need for soldiers on the front diminished over time the percentage of death sentences carried out in favor of saving manpower.[125]

It was not necessary to find more deaths, however, to condemn the treatment of German soldiers by their leaders in awarding so many deaths by execution, or to justifiably vilify a Nazi- system of *Wehrmachtjustiz* that was anything but just. "German military justice was imbued with the racist-eugenic tenets of the Nazi Party."[126] In this, the opportunity given Hans von Seeckt in 1921 to eliminate the death penalty for military offenses, which he dismissed as a naive idea, may have been the most fateful and blackly resonant through time of any decision he made in preparing the army for the war to come.[127]

In the armies studied here, the German court-martial system acted with the greatest severity, given the number of executions for military offenses noted earlier, if not the frequency of courts-martial in the Anglo-American armies. A soldier's relationships within his primary group, his squad or section, likely shielded its members from the worst possible effects on morale.[128] The internalization of the background authority over the soldier and the certainty of an external sanction beyond the primary group, as Samuel Stouffer stated, reinforced the unit norms for cohesive behavior. The Rass/Rohrkamp database provides statistical evidence that German soldiers were shielded from the worst effects of the Nazi justice system by keeping punishments for military offenses within the authority of the commanding officers. In this the leadership of the battalions and companies supported the cohesive behavior of the small infantry units.

Recognition

The German Army instituted a number of types and grades of awards for combat participation and valor during World War II. This was a reaction to the perception that the small number of awards and the difficulty in earning them during the Great War became detrimental to soldier morale. Enlisted soldiers were not eligible for the highest award, the *Pour le Mérite.* German awards for valor were sequential. The German Army reinstated the Iron Cross first and second class (*Eisernes Kreuz*) at the beginning of the war. The Knight's Cross (*Ritterkreuz*) became available after the Polish campaign for men who already possessed both Iron Cross awards.

Most awards were given to soldiers who survived their actions as living examples for their fellows. Additional medals were instituted in 1941 to allow for recognition of extraordinary behavior not rising to the level of the Knight's Cross, the German Cross in silver and gold. Awards included membership in the Honor Roll of the German Army for exceptional deeds not rising to the level of the German Cross or the Knight's Cross. Men holding the Iron Cross first class were eligible for the German Cross and the Honor Roll.[129]

To achieve one of the highest awards demanded individual valor and independent decision in an act of leadership. For officers, the bar was set higher still and demanded a continuous history of such actions to merit consideration; "mere bravery, however exceptional, was not enough," and "for a regimental commander to personally lead his men in a successful counterattack with machine gun and hand grenades constituted 'a self-evident duty.'"[130] Decisive battle valor with pivotal meritorious leadership was the requirement for the Knight's Cross.[131]

The German Army went to extraordinary lengths to recognize the actions of the infantryman. The badges awarded for combat participation included:

- Wound badge (*Verwundetenabzeichen*): In three classes, black (one to two wounds), silver (three to four wounds), and gold (five or more wounds, also presented to next of kin for all killed in action).
- Infantry assault badge (*Infanterie-Sturmabzeichen*): In silver for nonmotorized infantry for three individual infantry assault actions on three separate occasions and awarded by the regimental commander; in bronze for motorized infantry and mechanized infantry (*Panzergrenadier*).
- Close combat clasp (*Nahkampfspange*): In bronze for fifteen days of close combat, silver for thirty days, and gold for fifty days. Hitler reserved to himself the presentation of the gold award and it was accompanied by the award of the German Cross in gold. The award in gold allowed the recipient one year in the *Ersatzheer,* or in school, and the wearer received the same benefits as a Knight's Cross recipient; 403 gold clasps were awarded during the war.
- Tank destruction badge (*Sonderabzeichen für das Niederkämpfen von Panzerkampfwagen durch Einzelkämpfer*): Full title was "special badge for the single-handed destruction of a tank." These were braids worn on the sleeve with a tank motif, silver

for one tank and gold for five. To earn this badge an individual infantryman had to destroy a tank with a hand-held weapon, such as a *Panzerfaust,* magnetic mine, or satchel charge.

- Air force ground combat badge (*Erdkampfabzeichen der Luftwaffe*): For *Luftwaffe* field divisions and parachute units serving as infantry; direct equivalent of the infantry assault badge.
- Parachutist badge (*Fallschirmschützenabzeichen*): Not an infantry badge of the army, but as parachute soldiers fought on the ground as infantry, it was a unique ground combat badge after the assault on Crete in 1941.[132]

The award of the Iron Cross was the largest by number for battle conduct. By 1944, the leadership sought to prevent the lessening of the tenor of the Iron Cross, reminding division commanders of their duty to preserve the worth of the award. Noting data for all divisions as of June 30, 1944, army headquarters provided new direction. The average ratio between the Iron Cross first class and second class in 1918 was one to twenty-three, with the average since 1939 at one to just over nine. A ratio of one to eight was given as the standard, but exceptions were allowed for heavy combat and long service on the front. The 253rd Infantry Division, with 1,399 days' active service at the front, and 1,095 days on the line in the Soviet Union (by June 30, 1944), was 36 percent above the average number of days for divisions in the east. The division awarded the Iron Cross at the higher ratio of one to almost six, awarding 10,270 Iron Cross second class, and 1,820 Iron Cross first class. The division suffered 16,644 wounded and killed during thirty-six months on the Eastern Front.[133]

German Army awards were worn on the combat uniform. This was a powerful tool for building leader confidence in the infantry small unit and supporting the bonds essential for the maintenance of their cohesive behavior. One had no doubt of the combat capability of unit leaders by looking at their awards.

Replacements and Physical Wounds

The policy of the German Army was to return wounded recovered soldiers to their original regiments. Those whose recovery to full war-capable physical health was less than two months remained in a hospital behind the field army for eventual return to their units.[134] Those with longer recovery times returned to a hospital in their home

Wehrkreis. During convalescence and following release from the hospital, they stayed in the local area at home, and were placed on the staff of a training unit. When deemed fully war-capable physically again, they received assignment to a recovered-wounded march company (*Genesenen-Marschkompanie*) and returned to their field unit.[135]

As a result of the tremendous casualties incurred during the first few months of the Russian invasion, the General Army Office published a directive on September 16, 1941, requiring all wounded recovered soldiers to be returned to their original units as an essential element of unit cohesion.[136]

Replacement-army units responsible for recovering soldiers were admonished to provide well-thought-out daily and weekly plans that were tailored to the capabilities of the individual wounded. Important was the ability to allow for the development of local relationships in the area to speed the process: "The mission of the recovering-wounded units is the development of the wounded to fully war-capable physical health."[137]

Six years of war experience led to the creation of statistical planning models predicting the approximate number and time of the return of wounded to their units in aggregate. The model predicted that for every 1,000 battle-wounded men, or men wounded in accidents, 40 percent returned within three months, 60 percent within six months, and 70 percent within one year. For soldiers suffering frostbite, 40 percent returned in three months, 70 percent in six months, and 85 percent in twelve months. Men admitted to hospital or unit medical aid station for treatment of illness returned at the rate of 75 percent in one month, and 90 percent in three months.[138] The 253rd Infantry Division, for example, received a replacement march company of recovered wounded and trained recruits numbering around 200 men every three weeks on average.[139] The German Army's policy of sending recovered wounded soldiers to their original units kept faith with its stated concern for the development of cohesive behavior among the soldiers. Alone among the three armies studied here, the *Heer* related its personnel actions to the development of cohesive fighting units.

Psychological Wounds

In contrast to the British and US practice, the German Army conducted limited mental screening on induction, while recognizing the requirement for treatment in the field for mental combat trauma. In general, the German Army did not psychologically screen combat soldiers to identify and remove men who might be or become unsuited mentally for combat, relying instead on the collective interview process at mustering. The interview and physical processing determined to the greatest extent possible the traditional elements of character, physical and mental aptitude for handling the stress of combat, and the ability to work well with others.

On the other hand, in 1926, psychologists succeeded in gaining recognition for their view of the ability of observation and testing to inform the army of the character and capability of men to withstand the rigors of combat. From 1926 to 1942, the *Wehrmachtpsychologie* organization provided this service for limited numbers of men to assist in determining their suitability both for service and for the specialized task for which they were chosen. Unlike their British and American counterparts, German psychologists had access to a limited number of soldiers. *Wehrmachtpsychologie* personnel tested 1.2 million men (of nearly 18 million serving) including officer candidates, doctors, veterinarians, pilots and air crew, drivers, radio operators, as well as NCO candidates in the army, navy, and air force.[140]

As in the British and US systems, the initial response to psychological trauma was to treat as far forward in the field as possible. All cases of "battle fatigue" received rest, sleep, and military exercises for up to eight days behind the line. Soldiers not responding in the field went to rear-area hospitals for more active treatment with a variety of interventions such as electric shock, drugs, and other suggestive techniques. German records indicated 85 percent returned to their units. Those who did not went to prison or mental institutions.[141]

Overall the reported incidence of psycho-neuroses in the German Army was less than in either the British or US Armies. The author Martin van Creveld provided German data to suggest, in the few instances recorded, that a figure of around 2 percent of strength or lower was indicated as the incidence of psycho-neuroses in the soldiers. He concluded that "the connection between the Wehrmacht's organization (recruitment by land of origin, replacements on the same basis and in closed groups, the Feldersatzbattalion) and the low number of psychiatric casualties is therefore quite clear."[142]

Creveld provided an analysis by a Professor Panse, visiting psychiatrist to *Wehrkreis* VI, Münster, who described in 1945 the reasons for the army's success in reducing these types of casualties:

> In my opinion, the systematic attack on psychogenic disturbances is at least partly responsible for the fact that not even in the sixth year of war do such cases constitute a large scale problem. The very fact that the troops know that physicians are not helpless when faced with such cases probably prevents the appearance of many. The danger of a relapse, too, is very small for those who were correctly treated and concerns only constitutionally abnormal types.
>
> As far as *Wehrkreis* VI is concerned, a rise in the number of cases cannot be shown even in the sixth year of the war. A few more cases were brought to our attention in the last few months, it is true, but in my opinion this is simply the result of better diagnosis by the troops.[143]

This opinion is difficult to verify. British medical officers reported in Italy their belief that German medicine in general was far below the standards of the British and US medical services, as in the availability of whole blood. Antibiotics such as penicillin were not obtainable for wounded German soldiers.[144] Despite this, they observed, during the early part of the campaign to capture Caen in Normandy, there were few cases of "exhaustion amongst the Germans, notwithstanding the severity the Allied bombardment." Brigadier Glyn Hughes was puzzled by this in view of the fact that "our shelling and aerial attacks were much more intense than anything they produced. We were certainly up against the flower of their troops. . . . [T]here is no doubt that they made their troops psychologically prepared for war; desertion whatever the cause was punished by death. . . . Interrogation of captured German doctors, and direct observation of [prisoners of war] immediately after capture, showed that it [exhaustion] was practically unknown in both Panzer, *SS*, and *Wehrmacht* troops alike."[145]

Contrary to this were Allied medical reports concluding that German methods of treatment, and reporting of these cases, were characterized by a reluctance among German doctors to identify psychological cases; doctors rather treated them as organic physical complaints. This contributed to the belief that the mental state of the German Army deteriorated as the campaign became a general Allied advance following the breakout from the Normandy bridgehead.[146] This latter view appeared countered in its turn by combat reports of the intensity of German resistance during Operation Market Garden in September 1944. The fighting at Arnhem bore a greater resemblance to the fighting around Caen than the general German retreat of July and August 1944.[147]

If cases of desertion and AWOL during battle were an indicator of the psychological breakdown of soldiers, the numbers indicate this was a problem of significance for all armies studied. Of 210,029 British Army court-martial convictions during World War II, 106,064 were for unauthorized absence (75,157), cowardice (167), and desertion (30,740), and of more than 99,000 men struck from the rolls for absence (AWOL), 22,000 were still absent at the conclusion of the war.[148] The US Army convicted over 81,000 in courts-martial for these offenses.[149] The German Army's court-martial convictions for these offenses totaled 81,128—for desertion (12,895), cowardice (2,245), and unauthorized absence (65,988).[150] The total of 268,192 soldiers across the three armies indicated evidence of psychological trauma, and the armies sought both medical and psychiatric treatment in addition to punitive methods to keep combat soldiers in the line of battle.

The Strength of the German Riflemen

In the initial organization of the infantry from 1939 to 1943, the *Sollstärke,* or table of organization, strength was between 15,019 and 17,901. The number of frontline riflemen in nine battalions was 5,710. The riflemen stood at 34 percent of the total strength of the division.[151] The new division organization of 1943 provided for 13,463 soldiers[152] and reduced the infantry battalions to seven, with six in three regiments with a single fusilier battalion. The frontline rifle strength by percentage fell to 24. In addition, the new (1943) organization included 2,000 foreign "helpers," or *Hilfswillige,* former Russian prisoners of war, to help with supply support. The 1943 reorganization acknowledged the condition of the *Ostheer,* in particular, with a shortage of over 600,000 soldiers by the end of 1943. Officer strength in the infantry companies of the battalions was cut to two for identical reasons—no capability to make good the losses. All divisions retained their field replacement training battalion.[153] In contrast, Russian infantry divisions had a total strength of 9,375, with 4,630 frontline infantry in nine infantry battalions, with 49 percent in first-line infantry battalions.[154] The ninety-nine infantry divisions attacking the Soviet Union with the army groups on June 22, 1941, were at or near their full strength.[155] Of the 3 million men in the attack, fewer than 400,000 riflemen formed the anvil of the attack on the Soviet Union.

The infantry divisions increased after 1941, while the number of riflemen per unit decreased. The largest number placed into the field by the German Army occurred in mid-1944, with 284 divisions of all types in all theaters and the homeland.[156] In contrast to the 3 million of June 1941, in October 1943 the *Ostheer* deployed 2.5 million men, whose combat organizations included 151 infantry divisions and 26 Panzer divisions with 2,304 armored vehicles and 8,037 pieces of artillery. The infantry, reduced on average to around 50 percent or less, numbered some 200,000, or half the infantry present on June 22, 1941. The largest number of infantry divisions deployed in a single theater was the 151 in the *Ostheer* between July and October 1943.

These infantrymen faced a Red Army in October 1943 of 5.5 million men, 536 infantry and 324 tank/mechanized division equivalents, 8,400 armored vehicles, and 20,770 pieces of artillery. German infantrymen alone were outnumbered 200,000 to 2.5 million in the Red Army infantry formations. Long odds indeed.[157]

Demographics: The Face of the German Infantryman

The purpose of knowing the demographics of the German soldier is to put a face on the individual. The data tell the story of the infantryman: that the soldier's father was a metal worker; he had four brothers and sisters; he attended a *Volksschule;* he worked as a salesman before mustering into the army; he married and had a son while on active service; he entered service in the same town he was born in; he became an NCO and attended an NCO school in an occupied Belgium; he was wounded three times and returned to duty in his unit after each recovery; he was decorated for bravery; and he was a prisoner of war in the Soviet Union. All of this allows the soldier to come forward from an otherwise faceless mass—to matter as an individual. Knowing he was a Catholic or a Protestant like the vast majority of his Anglo-American opponents, that he was the same height and weight on average, that he left home at seventeen or eighteen years of age and served until death or wounding, allows a more definitive comparison with his Anglo-American counterparts. There is also now an objective, contemporary record of what he believed about his service, his nation, his army, and his fellow soldiers, about war and death, about his leaders, and about his life. These two aspects of the infantryman within the social construct and contract of the infantry division combine to provide a window into the battle characteristics of the German soldier and the perceptions of his opponents.

The data aid in the resolution of certain persistent myths about the German combat soldier. One of those misconceptions was that by the end of the war, old men and boys were in the combat formations of the army. Photographs and movies depicted Hitler in particular with boy soldiers in the final death throes of the *Reich* in a surrounded city of Berlin. In fact, as the data here show, there was a clear line separating soldiers allowed to serve in combat infantry formations, and soldiers whose physical and age qualifications placed their service in the rear areas and support services.

The US Office of Strategic Services reports for the period of June through August 1944 confirmed this. The office attributed the larger number of older soldiers captured as indicative of a decision by the German Army to place older, less fit soldiers in the infantry divisions.[158] For example, for the 253rd Infantry Division, fourth wave, *Wehrkreis* VI, Rheinland-Westphalia, the percentage of eighteen- to thirty-three-year-old combat-capable soldiers in the infantry units of this division was 77 in 1944 and 86 in 1945, with similar percentages in each year of the war. The eighteen- to thirty-three-year-old age groups represent the primary age cohort for infantry combat in all armies studied here.

For example, in the support units of the 253rd Infantry Division, more than 60 percent were in the age groups 1897–1910 in 1944, and more than 50 percent in 1945. These year groups represent the thirty-four- to forty-seven-year-old cohort. This confirms again the intent to place those less qualified by age and fitness in support organizations not required for combat in the front line as *Fechtende Truppe.* In fact, there were far more supporting soldiers than infantrymen; 5,600 supporting arms and support soldiers in a division reduced to half strength at 8,000 men, supporting 2,400 infantry soldiers on average, in 1944. The large number of stationary defense divisions all along the coast of France, and all without vehicle transport, contributed to the larger percentage of older soldiers captured during the headlong retreat to the German border from July to September 1944. These older, stationary divisions by design had a higher proportion of older men by army policy. These factors rather than a new mobilization policy account for the percentage of older soldiers.

A further sample taken across the entire *Wehrmacht* for company-sized units from the *Wehrkreis* VI, Rheinland-Westphalia region of Germany, demonstrates and validates this same trend across all combat arms units of company size.[159] The larger sample demonstrates the higher presence of the younger age groups for the first-quarter

1944 data than for the 253rd Infantry Division data. The bulk of these younger soldiers in the sample were in the 1922–1925 age groups, with none from 1926 in the first-quarter 1944 column. This was consistent with the October 1944 arrival of this year-group in their divisions. In the sample of combat-capable soldiers from fourteen frontline companies, the percentage of eighteen- to thirty-three-year-old soldiers was 81 in 1944.

Height and weight data carried the same pattern forward in each set of data. Average height and weight for the *Wehrkreis* VI combat arms soldiers was 170 centimeters (5 feet 7 inches) and 64 kilograms (141 pounds), and for the 253rd Infantry Division the averages were 170 centimeters (5 feet 7 inches) and 66 kilograms (146 pounds). Both sets of data showed a higher weight of 70 kilograms (154 pounds) for the older year-groups and lighter weights for the youngest age groups.[160] In comparison, US Army infantry soldiers averaged 170 centimeters (5 feet 7 inches) and 64 kilograms (145 pounds), and this was comparable with British Army infantry.

The US Office of Strategic Services reports indicated a larger percentage of the 1925 and 1926 year-groups than these datasets. These year-groups began to arrive in their units in April and October 1944. The Office of Strategic Services put the percentage of soldiers in year-groups 1921–1926 at 38, confirming the percentages in the company-level data across time, while the 253rd Infantry Division data indicated 15 percent based upon its sample of records. When the 253rd Infantry Division replacement and training regiment went into active service upon the US Army's capture of Aachen, its replacements ceased.

The numbers for this age group captured in Italy were even higher, at over 50 percent. This indicated the more stationary defensive battle resonant with infantry combat in mountain warfare typical of the Italian theater of operations, and the difficulty of retreat from close combat in the mountains.[161] With over 30 percent of the combat arms in these youngest year-groups, ages eighteen to twenty-three, in late 1944 the Office of Strategic Services data conform to the tally for these age groups present within the combat infantry formations of the German Army at the end of the war. Further, for the 253rd Infantry Division, the ages of the support service soldiers indicated no members in the year-groups 1921–1926, commensurate with German Army policy that placed the less physically capable men in the support echelons.

The soldiers of the 1925 and 1926 year-groups were nineteen- and eighteen-year-olds respectively in 1944. The presence of large numbers of these young combat soldiers was consistent with the manpower of both the US and British Armies at this time of the war, and does not support the hypothesis the German Army was manned by old men and boys, irrespective of the photographs of Hitler and very young men in Berlin in 1945.

The datasets highlight the regional foundation for building and manning combat divisions. This was a powerful tool for creating and sustaining cohesion. For the sample data of fourteen combat arms companies, the average regional affiliation, defined as born and inducted within the same region and *Wehrkreis,* was 93 percent. Some infantry units, including the First Company, 106th Reconnaissance Battalion, had 97 percent born and inducted in the region. The regional affiliation data for the 253rd Infantry Division noted 89 percent throughout the war from *Wehrkreis* VI, Rheinland-Westphalia, its home region, and the full dataset for regional affiliation in sixty-eight company-sized units across the *Wehrmacht* was 86 percent.[162] The data suggest the effect of a policy of regional recruitment succeeded as a powerful cohesive determinant in serving with men in a combat unit from their home area.

The data further indicated the very small percentage of soldiers with formal affiliation in the Nazi Party at 401 of over 18,536 records. The largest membership, at over 60 percent, was the Hitler Youth, required of every young person who turned fourteen from 1936. Soldiers in the full *Wehrkreis* VI database identified themselves as 56 percent Catholic, 38 percent Protestant, and 6 percent other or no religion. Over 76 percent were single. Illiteracy was not an issue for the German soldier, and in the full database under the section on school attendance there were 169 different types of school entries; the most common was *Volksschule,* with 9,331 entries, and 86 for *Abitur,* and for 302 *Gymnasium.*[163] Overall, the soldiers serving from this region evidenced a stable apportionment of population with a dominant working-class background. There was a strong social affinity supporting the cohesion of these soldiers as they came together into the army's small units.[164]

The Anschauung: *The Voice of the German Infantryman*

Felix Römer researched "the files of the US interrogation centre Fort Hunt, where over 3000 German soldiers were interrogated and wiretapped during the Second World War. The files contain the eavesdropping transcripts and biographical data that reveal both mentalities and socio-cultural backgrounds."[165] Study into the attitudes and experiences of the soldier allowed a limited view from letters and memoirs. These eavesdropped conversations allow a contemporary and candid window into what the German soldier thought of himself in relation to his service during World War II. When combined with the demographic data outlined earlier, a far richer and nuanced picture of the German soldier comes into focus, making the comparison of these soldiers with their Anglo-American counterparts more powerful.

The characteristics of the soldiers held at Fort Hunt were similar to the demographic data compiled from the Rheinland-Westphalia region of Germany earlier. From the data available of 730 soldiers who completed morale questionnaires, more than half of whom served in the *Heer,* 60 percent were rank-and-file soldiers, 30 percent were NCOs, and 10 percent were junior officers. The age groups matched the percentages for the combat units of the 253rd Infantry Division and the larger *Wehrkreis* VI data, with 41 percent born between 1911 and 1920, and 38 percent between 1921 and 1929. Half were single as opposed to 76 percent in the larger sample. These soldiers were from a working-class background, with Catholics representing 40 percent of the soldiers and Protestants 50 percent.[166]

Motivation and ideology. In confirmation of the Shils and Janowitz study, the Fort Hunt transcripts revealed an acceptance of Hitler among the soldiers at 60 percent; while "no above average resistance towards the *Nazi* regime emerged . . . almost 40 percent responded with rejection." It appeared their loyalty toward Hitler and the regime "was based to a considerable extent on their participation in the economic upsurge. . . . [A]fter years of unemployment the workers didn't ask where the work came from."[167] Römer noted their "concept of patriotism corresponded to rather traditional nationalistic views than to National Socialist theories. . . . The numerous Wehrmacht soldiers who credibly insisted on being 'non-political' personified the minimal impact of the interpretive culture of their social milieu. In the Wehrmacht the ideological indifference of the workers eased their collectivization as soldiers. The less the soldiers associated their deployment in the National Socialist armed forces with political substance, the easier it was for their working class ethos to be transferred from civilian life to the everyday life of the military."[168]

The racial implications of National Socialism appeared to be understood, even if the soldiers did not follow or prescribe as individuals to this portion of the ideological milieu of the Nazi worldview. Römer concluded that soldier attitudes concerning the racial policies of the Hitler regime were individual, and built upon a multitude of factors including his history, the attitudes of his family, and the experience of battle within his small unit. These differences appeared in the transcripts of members of the *Waffen-SS* and in the experiences of the soldiers on the Eastern as opposed to the Western or Italian fronts.[169]

The answer to the cliché of the fanatical (Nazi worldview) *Weltanschauungs*-inspired warriors, however, is that of a small minority of soldiers were such fanatics. A through-and-through political army the *Wehrmacht* was not, but it possessed a political backdrop and was in more than a few places interspersed with political soldiers.[170]

Kameradschaft: *cohesion, combat, and the small unit.* The view of the German soldier of his service from the evidence of the Fort Hunt transcripts was positive and internalized both the canons of the armed service and a bond between the soldiers in battle. The soldiers believed in themselves and their capabilities in comparison to other soldiers; it was noted that "the morale and fighting spirit of the German infantry are incomparable. No other group of soldiers could accomplish what they have done"; and again, "the Germans are excellent soldiers."[171] The eavesdropped conversations left no doubt of the soldiers' view of themselves, their experience of combat, and their admiration for their fellows.

These feelings extended throughout the transcripts to the soldiers' belief in the members of their small units, their leaders, and their battle experience and skill. Exemplary of these tendencies was twenty-four-year old Michael Schu, an unmarried private from Saarbrücken who deployed with the Third Armored Grenadier Division in Italy, where he was captured in May 1944: "His recollections of the fighting, during which he had been wounded and been decorated with the Assault Badge (*Sturmabzeichen*), were dominated by his admiration for the strength of his own unit: 'Man, how our tanks slammed in.' . . He talked proudly about an infantry attack in which he had participated: 'When we arrived at our destination we let loose against the enemy and we screamed; the enemy thought we were an entire regiment, but we were only 50 privates.'"[172]

The transcripts confirmed that the squad (*die Gruppe*) and its members were the social and cultural center and focus in the daily life of the individual soldier. He was not consumed in the collective. It was this sense of the individual that appeared lost in the historiography of the soldier and the dialectic between structure and actor that reappeared in these transcripts from Fort Hunt.[173] The soldiers were not afraid of their officers, demanding of them the best leadership from the front. These soldiers gave no evidence of fear of an overwhelming and constricting judicial system. The soldiers appeared to act with an agency and will as independent actors, committed to and functioning within their squads, a condition at odds with a historiography emphasizing the constrictions and "group terror" of their small units.[174]

Conclusion

The tenacity of the German infantry in the final offensives of 1941 and 1942 and the retreat into defeat in 1944 lay first in the disproportionate selection of the most physically and mentally qualified to withstand the horrific conditions of the face-to-face infantry battle. Second, the German Army intended and practiced the institutional norms and policies that ensured the creation and sustainment of the primary group. As Shils and Janowitz noted, "It appears that a soldier's ability to resist is a function of the capacity of his immediate primary group (his squad or section) to avoid social disintegration. . . . He was likely to go on fighting, provided he had the necessary weapons, as long as the group possessed leadership with which he could identify himself and as long as he gave affection to and received affection from the other members of his squad and platoon."[175] Römer noted that "the shared pride in the abilities of the Wehrmacht and its martial virtues was deeply rooted throughout German society. It combined to form a collective military morale that vaulted traditional dividing lines in the German armed forces and contributed elementarily to the motivation of the troops. Such complex military cohesion surely didn't represent an exclusively German phenomenon. Yet comparisons with other nations indicate that soldierly morale in the Wehrmacht was particularly strongly developed."[176] Why was the German so hard to kill? He was selected, trained, and bonded within his small unit to a degree perceived as unique and troublesome by his adversaries. The basis of his tenacity was this bond.

Notes

1. Glantz, *Before Stalingrad,* 14; Müller-Hillebrand, *Die Blitzfeldzüge,* vol. 2, 111; Kroener, Müller, and Umbreit, *Organisation und Mobilisierung des deutschen Machtbereichs,* 874–875.

2. Ibid., 881.

3. Absolon, *Die Wehrmacht im Dritten Reich: Band V,* 126; Absolon, *Die Wehrmacht im Dritten Reich: Band VI,* 275–276.

4. Kroener, Müller, and Umbreit, *Organisation und Mobilisierung des deutschen Machtbereichs,* 936.

5. Absolon, *Die Wehrmacht im Dritten Reich: Band V,* 638; Müller-Hillebrand, *Der Zweifrontenkrieg,* 28. Müller-Hillebrand gave a figure of 228,000 frostbite cases for the winter months of 1941–1942.

6. Boog et al., *Der Globale Krieg,* 787.

7. Absolon, *Die Wehrmacht im Dritten Reich: Band VI,* 276.

8. Rass, *"Menschenmaterial,"* 419.

9. BAMA RH 15/269, *Chef d.Heeres Rüstung und Befehlshaber des Ersatzheeres, Anliegend werden Auszüge aus dem Bericht des Major Oehmichen . . . der 4. Armee vom 9.-24. 42,* April 21, 1942.

10. BAMA RH 26-6/44, *Anlagenband XXII zum Kriegstagebuch Nr. 6 der 6.Inf.Div vom 1.4.-12.31.1942,* reports dated April 1, April 10, and April 13, 1942.

11. Absolon, *Die Wehrmacht im Dritten Reich: Band VI,* 277; Boog et al., *Der Globale Krieg,* 870.

12. Müller-Hillebrand, *Der Zweifrontenkrieg,* 60.

13. Ibid., 62.

14. Boog et al., *Der Globale Krieg,* 792.

15. Ibid., 802.

16. Frieser et al., *Die Ostfront 1943/44,* 4; Overy, *Why the Allies Won,* 85.

17. Boog et al., *Der Globale Krieg,* 1029.

18. Beevor, *Stalingrad,* 396, 430.

19. Frieser et al., *Die Ostfront 1943/44,* 1109.

20. Glantz, *When Titans Clashed,* 136–139.

21. Boog et al., *Der Angriff auf die Sowjetunion,* 1098–1102.

22. Cooper, *The German Army,* 446.

23. Kroener, Müller, and Umbreit, *Organisation und Mobilisierung des deutschen Machtbereichs,* 928; Blank et al., *Die Deutsche Gesellschaft,* 530, 542.

24. Kroener, *Generaloberst Friedrich Fromm,* 615–624.

25. Müller-Hillebrand, *Der Zweifrontenkrieg,* 211.

26. Bernhard Kroener, *Generaloberst Friedrich Fromm,* 30–43, 270, 730, 866; Kroener, Müller, and Umbreit, *Organisation und Mobilisierung des deutschen Machtbereichs,* 919.

27. Bernhard Kroener, *Generaloberst Friedrich Fromm,* 423–427; Kroener, Müller, and Umbreit, *Organisation und Mobilisierung des deutschen Machtbereichs,* 916–920.

28. Kroener, *Generaloberst Friedrich Fromm,* 402; Kroener, Müller, and Umbreit, *Organisation und Mobilisierung des deutschen Machtbereichs,* 917.

29. BAMA RH 15/369, *OKH Oberkommando des Heeres (Chef H Rüst u BdE),* Az.23b 12/14 AHA/Ag/H (IIb)/Nr.5660/41 geh., Berlin, August 15, 1941, *Personelle Überprüfung der Kdo.-Behörden, Stäbe, Dienststellen und Einheiten des Ersatzheeres,* 82; Kroener, Müller, and Umbreit, *Organisation und Mobilisierung des deutschen Machtbereichs,* 887.

30. Kroener, Müller, and Umbreit, *Organisation und Mobilisierung des deutschen Machtbereichs,* 890; Müller-Hillebrand, *Der Zweifrontenkrieg,* 51, 60; BAMA ZA 1/1778, Personnel and Administration (Project 2b), P-006, Teil I, *Das Rekrutierungssystem der deutschen Wehrmacht in Frieden und im Kriege;* Anlage I, *Durchführung einer Musterung und Aushebung,* 1948, 98.

31. BAMA RW 15/62, OKW, B50c, AHA/S In (Org IIIb I), October 7, 1941, *Auszug aus der Allgemeinen Heeres-Mitteilung* no. 994, October 21, 1941, *Zahnbehandlung Dienstpflichtiger und Freiwilliger vor Einstellung in die Wehrmacht,* 539.

32. BAMA RW 15/62, *Wehrbezirks-Kommando Münster,* IIc Az.12 Mu/24 no. 117/42, *Musterung des Geburtsjahrganges 1924,* January 23, 1942, 1.

33. BAMA RW 15/62, *Wehrbezirks-Kommando Münster,* IIc Az.12 Mu/24 no. 117/42, *Musterung des Geburtsjahrganges 1924,* January 23, 1942, 1; BAMA RW 15/62, OKW, Az.12.i.11.20 AHA/Ag/E (1c), no. 22 560/41, *Durchführung der Musterung von Wehrpflichtigen des Geburtsjahrganges 1924,* December 30, 1941, 1–4; Absolon, *Die Wehrmacht im Dritten Reich: Band I,* 132.

34. BAMA RH 26 6/48, 6.Division, Abt.Ia., no. 170/42 g./Az.Ausb., Div.Gefechtstab, April 1, 1942, *Ausbildung des Ersatzes,* 1.

35. BAMA RH 15/369, *OKH AHA/E Az.12.d E (IIb),* no. 4474/41 geh., Berlin, September 1, 1941, *Übersichten über die Ersatzlage,* 34; BAMA RH 15/369, *Wehrkreiskommando III,* Abt.Ib/Erg., no. 08610/41 geh., *Verbrauchsabrechnung der Übersicht Stärke der Ersatzeinheiten nach dem Stande vom 15.Juli 1941: Mannschaften,* 46; BAMA RH 15/369, *Oberkommando des Heeres (Chef H Rüst u BdE),* Az.23b 12/14 AHA/Ag/H (IIb), no. 5660/41 geh., Berlin, August 15, 1941, *Personelle Überprüfung der Kdo.-Behörden, Stäbe, Dienststellen und Einheiten des Ersatzheeres,* 82; BAMA RH 15/369 *Oberkommando des Heeres (Chef H Rüst u BdE),* AHA/Ia(II), no. 13346/41 geh., Berlin, June 11, 1941, 130–131.

36. Kroener, Müller, and Umbreit, *Organisation und Mobilisierung des deutschen Machtbereichs,* 800.

37. BAMA RH 53-7/750, OKW, no. 8100/42 geh. AHA/Ag/E (VZ), December 23, 1942, *Durchführungsbestimmungen zum Führerbefehl vom 19.12.42.,* "Ur 43 Tausch"; Absolon, *Die Wehrmacht im Dritten Reich: Band VI,* 295.

38. Absolon, *Die Wehrmacht im Dritten Reich: Band VI,* 295–296.

39. Müller-Hillebrand, *Der Zweifrontenkrieg,* 254.

40. Kroener, Müller, and Umbreit, *Organisation und Mobilisierung des deutschen Machtbereichs,* 480.

41. Ibid., 769–774.

42. Ibid., 832–833.

43. Müller-Hillebrand, *Der Zweifrontenkrieg,* 134; Kroener, Müller, and Umbreit, *Organisation und Mobilisierung des deutschen Machtbereichs,* 832–833; Absolon, *Die Wehrmacht im Dritten Reich: Band VI,* 267–268.

44. Müller-Hillebrand, *Der Zweifrontenkrieg,* 71–72; Kroener, *Generaloberst Friedrich Fromm,* 527–528.

45. BAMA RH 53-7/751, *Der Chef der Heeresrüstung und Befehlshaber des Ersatzheer,* AHA Ia(I) no. 4167/42 g.Kdos., *Betr.: Feld-Ausbildings-Divisionen,* Berlin, August 31, 1942, Anlage 4, p. 2.

46. Müller-Hillebrand, *Der Zweifrontenkrieg,* 149.

47. BAMA RW 15/9, *W.E.I (Wehrersatzinspektion) Münster,* II/IIA Az. 12 i 10.20, April 19, 1943, *Zuweisung brauchbaren Ersatzes, für die verschiedenen Waffengattungen,* location: *W.M.A.* [*Wehrmeldeamt*] *Landkreise Bielefeld, Wehrbezirksoffizier Bielefeld 2, Wehrbezirkskommando Bielefeld,*

Wehrersatzinspektion Münster, Wehrkreis VI; Regierungsbezirk: Minden; Provinz: Westfalen; Land: Preußen.

48. Rass, *"Menschenmaterial,"* tab. A2, "Personalersatz für die 253.Infanteridivision," 420–421.

49. Müller-Hillebrand, *Der Zweifrontenkrieg,* 60.

50. Absolon, *Die Wehrmacht im Dritten Reich: Band VI,* 776–777.

51. BAMA RW 15/81, *Wehrkreiskommando* VII and various headquarters including *Wehrersatz-Inspektion München,* on the dates indicated. For *Totaler Krieg: OKH,* no. 8050/44 geh., November 11, 1944.

52. Glantz, *The Battle of Kursk,* 229–237, 241–249.

53. Sokolov, "The Cost of War," 187; Korol, "The Price of Victory," 425. Korol quotes Major General J. R. Deane (head of the US military mission in Moscow): "From October 1941 and to war's end, more than 15 million tons of supplies were dispatched to Russia, including 427,000 trucks . . . 22 million shells, 5.5 million pairs of army boots, etc."

54. Glantz, *When Titans Clashed,* 136–139, 53, 66–70, 75.

55. Burkhart Müller-Hillebrand, *Der Zweifrontenkrieg,* 170–172.

56. Sokolov, "The Cost of War: Human Losses for the USSR and Germany," 185–186; Korol, "The Price of Victory," 422.

57. Sokolov, "The Cost of War," 187; Korol, "The Price of Victory," 423; Frieser et al., *Die Ostfront,* 204.

58. Sokolov, "The Cost of War," 182, 186.

59. Frieser et al., *Die Ostfront,* 204.

60. Ibid., 648.

61. Müller-Hillebrand, *Der Zweifrontenkrieg,* 21, 41, 214–215; Weigley, *Eisenhower's Lieutenants,* 264.

62. Boog et al., *Der Zusammenbruch des deutschen Reiches,* MGFA Map 05579-03, MGFA Map 05581-02, 391.

63. Frieser et al., *Die Ostfront,* MGFA Map 0489-22, 572.

64. Keegan, *The Second World War,* 394–395.

65. BAMA RH 14/5, *HOKW 451937, Ch H Rüst u BdE, Generaloberst Fromm, Fernschreiben Telegram, 0021 Hours, 21.7.1944* [Fromm message indicated he had had the leaders of the conspiracy shot]; BAMA RH 14/5, *HOKW 453217 Heinrich Himmler, Fernschreiben Telegramm, 0431 Hours, 21.7.44* [Himmler message nullifying the message from Fromm at 0021 hours and confirming he was in command of the *Ersatzheer*]; Keegan, *The Second World War,* 394–395, 439.

66. Absolon, *Die Wehrmacht im Dritten Reich: Band VI,* 588.

67. BAMA RW 15/81, *Wehrkreiskommando* VII and various headquarters including *Wehrersatz-Inspektion München,* on the dates indicated. For *Totaler Krieg: OKH,* no. 8050/44 geh., November 30, 1944.

68. BAMA RH 11 1/28, Middeldorf, Major, i.G., *Bericht über die Reise Heeresgruppe B,* OKH Gen.St.d.H./Ausb.Abt.(II), no. 1464/44 g.Kdos., H.Qu. OKH, November 12, 1944; Müller-Hillebrand, *Der Zweifrontenkrieg,* 168–170, 175, 237.

69. Rass, *"Menschenmaterial,"* 60–61; Müller-Hillebrand, *Der Zweifrontenkrieg,* 71–72.

70. Absolon, *Die Wehrmacht im Dritten Reich: Band VI,* 21–25.

71. BAMA RW 15/126, *Der Chef der Heeresrüstung und Befehlshaber des Ersatzheeres,* AHA/Stab II (1) no. 418/45 geh., Berlin, January 3, 1945, *Verwendung junger Jahrgänge,* 153–154; BAMA RW 15/126, *Vortragsnotiz Einstel-*

lung des Jahrganges 28, Berlin, January 9, 1945, 155; BAMA RW 15/126, AHA/Stab II (1), December 14, 1944, *Notiz, Einstellungs-Übersicht für die Verplanung des Jg. 27/28,* 147.

72. BAMA RW 15/126, OKW, no. 2884/45 geh.Kdo., *Zugänge in das Heer im 1. Vierteljahr 1945, insbesondere sog. Goebbelsaktion,* March 30, 1945.

73. BAMA RW 15/126, OKW, AHA/Stab II (1), no. 22884/45 geh.Kdos., Berlin, March 30, 1945, *Zugänge in das Heer im 1.Vierteiljahr 1945, insbesondere sog.Goebbelsakte,* 133.

74. BAMA ZA 1/1785, Personnel and Administration (Project 2b), P-006, Teil I, *Ausbildung und Verwendung der Unteroffiziere und Mannschaften im deutschen Heer,* 1948, 37.

75. BAMA RW 15/126, OKW, *Abschrift aus* no. 3053/45 g.K. AHA /Ia, April 27, 1945, *Unterstellung der Ausbildungseinheiten des Ersatzheeres unter das Feldheer,* 217; Absolon, *Die Wehrmacht im Dritten Reich: Band VI,* 873.

76. BAMA RW 15/126, *Wehrbezirkskommando Passau IIc,* April 25, 1945, *Organisation des Ersatzes für die Front: Erfassung und Einberufung von Wehrpflichtigen.*

77. Deggendorf, Kötzting, Passau, Straubing, Winterberg, Wolfstein, and Zwiesel.

78. BAMA RH 15/126, *Wehrbezirkskommando Passau IIc,* April 24, 1945.

79. Bidermann, *In Deadly Combat,* 273–274. *Kettenhund:* literally a "chained dog" was German slang for a military policeman. The term derived from a metal gorget suspended around the neck from an aluminum chain. This field badge was the equivalent of the MP (military police) arm bands of their US counterparts.

80. Hitler, *Mein Kampf,* 654.

81. Overmans, *Deutsche Militärische Verluste im Zweiten Weltkrieg,* 265, tab. 52.

82. Weinberg, "Dokumentation," 445, 451.

83. Ibid., 445.

84. Ibid., 448.

85. Castillo, *Endurance and War,* 10; Bartov, *The Eastern Front,* 76.

86. Förster, "Der 'Führerbefehl,'" 596; Zoepf, *Wehrmacht zwischen Tradition und Ideologie,* 203.

87. Zoepf, *Wehrmacht zwischen Tradition und Ideologie,* 50, 371–374.

88. Quinnett, "The German Army Confronts the NSFO," 62.

89. Thamer, "Die Erosion einer Säule," 434–435.

90. Absolon, *Die Wehrmacht im Dritten Reich: Band VI,* 282, 782.

91. BAMA ZA 1/1778, Personnel and Administration (Project 2b), P-006, Teil I, *Das Rekrutierungssystem der deutschen Wehrmacht in Frieden und im Kriege;* Anlage I, *Durchführung einer Musterung und Aushebung, Beilage XI, Fehlertabelle,* 1948, 177; *Vorschrift über wehrmachtärtzliche Untersuchungen im Kriege: Ärztliche Anweisung zur Beurteilung der Kriegsbrauchbarkeit bei Kriegsmusterungen* [HDv 252/4.LDv 399/4], April 1, 1944.

92. Absolon, *Die Wehrmacht im Dritten Reich: Band VI,* 282.

93. BAMA ZA 1/1778, Personnel and Administration (Project 2b), P-006, Teil I, *Das Rekrutierungssystem der deutschen Wehrmacht in Frieden und im Kriege;* Anlage I, *Durchführung einer Musterung und Aushebung,* 1948, 32; Absolon, *Die Wehrmacht im Dritten Reich: Band VI,* 283.

94. BAMA ZA 1/1778, Personnel and Administration (Project 2b), P-006, Teil I, *Das Rekrutierungssystem der deutschen Wehrmacht in Frieden und im Kriege;* Anlage I, *Durchführung einer Musterung und Aushebung,* 1948, 158.

95. Overmans, *Deutsche Militärische Verluste im Zweiten Weltkrieg,* 333, tab. 72a, and 334, tab. 73a.

96. BAMA RH 17/374, *Lehrstab Maria ter Heide, Ausbildung im Nahkampf,* September 1943, Oblt. Henke; *Zusammenstellung der Gruppenaufgaben,* April 28, 1943, *von Oberst Hegwein,* 10.

97. BAMA RH 26 6/48, 6.Division, Abt.Ia Nr.170/42/Az. Ausb., *Ausbildung des Ersatzes,* April 1, 1942.

98. BAMA RH 26 6/48, *Zug- und Unterführerlehrgange, 6.Division, Ausbildungsplan für den 3, Unterführerlehrgang vom 8.1.1943–30.1.1943,* January 6, 1943.

99. BAMA RH 26 6/48, Anlage zu 6.Div.Abt.Ia Az.Ausb., no. 15/43 geh., January 6, 1943, *Durchgeführte und vorgesehene Lehrgänge der Infanterie.*

100. Glantz and House, *When Titans Clashed,* 136–139.

101. BAMA MSG 1/2527, Otto Gunkel, *Aus Meinem Leben,* 19; Müller-Hillebrand, *Die Blitzfeldzüge,* 166.; Müller-Hillebrand, *Der Zweifrontenkrieg,* 222.

102. BAMA MSG 1/2527, Otto Gunkel, *Aus Meinem Leben,* 20. Translation for "dirt-bunnies" (*Sandhasen*) or "stubble-jumpers" (*Stoppelhopser*); Houlihan, *Kriegsprache.*

103. USAMHI, MS # B-068, Denkert, *3rd Panzer Grenadier Division (Ardennes),* 5–6.

104. Rass, *"Menschenmaterial,"* 60–61.

105. USAMHI, MS # B-046, Denkert, BG, *Operation of the 3rd Panzer Grenadier Division in the Ardennes Operation,* May 1946, 8, emphasis in original.

106. BAMA MSG 1/2527, Otto Gunkel, *Aus Meinem Leben, Die Zeit vom September 1943 bis November 1946: Kriegsdienst und Kriegsgefangenschaft,* 8.

107. Ibid., 9; Absolon, *Die Wehrmacht im Dritten Reich: Band VI,* 102.

108. Absolon, *Die Wehrmacht im Dritten Reich: Band VI,* 107.

109. Bidermann, *In Deadly Combat,* 151–155.

110. Absolon, *Die Wehrmacht im Dritten Reich: Band VI,* 373–374.

111. Ibid.

112. BAMA ZA 1/2003, Personnel and Administration (Project 2b), Teil IV, *Stellungnahme zu "Erziehung des Offiziers,"* 12–13.

113. Müller-Hillebrand, *Der Zweifrontenkrieg,* 255; van Creveld, *Fighting Power,* 152–155.

114. Römer, *Kameraden: Die Wehrmacht von Innen,* 313.

115. Römer, "Milieus in the Military," 136.

116. Römer, *Kameraden,* 314.

117. Ibid., 315.

118. Bartov, *Hitler's Army,* 104.

119. Römer, "Milieus in the Military," 136; Shils and Janowitz, "Cohesion and Disintegration in the Wehrmacht," 292–294.

120. Ibid.

121. Rass and Rohrkamp, *Deutsche Soldaten,* 270–273.

122. Rass, *"Menschenmaterial,"* 307.

123. Overy, *Russia's War,* 160; Messerschmidt, *Die Wehrmachtjustiz,* 172.

124. Rass/Rohrkamp database derived from analysis of individual records in the *Tabelle Strafen,* or Punishment Tables.

125. BAMA RH 14/55, *Kriminalität im Heere, Beschleunigte Auswertung, Personen,* Ag HR Wes/HR (II 2b), *Jahresübersicht über die verurteilten Soldaten und Wehrmachtbeamten des Heeres,* 1945, 7–12; Messerschmidt and Wüllner, *Die Wehrmachtjustiz im Dienste des Nationalsozialismus,* 77–89; Wüll-

ner, *Die NS-Militärjustiz und das Elend der Geschichtsschreibung,* 852; Messerschmidt, *Die Wehrmachtjustiz,* 398, 420, 453.

126. Bryant, "American Military Justice from the Revolution to the UCMJ," 11.

127. Von Rabenau, *Seeckt,* 489.

128. Römer, "Milieus in the Military," 136; Shils and Janowitz, "Cohesion and Disintegration in the Wehrmacht," 284.

129. Van Creveld, *Fighting Power,* 109–111.

130. Ibid., 111.

131. Absolon, *Wehrgesetz und Wehrdienst,* 261.

132. Ibid., 261–275; Williamson, *Infantry Aces of the Reich,* 15–26.

133. BA ZNS, *Oberkommando des Heeres,* PA/ P 5(b) 1.Staffel/Az. 29a12, 732/44 g.Kdos, *Eiserne Kreuze,* May 16, 1944; BA ZNS, *Ordensabteilung,* P5 (10), *Verleihungstand,* December 31, 1943; BA ZNS, *Ordensabteilung,* P5.(10), *Verleihungsstand,* June 30, 1944; BA ZNS, *Der Kommandierende General des XVII: Armeekorps, Zuweisung von Tapferkeitsauszeichnungen, Kampf Gefechtsstand,* November 11, 1943.

134. Absolon, *Die Wehrmacht im Dritten Reich: Band V,* 62.

135. BAMA ZA 1/1784, Statistics Systems (Project 4), Teil I, *Das statistische System,* 25.

136. BAMA RH 15/369, *Allgemeines Heeresamt,* 1a, *Vortragsnotiz,* Betriff: *Zuführung der Genesenen zu ihren alten Feldtruppenteilen,* Berlin, September 16, 1941.

137. BAMA RH 14/7, *Chef d. Hr. und B.des Ers.H., Ausbildung bei Marschbataillon,* Teil II, *Ausbildung in den Genesendeneinheiten,* Berlin, September 16, 1942.

138. BAMA ZA 1/1784, Statistics Systems (Project 4), Teil I, *Das statistische System,* 166.

139. Rass, "Das Sozialprofil von Kampfverbänden," 673.

140. Fritscher, *Dokumente Zur Deutschen Wehrpsychologie,* 120–121.

141. Van Creveld, *Fighting Power,* 93.

142. Ibid., 94.

143. Ibid.

144. Harrison, *Medicine and Victory,* 168–169.

145. Ibid., 252.

146. Ibid., 253.

147. Ibid., 260–261.

148. McPherson, *Discipline,* appendix 1(a); Glass, *The Deserters,* 307.

149. Rose, "The Social Psychology of Desertion from Combat," 614; US Department of Army, *Army Almanac,* 625, 634–633; Glass, *The Deserters,* 203.

150. BAMA RH 14/55, *Kriminalität im Heere, Beschleunigte Auswertung, Personen,* Ag HR Wes/HR (II 2b), *Jahresübersicht über die verurteilten Soldaten und Wehrmachtbeamten des Heeres,* 1945, 3.

151. Kroener, Müller, and Umbreit, *Organisation und Mobilisierung des deutschen Machtbereichs,* 960.

152. Müller-Hillebrand, *Der Zweifrontenkrieg,* 138. Infantry Division no. 44, beginning in April 1944, reduced by a further 1,000 soldiers to 12,769, with six battalions of infantry.

153. Kroener, Müller, and Umbreit, *Organisation und Mobilisierung des deutschen Machtbereichs,* 958–961.

154. Ibid., 960.

155. Ibid., 874–875.

156. Müller-Hillebrand, *Der Zweifrontenkrieg,* 152.

157. Kroener, Müller, and Umbreit, *Organisation und Mobilisierung des deutschen Machtbereichs,* 953.

158. USAMHI, US Office of Strategic Services, Research and Analysis Branch, *Nationality and Age of German Armed Forces Prisoners Captured in Northern France to 10 Sept. 1944,* 7. The office based its calculations on a 3.3 percent sample of 150,000 prisoners of war, 88 percent of whom were German soldiers of all ages.

159. Rass and Rohrkamp, *Deutsche Soldaten,* 69. All data used by permission of the author and the *Deutsche Bundesarchiv.* The principal sources of the data were the following documents: *Erkennungsmarkenverzeichnisse,* or unit morning reports; *Personalunterlagen der Bundesarchiv-Zentralnachweisstelle (BA-ZNS) Kornelimünster,* as the surviving personnel documents such as the service record (*Wehrstammbuch*); soldier personnel records that went with him, including a pay record (*Soldbuch/ Gebührniskarte*), health record (*Gesundheitsbuch*), and the *Wehrpass,* primarily used between mustering and enlistment; *Nachkriegskarteien zur Klärung von Kriegsschicksalen,* or various records for displaced persons following the war; *Vermisstenbildliste,* or a list of missing soldiers compiled from multiple sources following the war; and *Heimkehrerkartei,* or a repatriation card including for returning prisoners of war.

160. Rass, "Das Sozialprofil von Kampfverbänden," 680, 691.

161. Ibid.

162. Rass and Rohrkamp, *Deutsche Soldaten,* 197–218. The full database included 18,536 records for sixty-eight company-level organizations from the *Wehrkreis* VI, Rheinland-Westphalia region.

163. Rass and Rohrkamp, *Deutsche Soldaten,* 199. In addition, the database records the *Volksschule* at different levels of achievement, other school types and forms, and other qualifiers. The interviewers of the drafted German citizen had a very clear understanding of the intellectual accomplishments of the soldiers before selection for the infantry and other combat arms.

164. Rass, "Das Sozialprofil von Kampfverbänden," 686.

165. Römer, "Milieus in the Military," 125.

166. Ibid., 132.

167. Römer, "Milieus in the Military," 133–134; Römer, *Kameraden,* 94–95, 110.

168. Römer, "Milieus in the Military," 144–145. See also Geary, "Working-Class Identities in the Third Reich," 46. Geary argues "that working-class behavior in Germany between 1933 and 1945 cannot be understood without reference to specific context and that in any case it has never been appropriate to generalize about 'the German working class.' In fact, working-class identity in Germany had always been fragmented and had become more so as a consequence of mass and long-term unemployment before the Nazi seizure of power."

169. Römer, *Kameraden,* 408–409, 432–433, 465–466.

170. Ibid., 471.

171. Römer, "Milieus in the Military," 136.

172. Römer, *Kameraden,* 314.

173. Ibid., 478

174. Römer, *Kameraden,* 478–480; Kühne, "Gruppenkohäsion und Kameradschaftmythos," 540.

175. Shils and Janowitz, "Cohesion and Disintegration in the Wehrmacht," 281; Cooley, *Social Organization,* 23.

176. Römer, "Milieus in the Military," 136.

PART 4

Conclusion

11

The Anvil Forged

THIS WORK BEGAN WITH TWO THOUGHTS THAT ENCAPSUlated the goal of a soldier's training and the key factor that enabled him to risk his life:

> Combat is the end toward which all the manifold activities of the Army are oriented. . . . The role of the combat soldier may well be considered the most important single role for the understanding of the Army.[1]

> I hold it to be one of the simplest truths of war that the thing which enables an infantry soldier to keep going with his weapons is the near presence of a comrade. The warmth which derives from human companionship is . . . essential. . . . It is that way with any fighting man. He is sustained by his fellows primarily and his weapons secondarily.[2]

The intent here has been to determine the substance behind the perception, both during and after World War II, that the German infantryman possessed a singular advantage in tactical skill, in the opinion of his adversaries. The path to that determination led through the history of the Anglo-American and German armies in modern times, seeking evidence to illuminate the differences in the selection and preparation of their infantry.

Classification for Combat Duty and the Infantry

The principal factor demonstrated was the manner and method of classification and selection, and on this all other variables appeared dependent. Statistical evidence was sought wherever such existed to support factually, in addition to narratively, the power of this element. The principal lens for viewing this cause was the determination of the US Army Ground Forces, founded upon their statistical analysis of soldier characteristics, that the two factors most correlated in the nature of soldiers at induction were physical health (as either physicality or robustness) and intelligence (as measured by test scores). An important assumed aim of all armies was the development of cohesive fighting units that such selection supported.

From this viewpoint, the analysis of the classification systems of each army became differentiated. The single most important finding here, one that played the greatest role in delineating the characteristics of the cohesive behavior of infantrymen, was the manner of assignment of men of high physicality and high intelligence into the infantry, in particular, and into the combat arms in general. The strongest demarcation between the armies was the Anglo-American method for the disproportionate assignment of men with lower intelligence test scores and lower physical robustness to the infantry. This was the result of a classification scheme that prioritized civilian work skills and intelligence test scores above physicality. This resounded throughout the war in the concerns of the Anglo-American general officers and other leaders in the steadfastness of the infantry in the close combat required in offensive warfare to dislodge the German infantry. The Anglo-American reliance on overwhelming firepower rather than on the tactical prowess of their advancing infantry was a direct result of this selection method.

The German Army classification and assignment method resulted in the disproportionate assignment of men of the highest physicality, and its correlation of higher intelligence, to the infantry. Based first on the presumed requirements of combat on the modern battlefield rather than civilian skill, the German system did not use either intelligence tests or modern psychological/scientific methods, relying first on physical robustness. With near zero illiteracy in the German population, the multiperspective interview process combined with the doctor's determination of combat-capable physicality, assigned the largest majority of such men to the infantry. German fidelity to fielding infantrymen believed capable of withstanding the mental and physical challenge of battle provided the first large distinction between the armies. From this difference, all other constituent elements were subordinated as dependent variables in the ultimate aim of the development of cohesive fighting small units.

In social science terms, this factor of selection has causal priority over other factors in determining the success or failure of the appearance of cohesion among the men assigned to the infantry squads, platoons, and companies that bore the vast majority of casualties inflicted and received by the armies studied here. In other words, no leader, no method and manner of training, no organizational scheme, no disciplinary regimen, no political ideology, not even victory or defeat, overcame the differences, the capabilities, the quality, and the character of the infantry decided upon at selection.

Cohesion and the Primary Group

To win a war with the political object of final victory—the complete overthrow of an enemy's ability to resist—the ground forces of a nation, its army, must accomplish the tasks of destroying an enemy's physical force and its will to fight, and then occupy the territory of its homeland. Battle is the principal method to achieve these aims, and the combat soldier is the means.[3] If that which most enables the infantryman to succeed is the near presence and warmth of his fellows, then the purpose of the selection and training of the infantry is to develop the companionship of the small unit. The goal is to build his primary social group in preparation for battle. The development of the alertness, the agency,[4] of the individual within the community structure of his primary group, his sense of obligation to his fellows, and his acceptance of the norms of caring for his family in arms are the requirements that define the term *cohesion.*

The German Infantry

This study reinforces the conclusions reached by Shils and Janowitz in their 1948 article, providing an explanation for the observed resilience of German infantry soldiers in the face of overwhelming odds and defeat against the Soviet Red Army after 1942 and the Anglo-American armies in Italy and northwestern Europe in 1944–1945.

The comparison related here between these armies goes beyond what Shils and Janowitz observed. The statistical evidence for the German infantry soldier, when added to the cohesion policies of the German Army, demonstrated how the regionally recruited citizen-soldier found a home for his service in the division organization in which his time began and in which, barring its destruction, or his own death or service-ending injury, he remained through the exigencies of the war.

The key departure in this work from historical narratives of the German infantry is the delineation of the experience of the combat soldier from two points of view. The first is Christoph Rass's data from the social history of the 253rd Infantry Division in the Soviet Union in the course of the war; the second is Rass's work with René Rohrkamp in producing a statistically relevant demographic database of over 18,000 soldiers of the wartime *Wehrkreis* VI in the Rheinland-Westphalia region of Germany. Alongside these statistical data lay Felix Römer's presentation of data from the transcripts of eavesdropped conversations of German prisoners of war held at Fort Hunt near Washington, DC, between 1942 and 1945. The Römer work highlights heretofore inaccessible information on soldier attitudes in unforced, eavesdropped recordings done by US Army intelligence. Together these two sets, data and transcripts, provide a more complete picture of the German soldier from an individual and small-group perspective. This contrasts with a more typical historiographic treatment of the German soldier that relies upon an examination of the words and materials of the Nazi elites and the structures of the National Socialist state.

Römer demonstrated a powerful link between primary bonds, based on the sense of community in the small units, and a secondary cohesion focused on commitment to the ethos and norms of the military as a uniquely German institution. In this, one found confirmation of the Shils and Janowitz thesis associated to primary gratifications for solidarity, bonding, and the reward of close comradeship in the small unit. "The soldiers foremost sought to emulate their common values and norms in order to live up to their comrade's expectations, to maintain their positive self-perceptions and to gain social esteem."[5] Taken together, the statistical demographics and the soldier narratives complete a picture of the German soldier. This counters a narrative denying the intent and effect of the bonding, *Kameradschaft;* the soldiers themselves believed this to be an integral and essential part of their self-worth, as well as their best chance for survival.

Omer Bartov proposed a narrative that denied these effects. He postulated that the primary groups, if such existed, were destroyed in 1941 and never reconstituted. He stated that the principal reason for the wartime cohesion of such primary groups was based on fear of officers, fear of an overwrought and deadly military judicial system, and the brutalization of the soldiers by the intensity of the Nazi propaganda for a racial war of extermination against the Soviet Union, and overall for soldiers in all theaters.[6] The evidence presented here counters all three of these contentions.

Rass, in his study of the 253rd Infantry Division, provided evidence of the existence of a powerful social support milieu in all divisions as the home to the soldier in his small units throughout his service. The statistical evidence for the return of wounded soldiers to their original units undermines the Bartov contention that the primary groups were destroyed and never reconstituted. If more than 70 percent of wounded soldiers returned to their home units within six months of wounding, then the primary groups were in a continual process of reforming, reconstitution, and regeneration as these combat veterans reintegrated into their small units. Through the course of the war, the German Army never ceased to identify, train, and assign leaders for the small combat units of the divisions.

The narratives reported from Fort Hunt directly contradict the notion that the soldiers feared their officers, noting the exact opposite—a powerful respect for the abilities of both their officers and NCOs was the norm for the German soldier. Nowhere in the Fort Hunt transcripts is there evidenced fear of the military judicial system across any service represented during 1942–1945. Rass provides statistical evidence of the relatively limited disciplinary interactions of the German soldier beyond his commanding officer, especially in comparison to the numbers of courts-martial awarded by the British and US Armies. As Rass noted for the 253rd Infantry Division, up to 70 percent of the sentences awarded by court-martial were reduced and the soldier quickly returned to his unit. Despite the executions of German soldiers, a strong contrast with the Anglo-American armies, they did not evidence the fear that Bartov presented as evidence.

The final contention that the intensity of Nazi propaganda succeeded in cowing the soldiers into staying on the battlefield is again not present in the transcripts from Fort Hunt. Römer found that while there were many politicized soldiers, the Fort Hunt transcripts indicated that the majority were motivated by more traditional themes of nationalism and their duty as Germans and soldiers, and were not a through-and-through political army.[7] The evidence of the generally ineffective introduction of the NSFO as political officer in 1944 demonstrated the difficulty in politicizing the majority of soldiers, and was in consonance with the evidence from Fort Hunt that the soldiers believed themselves to be apolitical—supporters of the idea of Adolf Hitler as leader but ambivalent about the goals of the National Socialist state in the war.[8]

The German soldier benefited from his army's intent to create and sustain the bonds between soldiers deemed essential to his performance and survival. They were selected for their physical and mental capability to withstand the stress of the face-to-face battle of the infantry. The entire social, training, and combat milieu of the German Army found its expression in the powerful perceptions of the German soldiers of themselves, their individual prowess as combat soldiers, and their acceptance of the canons of the army. All of these led to their ability to hold together in the face of defeat through the formation of the bonds of their primary groups first, and secondarily from the army itself. The policies creating, then sustaining, cohesive behavior in the infantry small unit included the regional basis for recruiting the infantry divisions, the return of recovered wounded soldiers to their home unit, the formation of home-unit replacement formations combining both recovered wounded and trained recruits, the cohesive effect of selecting junior officers from the quality basis of the infantry soldiers, and the wearing of combat awards on the battle uniform, which instilled confidence in the soldiers of the capability and competence of the new officer. Officers selected during wartime were returned to their divisions and regiments upon completion of officer training, arriving in units where their reputation and expertise were known. In these lay the reality of the perception of German individual and small-unit tactical excellence observed by his foes.

The British Infantry

The outcomes that defined the quality of the British infantry in the course of the war surprised the leaders of the British Army. The metanarrative of World War I and its legacy of the tenacity and stamina and leadership characteristics of the small infantry unit in the mass infantry army was lost. The infantry was the destination for the vast majority of conscripted citizens, who, in their masses, provided the wide range of mental and physical characteristics that enhanced the ability of the infantry to sustain its battle capability and to find leaders. The interwar years enhanced the anxiety that the quality of the infantrymen during the Great War was no longer forthcoming. This enhanced the value and expectations for the regimental tradition.

The key lesson from World War I was the need for complete control of the manpower of the nation to ensure its best division between armed forces and industry. The regiment was lost in this. The regimental tradition and assumption, that the qualities needed in the mass army were sustained therein, did not survive the manpower shortages of the Great War. These lessons in manpower management were the foundation for the conscription of World War II. The science of selection and classification through psychological, intelligence, and mechanical aptitude testing ensured that the most capable mentally were siphoned away from combat units for the Royal Air Force, the Royal Navy, and the technical services. The British Army noted this relationship between physicality and intelligence, observing with emphasis that "it was known that the less fit men tended to be somewhat less intelligent."[9] This method denied the infantry the wide range of mental and physical characteristics that enhanced the ability of the infantry to sustain its battle capability and to find leaders as so essentially needed in 1914–1918.

In three stages the regimental tradition was systematically stripped of its assumed capability to instill the traditional, cohesive effects of its long-standing history and expectations of battle experience and leadership. In the end, the regiments were combined and subsumed in the general service corps methodology. Lost was the tradition, the metanarrative master plot that assumed that the regiment provided the cohesive bonding deemed essential for infantry close combat. The regiments did not hold their regional focus in recruiting; the army policy of returning recovered, wounded men to their original units proved difficult under the conditions of limited shipping and the continual disbanding of divisions and reassignments of weary infantrymen to other divisions. Replacements were individual, and again due to long transit times aboard ship and indifferent facilities to hold soldiers prior to their going up to the front, cohesion suffered. Officers and NCOs came from the same quality stock as the men of the infantry formations. Desertions were a significant problem for the infantry, and considered a key indicator of low morale in the infantry, with nearly 90 percent of deserters coming from their ranks. Only the US soldier was more litigated against than the British soldier. Formal discipline was pervasive. The final result of the scientific selection methodology was not conducive to creating and sustaining the cohesive behavior and bonding of the infantry small unit. The external, institutional cohesion intent was not met and strongly reduced the effect of attempts to create such bonding. The metanarrative of the regiment failed in the course of the war, and in failing surprised the leaders of the army.

The US Infantry

The American metanarrative, the master plot, that assumed native male excellence in battle, faced the same difficulties as the British regimental tradition. For the same reasons, the metanarrative failed to anticipate the effects of the scientific method of selection and classification of soldiers for the infantry small unit. The disproportionate assignment of the least–physically and mentally capable men for the infantry denied the assumptions that governed institutional understanding embedded in the history of the Revolutionary War, the Civil War, and World War I. The destination for the physically capable manpower of the nation in these prior wars was in the mass of the infantry army. The mass of qualified citizens brought with them a wide range of mental and physical characteristics that enhanced the ability of the infantry to sustain its battle capability and to find leaders. The methodology that sought first civilian skills, then intelligence, and only then physicality, denied the reality of these assumptions.

The US Army Ground Forces determined that the two characteristics that most closely correlated in the conscripted citizen were intelligence and physicality. The system the army created ensured that the mass of these most intelligent and physical men were assigned away from the combat arms in general and the infantry in particular. This denied the infantry the quality assumed in the metanarrative and diminished the institutional intent to create the opportunity to build the cohesion deemed essential in doctrine.

There was no regional or regimental intent for assignment to units. There was no intent to return recovered, wounded soldiers to the original unit; replacements arrived in combat as individuals after extensive travel and loss of physicality and training preparation. The infantry officers were selected from the same pool as the infantry soldiers. The US soldier was the most litigated against of all the armies studied here, with over 2 million courts-martial.

The final result of the scientific selection methodology was not conducive to creating and sustaining the cohesive behavior and bonding of the infantry small unit. The external, institutional cohesion intent was not met and strongly reduced the effect of attempts to create such bonding. The metanarrative assumption of a traditional American affinity for battle failed, and in failing surprised the leaders of the army, and was not overcome in the course of the war.

This study finds substantial agreement with Samuel Stouffer's hypothesis that primary-group battle cohesion of US soldiers was the result of the experience of combat itself:

> The rigid and complexly hierarchical Army organization, with its accompanying set of formal rules, was the Army's main answer to the stress and confusion of battle. The soldier was not an individual atom in the tide of warfare; he was an integral part of a vast system of discipline and coordination. The chain of command was implemented by stringent sanctions for failure to conform. Men faced combat in tightly organized formal groups, and were held in those groups by the ultimate sanctions their society wielded, including the power, almost never used, of punishment by death. Thus, the individual in combat was simultaneously guided, supported, and coerced by a framework of organization.[10]

The tenor and substance of the preceding passage might apply to the British soldiers as allies, or the German infantryman who faced these men in battle. All of these soldiers learned the experience of combat in the most stressful manner—in the life-or-death struggle of combat itself. The victory of the Anglo-American infantry was a tribute to its ability to adapt to the conditions of killing face-to-face, despite the manner and method of their selection or training.

Perception, Reality, and Consequences

It is a conclusion of this study that if combat itself and the complex hierarchy of military organizations built a form of cohesion in the Anglo-American armies, that German Army policy and practice intended to build and sustain cohesive behavior in small units in combat enhanced this concept beyond that of the Anglo-American armies. The resulting perception of the prowess of the individual and German infantry small unit was based upon the outcome of the interaction of these factors: that the German Army infantry selection had as its foremost tactical goal the development and sustainment of the close association of the members of their small infantry units in combat, the primary group, based on a combination of factors that when taken together provided the framework on which perception and reality coincided.

Did it matter that the bonding of soldiers within their primary groups to enhance their effectiveness in battle was the policy of the German Army, or that it produced strong, cohesive small infantry combat units throughout World War II? The German Army was destroyed, Germany was occupied by the armies of the Western Allies and the Soviet Union, and the will of the German people to continue the struggle ended, fulfilling all of the conditions for victory outlined by Carl von Clausewitz. Perhaps one can speculate on the course of the war if the German Army had not been so successful in putting together such battle-worthy small units. Would the war have lasted six years at the cost of the lives of over 11 million Russian soldiers, as well as more than 144,000 British Army deaths, and nearly 235,000 US Army battle deaths? The German Army destroyed over 4 million of its own soldiers in addition to the Allied losses.

The key fallacy of much historical analysis of the war was the superiority of the German technical means of war. From the beginning, the bulk of the German Army was horse-drawn and foot-mobile. The advantages of German operational and tactical superiority displayed against the British and French Armies in 1940, and in the first year of the war against the Soviet Union, were based on a very small number of mobile divisions in comparison to the totality of the army. The *Blitzkrieg* legend was mythology as a national policy of war. Rather, as a philosophy of warfare, of movement on the battlefield, the concept of *Bewegungskrieg* returned mobility and decisiveness to the armies of World War II. Once the world war commenced with the attack against the Red Army, the German Army was unable to match the mobility, manpower, and firepower of the greatest industrial powers of the mid–twentieth century in the Soviet Union, the United Kingdom, and the United States. The German Army, stripped of its air force by the Western Allies and unable to match the motor mobility of the mass of Russian divisions or the mobile British and US Armies, fought from the end of 1942 until its capitulation as a World War I army.[11]

The cohesive battle performance of the German infantry, as for all of its combat units, was the determining factor in the length and deadliness of the war. The remarkable tenacity of the German soldier was the result of the policy intent of the German Army prior to the overlay of National Socialism, not an outcome of Nazi ideology. The ability of the German Army to reconstitute cohesive battle formations time and again throughout the war demanded overwhelming mass and firepower as the price of their defeat. Destroying German soldiers in their positions with the weight of metal thrown by the Allies, both ground and air, when coupled with the Allied superabundance of motor mobility and mechanization, was the key to defeating German cohesion.

The German Army bled to death in Russia. The mistakes of that campaign destroyed the ability of the German soldier to find victory. The German general officers bear the blame for the destruction of so many men for so little gain, and in so murderous and evil a cause. They betrayed their professional training and warfighting expertise to the destruction of their soldiers and their nation.[12] To the Red Army must go the largest share of the cause of the destruction of Germany through the devastation of the German soldier on the ground. In this the Red Army owes thanks to the Anglo-American air forces for the destruction of the *Luftwaffe,* thereby destroying the ability of the German Army to conduct the modern mechanized, decisive, combined arms warfare of movement it introduced to the world against France in 1940. Only after five years of war did the Western Allies land their main effort against the German Army in northwestern Europe, and as in World War I the German Army that faced the Americans and British in northwestern Europe was worn down by years of fighting. The war against the Western Allies was the German Army's economy-of-force theater of war. Never able to match the mobility of the Allies, the loss of its air force proved decisive in reducing the German divisions of World War II into foot-mobile, horse-drawn apparitions of their World War I forebears, a fact the German general officers could not overcome.

US and British policy disproportionately provided the most physically robust and intelligent men to their naval and air forces and technical ground forces. Given the limited number of ground divisions of the Allied armies, this preference for the other services provided the basis for the key Allied contribution to victory. The concentration in the US Army on the quality of its air arm, and the British on the Royal Air Force, had the effect of destroying the German air force and hastening the reduction of all German infantry to a World War I status. Through the destruction of the German air force, the Allies brought the full weight of their firepower superiority to bear without interference. In this, the national focus of the United States and Britain on quality airmen proved both beneficial and decisive. In combination with the extraordinary weight and accuracy of Allied artillery fires, tactical and strategic airpower destroyed that portion of the German Army arrayed against them on the continent. There were never enough British or US ground forces to destroy the German Army; the Red Army was the arbiter of the ground defeat of Germany, and its war there constituted its theater of main effort; the Anglo-American air forces made possible the final destruction and defeat of Germany by the ground forces of the Allied and Soviet armies.

Judgments made following the war that the Anglo-American general officers were stolid, unimaginative, and unthinking in devotion to mass fires as opposed to their more nimble-minded, maneuverist, and technically superior opponents in the German general officers are not supported by this evidence. The Anglo-American general officers fought with full knowledge of the capabilities of the citizen-soldiers and the infantry they built. They knew the nature of the anvil they forged. The report cards provided for each infantry division on departure from the United States made certain that the training status and apparent combat capability of these divisions were known quantities. Given this knowledge, the Anglo-American general officers acted correctly to employ the key advantage they possessed, massive ground and air fires, rather than any expectation of infantry or armored maneuver to destroy their opponents. In this determination, the Anglo-American soldier was spared the butcher's bill that his German counterpart endured, regardless of selection, training, and small-unit leadership. It remained for the German generals to demonstrate their failure to adapt to the loss of their air force and the concomitant loss of their ability to conduct combined arms maneuver. If one seeks to find where the German Army excelled in World War II, it was more likely in its infantrymen than in its general officers. The manner and method of the selection of citizen-soldiers for the infantry provided perhaps a greater lesson than any given by German general officers, whose methods cost the lives of over 4 million of their own men in defeat.

The Future

The modern volunteer methods now allow for greater emphasis on quality selection for smaller armed forces, including the infantry. Combat soldiers today fight as the elite formations of World War II, with definitive tours of battle as units, followed by periods of refreshment prior to entering combat again. In this way, one of the possible lessons from this study is already in effect. The co-relation of physicality and intelligence test scores will never again have the discriminating impact upon selection as for the post–World War I generation in the effects of the worldwide Great Depression. Food today in the Western democracies is cheap and abundant. The task of parenthood is not whether or not one's child can be fed and housed and go to school, as for the interwar generation. Intelligence tests and psychological evaluation are ubiquitous as well. As long as the force remains small and volunteer, the dangers of too great a reliance on manpower quantity over quality remain low. Today's Anglo-American infantry fights with the combination of World War II German Army emphasis on quality infantry soldiers, and the Allied excellence and super-abundance of air and ground fires—the unequaled ability to deliver "thunderbolts from heaven." In combination, these factors represent the joining of the two most successful forms of warfare from the experience of ground combat during 1939–1945. A force that possesses high-quality manpower, overwhelming mobility, and massive, agile firepower, even if limited in size compared to the forces of World War II, will always be deadly to its enemies.

The King with half the East at heel is marched
from lands of morning;
Their fighters drink the rivers up, their shafts
benight the air.
And he that stands will die for nought, and home
there's no returning.
The Spartans on the sea-wet rock sat down and
combed their hair.
—A. E. Housman, *The Oracles*

Notes

1. Stouffer, *The American Soldier,* 59.
2. Marshall, *Men Against Fire,* 42.
3. Von Clausewitz, *On War,* 95, 596.

4. Emirbayer and Mische, "What Is Agency?" 962–963.

5. Römer, "Milieus in the Military," 137.

6. Bartov, *Hitler's Army,* 59

7. Römer, "Milieus in the Military," 144–145; Stargardt, *The German War,* 17–18.

8. Quinnett, "The German Army Confronts the NSFO," 62; Thamer, "Die Erosion einer Säule," 434–435.

9. Ungerson, *Personnel Selection,* 47.

10. Stouffer, *The American Soldier,* 96–97.

11. Frieser, *The Blitzkrieg Legend,* 29.

12. Craig, *The Politics of the Prussian Army,* 503.

Appendix 1: Comparative Ranks

Table A1.1 Comparative Army Ranks: Germany, the United Kingdom, and the United States

Germany	United Kingdom	United States
Grenadier/Schutze	Private	Private
Obergrenadier/Oberschutze	Private	Private
Gefreiter	Lance corporal	Private first class
Obergefreiter	Corporal	Corporal
Unteroffizier	Corporal	Sergeant
Unterfeldwebel	No equivalent	No equivalent
Feldwebel	Sergeant	Staff sergeant
Oberfeldwebel	No equivalent	Technical sergeant
Stabsfeldwebel	Staff sergeant	Master sergeant
Hauptfeldwebel	Warrant officer second class (company sergeant major); warrant officer first class (regimental sergeant major)	Sergeant major
Fahnenjunker[a]	Officer cadet	Officer candidate
Fähnrich (ensign)	No equivalent	No equivalent
Oberfähnrich (senior ensign)	No equivalent	No equivalent
Leutnant	Second lieutenant	Second lieutenant
Oberleutnant	First lieutenant	First lieutenant
Hauptmann	Captain	Captain
Major	Major	Major
Oberstleutnant	Lieutenant colonel	Lieutenant colonel
Oberst	Colonel	Colonel

continues

Table A1.1 Continued

Germany	United Kingdom	United States
Generalmajor	Brigadier	Brigadier general
Generalleutnant	Major general	Major general
General der Infanterie	Lieutenant general	Lieutenant general
Generaloberst	General	General
General-feldmarschall	Field marshal	General of the army

Sources: For Germany, Absolon, *Die Wehrmacht im Dritten Reich: Band I;* Absolon, *Die Wehrmacht im Dritten Reich: Band III;* Absolon, *Die Wehrmacht im Dritten Reich: Band V,* 157–161. For the United Kingdom, Forty, *British Army Handbook, 1939–1945,* 188–191.

Notes: All UK and US ranks are approximately equal to the listed German rank in responsibility.

a. *Fahnenjunker, Fahnenjunkergefreiter, Fahnenjunkerunteroffizier, Fähnrich,* and *Oberfähnrich* all fell under the title *Offizieranwärter* (officer aspirant), and these were all steps in the progression of an officer aspirant as he moved from his enlisted grade to finally the position of *Leutnant* or second lieutenant–equivalent.

Appendix 2: Small-Unit Organizations

Figure A2.1 British Rifle Company

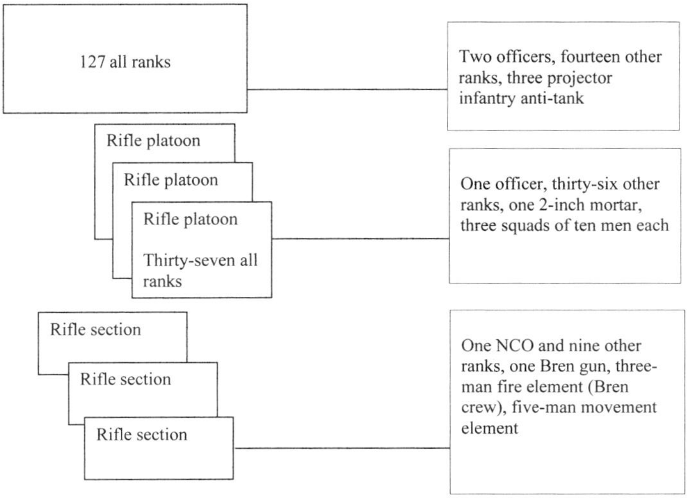

Sources: TNA PRO WO 32/10899, appendixes F and F2 to amendment no. 1 to 20/Gen/6166 (SD 4), May 1944, "Review of Army Manpower," June 7, 1944; "Reorganization of Ground Troops for Combat," *Study no. 8,* 1946, US Army Ground Forces, Ground Hist. Sec. 1942–1947, RG 337.6.1, NACP, annex 1, 71; BAMA RH 2/1331, *Kriegstärkenachweisungen (Heer)*, KStN 131c/b and 131n; Tessin, *Formationsgeschichte der Wehrmacht,* p. 151.

Figure A2.2 US Army Rifle Company

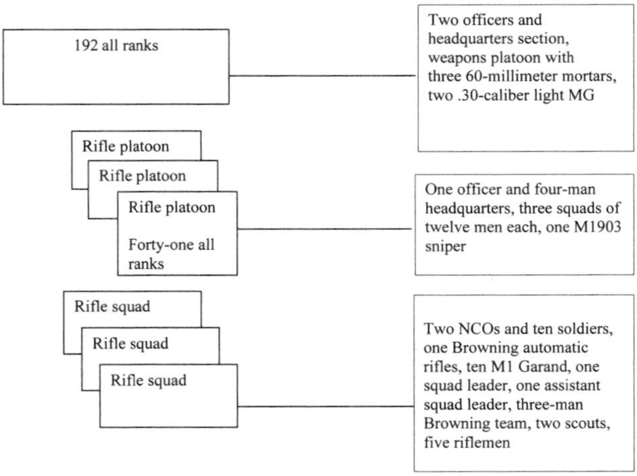

Figure A2.3 German Rifle Company

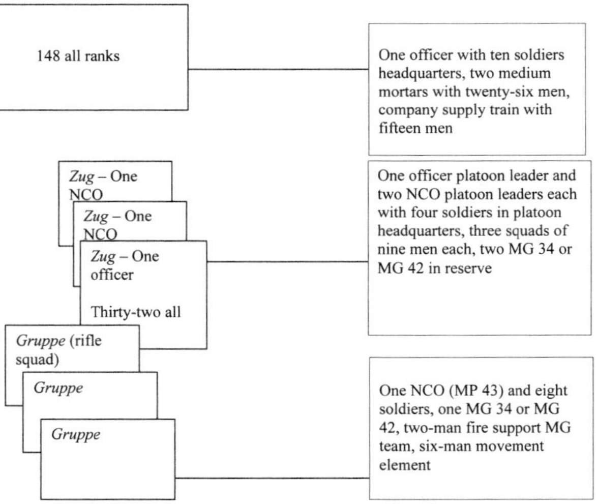

Sources: TNA PRO WO 32/10899, appendixes F and F2 to amendment no. 1 to 20/Gen/6166 (SD 4), May 1944, "Review of Army Manpower," June 7, 1944; "Reorganization of Ground Troops for Combat," *Study no. 8,* 1946, US Army Ground Forces, Ground Hist. Sec. 1942–1947, RG 337.6.1, NACP, annex 1, 71; BAMA RH 2/1331, *Kriegstärkenachweisungen (Heer)*, KStN 131c/b and 131n; Tessin, *Formationsgeschichte der Wehrmacht,* p. 151.

Appendix 3: Courts-Martial

Table A3.1 Courts-Martial: Germany, the United Kingom, and the United States

Germany	United Kingdom	United States
Wehrmachtsgerichtbarkeit 1934[a]	Army and Air Force Acts (annual) 1939–1945[b]	Articles of War 1920[c]
Kriegsgericht and ***Feldkriegsgericht:*** **Convening authority:** Division commander and above and the *Wehrkreis* commanders. **(Equivalent to a general court-martial)** At least three officers presiding. If the sentence awarded contemplated death, or more than fifteen years or life in prison, then five officers were required, authorized to award death and penal servitude. Authorized to try any person subject to military law including officers. (*Feldkriegsgericht* referred to court-martial proceedings held in the field of operations rather than a unit location in the homeland.)[d] ***Oberkriegsgericht:*** **Convening authority:** Division commander and above and the *Wehrkreis* commanders. **(Equivalent to a general court-martial and appeal authority for the *Kriegsgericht*)** Five officer judges presiding. Authorized to try any person subject to military law including officers. (This court went out of use after 1940.)[e]	**General court-martial convening authority:** The king or general officer commanding-in-chief of a command or field army. At least five officers presiding, authorized to award death, penal servitude, and imprisonment. **Field general court-martial convening authority:** General officer commanding a division or higher formation. At least three officers presiding, authorized to award death, penal servitude, and imprisonment, but punishment award authority diminished if only two officers presiding. **Punishments:** • **Officers:** Penal servitude for a term not less than three years; imprisonment, with or without hard labor, for a term not exceeding two years; cashiering; dismissal; forfeiture of seniority in rank; severe reprimand or reprimand; stoppages. • **Enlisted** (including warrant officers): Penal servitude for a term not less than three years; imprisonment, with or without hard labor, for a term not exceeding two years; detention for a term not exceeding two years; discharge with ignominy. • **Noncommissioned officers:** Reduction to the ranks or to a lower grade, or forfeiture of seniority in rank; severe reprimand or reprimand;	**General court-martial** (Articles 8 and 12): **Convening authority:** The president of the United States, commanding officer of a territorial division or department, superintendent of a military academy, or commanding officer of an army, army corps, division, or separate brigade. No less than five officers may try any person subject to military law for any crime or offense made punishable by the Articles of War, including capital offenses.

continues

Table A3.1 Continued

Germany	United Kingdom	United States
	forfeitures, fines, and stoppages; field punishment (if on active service); forfeiture of all ordinary pay for a period not exceeding three months (if on active service).	
Reichskriegsgericht:	No equivalent.	No equivalent.
Convening authority: *Der Führer* and *Reichskanzler,* or *Reich* Senate president in Berlin with general or admiral with authority of a commanding general, authorized to award death or penal servitude to life. Five judges presiding, including one general or admiral appointed by the Senate. Authorized to try any person subject to military law including officers. (Appeal authority for the *Oberkriegsgericht* and in wartime the *Kriegsgericht* and *Feldkriegsgericht.*)		
Fliegenden Standgericht (Flying Court). Instituted March 9, 1945, by Hitler. Allowed for summary execution of soldiers without appeal if found unwounded away from their unit without proper authority.[f]	No equivalent.	No equivalent.
Civil offenses: As in the case of the British Army, but with larger specificity. Military offenses were strongly segregated from civil offenses, and the offenses reserved to prosecution by civil authority were far more extensive than those pertaining to the British Army, but included the offenses listed here for the United Kingdom and an entire taxonomy of civil offenses besides. All cases of treason, however, charged to a member of the *Wehrmacht* remained with the military court system. The *Kriminalstatistik*, or criminal statistics, maintained until February 1945, listed no	**Civil offenses:** Laws of the United Kingdom sharply differentiated between civil offenses and military offenses, with only a few civil offenses specifically reserved for civil court jurisdiction. But no court-martial could try the offenses of treason, murder, manslaughter, treason-felony, or rape committed in the United Kingdom. Nor could a court-martial try those five excepted offenses if committed in any place other than the United Kingdom and Gibraltar, unless the offense was committed when the offender was	No equivalent restrictions to German or British law for active service persons. The army tried and executed 142 persons during the war—106 for

continues

Table A3.1 Continued

Germany	United Kingdom	United States
crimes of murder, for example, among other civil offenses.	on active service, or the place where the offense was committed was more than a hundred miles distant from a city or town that had a civil court of competent jurisdiction.	murder, 35 for rape, and 1 for the military offense of desertion.[g]
No equivalent.	**District court-martial:**	No equivalent.
	Convening authority: Not usually below the rank of brigadier. At least three officers presiding. Not authorized to try officers or to award more than two years' imprisonment with or without hard labor.	

Sources and notes: a. Altrichter, *Der Reserveoffizier*; Absolon, *Das Wehrmachtsrafrecht im 2*, xiii–xvi. In wartime the *Kriegssonderstrafrechtverordnung* KSSVO 1938 (Special Wartime Criminal Law) and *Militärstrafgesetzbuchs* MStGB 1940 (Military Criminal Statutes), and *Wehrmachtgerichtsbarkeit* 1934 (Armed Forces Criminal Courts Jurisdiction), were the principal governing documents for the *Wehrmacht,* and these received occasional updates during the war, becoming especially more strident as the defeat became more certain for Germany and as a result of the attempt on Hitler's life on July 20, 1944. In addition the references for these regulations included *Das Wehrmachtstrafgeseztbuch für das Deutche Reich (W.St.G.B.) für soldatische Vergehen und Verbrechen,* and *Das Strafgesetzbuch für das Deutsche Reich (R.St.G.B.) für bürgerliche Verbrechen, Vergehen und für Übertretungen: W.St.G.B.* for soldier offenses against military law and *R.St.G.B.* for soldier offenses against civil law, ensuring that the soldier was subject to both, and providing a bright dividing line between military and civil offenses for which a soldier could or could not be tried, and which had jurisdiction.

b. Secretary of State for War, *Report of the Army and Air Force Courts-Martial Committee 1946,* 9–14, 36.

c. War Department, *The Articles of War,* 4–6.

d. Absolon, *Das Wehrmachtsrafrecht im 2,* 217–218. Regimental commanders could be authorized to conduct speedy trials in special cases, especially for the most serious crimes, provided permission from the division commander. The Rass/Rohrkamp database records no instances of regimental courts in the 605 court-martial cases presented; see Rass and Rohrkamp, *Deutsche Soldaten, 1939–1945.*

e. Absolon, *Das Wehrmachtstrafrecht im 2,* xvi.

f. Absolon, *Die Wehrmacht im Dritten Reich: Band VI,* 604; Absolon provided an example of 21 soldiers found guilty of desertion at a *Wehrkreis Kreigsgericht* and awarded death in Hamburg under this directive, by order of the *Wehrkreis X* commanding general Wetzel on March 27, 1945. Wüllner, *Die Ns-Militärjustiz Und Das Elend Der Geschichtsschreibung,* 852; Wüllner provided evidence of 122 soldiers awarded death in Berlin during early 1945, at least 58 under this directive. None of these awarded punishments appear in the archival source BAMA RH 14 55 Kriminalität im Heere, Beschleunigte Auswertung, Personen, Ag HR Wes/HR (IIb), *Jahresübersicht über die verurteilen Soldaten und Wermachtbeamten des Heeres,* 1945, 7–12, which recorded death sentences and execution only through the end of January 1945 and partially in February 1945 before the directive issued by Hitler on March 9, 1945. Given the manpower situation at the end of the war in Germany, the award of a death sentence did not necessarily mean a soldier would be executed. The percentage of executions against death sentences declined dramatically as the war came to its end. The award of harsh sentences was generally followed by returning the soldier to the front, where death was a very strong possibility and an example could be presented to the soldiers while maintaining in small measure the fighting manpower at the front.

g. Willis, "The United States Court of Military Appeals," 40.

Table A3.2 Sample General Court-Martial Offenses Tried, US Army, European Theater of Operations, July 1942–May 1945

Total general courts-martial cases tried, European theater	10,672
Total all theaters	66,993
Total individuals accused, officer/enlisted, European theater	1,013/11,106
AWOL (average sentence: 15 years)	3,857 cases 5,834 absences
Desertion (average sentence: 20 years)	1,963
Misbehavior before the enemy (average sentence: 35 years)	494
Misbehavior of a sentinel/post (average sentence: 7 years)	935
Willful disobedience (combat), striking an officer, drawing weapon against an officer, offering violence to an officer (average sentence: 15 years)	1,424
Theft/larceny (average sentence: maximum)	1,191
Murder (average sentence: 8 years) (35 received death, 48 life in prison)	290
Rape (average sentence: 8 years) (43 received death, 14 life in prison)	169
Mutiny (average sentence: maximum)	25

Source: General Board, European Theater, *Military Justice in Theater of Operations,* appendix 1, 59–61.

Table A3.3 Total Court-Martial Convictions, British Army (Other Ranks), September 1, 1939–August 31, 1945

Offense	1939–1940	1940–1941	1941–1942	1942–1943	1943–1944	1944–1945	Total
Desertion	552	2,724	4,322	5,113	5,719	12,310	30,740
Cowardice	0	5	26	24	41	71	167
AWOL	1,488	12,358	15,091	13,936	14,167	18,117	75,157
Disobedience, resisting escort, striking or violence to superior, threatening or insubordinate language, ill-treating soldier	1,653	3,777	3,742	4,492	5,409	5,728	24,801
Fraud, losing by neglect, injuring property, falsifying official document, enlisting after discharge with ignominy, false answer on attestation, fraudulent enlistment	826	5,553	5,711	1,951	1,366	1,127	16,543
Theft	639	2,186	3,230	4,166	3,842	4,536	18,599
Self-inflicted wound	22	10	19	8	40	238	337
Sleeping or leaving while on post, indecency, drunkenness, allowing to escape, escaping	1,263	2,860	1,974	2,277	2,123	2,699	13,196

continues

Table A3.3 Continued

Offense	1939–1940	1940–1941	1941–1942	1942–1943	1943–1944	1944–1945	Total
Mutiny	72	4	79	61	344	240	800
Offense against inhabitants (572 total), miscellaneous offenses military and civil	1,168	3,927	5,003	6,204	6,321	6,075	29,698
Total	7,683	33,404	39,197	38,232	39,372	52,141	210,029

Sources: TNA PRO WO 277/7; McPherson, *Discipline,* appendix I(a).

Table A3.4 Total Court-Martial Convictions, German Army (*Heer*), August 29, 1939–December 31, 1944 (summary by year for soldiers and civilians)

Offense	1939–1940	1941	1942	1943	1944	Total
Desertion	720	1,067	2,394	2,931	5,783	12,895
Cowardice	180	246	325	439	1,055	2,245
AWOL	7,632	11,106	13,795	16,310	17,145	65,988
Disobedience, refusal to obey, making threats, resisting orders, assault/battery, mutiny, revolt	4,116	6,483	6,628	8,448	11,121	37,396
Plundering, looting, profiteering	2,712	2,041	2,068	2,206	4,312	13,339
Military theft, misappropriation	7,464	12,144	25,858	26,005	21,936	93,407
Misuse of authority of one's office, assaulting a subordinate	720	1,150	1,061	1,385	1,097	5,413
Sleeping on watch, failure to keep proper watch	744	2,293	7,349	8,140	9,073	27,599
Undermining fighting power	600	912	2,057	3,340	5,677	12,586
Miscellaneous	18,072	25,525	27,428	34,223	39,566	144,814
Total	42,960	62,967	88,963	103,327	126,826	415,742[a]

Source: BAMA RH 14/55, Kriminalität im Heere, Beschleunigte Auswertung, Personen, Ag HR Wes/HR (II 2b), *Jahresübersicht über die verurteilten Soldaten und Wehrmachtbeamten des Heeres,* 1945, 3.

Notes: The title for civilians, under military legal disciplinary status, was *Wehrmachtbeamten*. They generally held officer or NCO rank for pay purposes, and were not in responsibility or authority over soldiers, nor in administrative and technical positions ranging from secretary to armament artificer to school master to forester to restaurant manager to master mechanic, and many more. The report does not distinguish between those convicted as soldiers or civilians. See Absolon, *Die Wehrmacht im Dritten Reich: Band V,* 206–222.

a. The archival sources do not include any indication of the severity of the offenses nor indications of the sentences and the length of time or commutations or reductions of same. Evidence for both exists in the Rass/Rohrkamp database within the records of the 2,456 individuals with disciplinary histories. See Rass and Rohrkamp, *Deutsche Soldaten, 1939–1945*; Rass, "*Menschenmaterial,*" 276–307.

Appendix 4: Non-Court-Martial Punishments for Commanding Officers

Table A4.1 US Army Commanding Officers, Non-Court-Martial Punishments

US Army: Article 104
"The commanding officer of any detachment, company, or higher command may, for minor offenses, impose disciplinary punishments upon persons of his command without the intervention of a court-martial, unless the accused demands trial by courts-martial."

Officers
"Except that in time of war or grave public emergency a commanding officer of the grade of brigadier general or higher grade may, under the provisions of this article, also impose upon an officer of his command below the grade of major a forfeiture of not more than one-half of such officer's monthly pay for one month."

Enlisted
- Admonition
- Withholding of privileges not exceeding one week
- Extra fatigue not exceeding one week
- Shall not include forfeiture of pay or confinement under guard

Source: US Department of War, *The Articles of War.*

Table A4.2 British Army Commanding Officers, Non-Court-Martial Punishments

By Commanding Officers and Detachment Commanders (such as regimental or battalion commanders)					By Company Commanders	
To Officers[a]	Summary Punishments to Noncommissioned Officers	Summary Punishments to Private Soliders	Minor Punishments to Noncomissioned Officers	Minor Punishments to Private Soldiers	To Noncommissioned Officers Below the Rank of Sergeant	To Private Soliders
Forfeiture of seniority in rank Severe reprimand or reprimand Deductions from ordinary pay	Deductions from ordinary pay	Detention not exceeding 28 days Fine (for drunkenness only) not exceeding 28 days; forfeiture of all ordinary pay (on active service only) for a period not exceeding 28 days	Severe reprimand or reprimand Admonition	Confinement to barracks (or camp) for a period not exceeding 14 days Extra guards and piquets for minor offenses or irregularities when on, or parading for, these duties Admonition	Reprimand Admonition	Seven days confinment to barracks/camp Extra guards or piquets Fines for drunkenness Admonition

Source: Secretary of State for War, *Report of the Army and Air Force Courts-Martial Committee, 1946*, 36–38.

Note: a. Only in cases where authority existed to punish officers below the rank of lieutenant colonel under Section 47 of the Army and Air Force Acts (annual).

Table A4.3 German Army Commanding Officers, Non-Court-Martial Punishments

Commanding Officer Position	Punishments Against: Officers	NCO with Swordknot	NCO Without Swordknot	Soldiers
Company, squadron battery, commanders	Simple reprimand Strict reprimand	Simple reprimand[a] Strict reprimand[b] Restrction to quarters up to two weeks[c] Mild arrest up to two weeks[d]	Simple reprimand Strict reprimand Extra duties Gate restriction returning to barracks before, at, or after taps up to four weeks Restriction to barracks or quarters up to two weeks Mild arrest up to two weeks Severe arrest up to ten days[e]	Reprimand Performance of duty outside the norm, including disciplinary physical exercise[f] Gate restriction returning to barracks before, at, or after taps up to four weeks Restriction to barracks or quarters up to two weeks Mild arrest up to two weeks Severe arrest up to ten days
Battalion commanding officers	As above including: Restriction to quarters up to five days	As above including: Restriction to quarters up to four weeks	As above including: Restriction to barracks or quarters up to four weeks Mild arrest up to four weeks Severe arrest up to three weeks	As above including: Restriction to barracks or quarters up to four weeks Mild arrest up to four weeks Severe arrest up to three weeks

continues

Table A4.3 Continued

	Punishments Against			
Commanding Officer Position	Officers	NCO with Swordknot	NCO without Swordknot	Soldiers
Regiment commanding officers	As above including: including: Restriction to quarters up to ten days	As above	As above	As above including: Reduction by one grade in rank[g]
Division commanding officers and above	As above including: Restriction to quarters up to four weeks	As above	As above	As above including: Reduction by more than one grade in rank

Source: Source: Altrichter, Der Reserveoffizier, 78–79, 82–83.

Notes: a. Against soldiers a simple reprimand was awarded by publishing in front of at least three members of the unit and in the order of the day. Against NCOs awarded through announcement in the order of the day and in the presence of a senior NCO.

b. Against NCOs a severe reprimand was published through announcement in front of the assembled NCO corps of the company.

c. Barracks or quarters arrest held the offender to his place of duty, but he was not allowed to leave his quarters building after working hours, including to any adjacent courtyard, and could not visit a canteen or friend's quarters.

d. Mild arrest was awarded as solitary confinement. Meals were taken with the unit. Allowed a firm bed, books, and writing materials, but no tobacco or alcohol. Exercise was permitted after one week under supervision.

e. Severe arrest was awarded in solitary confinement with a hard bed, bread, and water. These hard conditions stopped on the fourth day and then every third day. Books and outside exercise permitted only on good days, and use of writing material was denied in the worst cases.

f. Disciplinary exercise could only be given to soldiers with less than four years of service, could not be more than one-half hour and not excessive, and must be supervised by an officer.

g. Grade reductions could also include the withholding of pay for up to two months, but the pay was returned upon completion of the disciplinary period.

Appendix 5: Casualties

Table A5.1 German and Allied Killed in Action, September 1939–May 1945

Nation	Number Killed (including deaths by wounds)
Germany (army only, all theaters)	4,202,030[a]
United States (ground army, all theaters with marine corps ground and air)[b]	201,324[c]
United Kingdom (army, all theaters)	144,079[d]
Soviet Union (army and air force, includes missing and captured soldiers presumed dead)	11,285,057[e]

Sources and notes: a. Overmans, *Deutsche militärische Verluste im Zweiten Weltkrieg,* 335.

b. US Army Air Corps battle deaths (52,173) and US Navy (36,950) and Coast Guard (574) deaths not included in order to focus on ground casualties only through separation from other losses, where possible.

c. US Army Command and General Staff College (USACGSC) Library, Fort Leavenworth, Kansas, *Army Battle Casualties and Non-Battle Deaths in World War II, Final Report, 7 December 1941–31 December 1946,* Statistical and Accounting Branch, Office of the Adjutant General, War Department, June 1, 1953, 5; History Division Files, *Casualties,* Marine Corps University, Quantico, Virginia, Letter (AG-1309-ec) to Mr. John Kiernan, *Information Please Almanac,* from Colonel O. H. Wheeler, Assistant Director, Division of Public Information, November 7, 1947.

d. TNA PRO WO 277/12; Pigott, *Manpower Problems,* appendix E, 81.

e. Frieser et al., *Die Ostfront 1943/44,* 204. This number represents the official Soviet tabulation of irreversible military losses during World War II.

Table A5.2 British Army Casualties (less the Dominions), 1914–1918 and September 3, 1939–April 30, 1945 (all theaters)

	1914–1918			September 3, 1939–April 30, 1945		
Nature of Casualty	Officers	Other Ranks	All Ranks	Officers	Other Ranks	All Ranks
Killed (including deaths of wounds)	38,834	665,969	704,803	10,773	133,306	144,079
Wounded	81,178	1,583,772	1,664,950	17,749	221,826	239,575
Prisoners of war	6,226	150,811	157,037	8,096	143,980	152,076
Missing (including prisoners of war)	—	—	—	1,585	32,186	33,771
Total	126,238	2,400,552	2,526,790	38,203	531,298	569,501

Sources: TNA PRO WO 277/12; Pigott, *Manpower Problems,* appendix E, 81.

Table A5.3 US Armed Forces Casualties, April 1, 1917–December 31, 1918, and December 1, 1941–December 31, 1946

	Branch of Service	Number Serving	Total Deaths	Battle Deaths	Other Deaths	Wounds (not mortal)
World War I (1917–1918)[a]	Army	4,057,101	106,378	50,510	55,868	193,663
	Navy	599,051	7,287	431	6,856	819
	Marines	78,839	2,851	2,461	390	9,520
	Total	4,734,991	116,516	53,402	63,114	204,002
World War II (1941–1946)[b]	Army[c]	11,260,000	318,274	234,874	83,400	565,861
	Navy	4,183,466	62,614	36,950	25,664	37,778
	Marines	669,100	24,511	19,733	4,778	68,207
	Total	16,112,566	405,399	291,557	113,842	671,846

Source: Congressional Research Service, American War and Military Operations Casualties, "Lists and Statistics," updated May 14, 2008, 2–4.

Notes: a. Includes air service. Battle deaths and wounds not mortal include casualties suffered by US forces in northern Russia to August 25, 1919, and in Siberia to April 1, 1920. Other deaths cover the period April 1, 1917, to December 31, 1918.

b. Data are for the period December 1, 1941, through December 31, 1946, when hostilities were officially terminated by presidential proclamation, but a few battle deaths or wounds not mortal were incurred after the Japanese acceptance of the Allied peace terms on August 14, 1945. Numbers serving from December 1, 1941, through August 31, 1945, were: army, 10,420,000; navy, 3,883,520; marines, 599,693; total, 14,903,213.

c. Includes US Army Air Forces.

Table A5.4 US Army Battle Casualties by Type and Duty Branch, December 7, 1941–December 31, 1946

Type of Battle Casualty and Disposition	Total	Air Corps (including flight officers)	Other Branches Total	Infantry	Field Artillery	Corps of Engineers	All Other
Total	936,259	115,382	820,827	661,059	42,692	29,806	87,320
Deaths among battle casualties	234,874	52,173	182,701	142,962	9,585	7,691	22,463
Killed in action	192,798	45,520	147,278	118,376	6,979	5,920	16,003
Died of wounds and injuries	26,762	1,140	25,622	19,799	1,701	1,149	2,973
Declared dead	6,058	3,603	2,455	1,735	176	137	407
Died of other causes (nonbattle)	9,256	1,910	7,346	3,052	729	485	3,080
Other battle casualties[a]	701,385	63,209	638,176	518,097	33,107	22,115	64,857

Source: US Army Command and General Staff College (USACGSC) Library, Fort Leavenworth, Kansas, *Army Battle Casualties and Non-Battle Deaths in World War II, Final Report, 7 December 1941–31 December 1946,* Statistical and Accounting Branch, Office of the Adjutant General, War Department, June 1, 1953, 5.

Note: a. Includes wounded returned to duty overseas, returned prisoners of war, and wounded discharged from service.

Table A5.5 German Army Deaths on the Eastern Front, June 1941–December 1944

1941	302,000
1942	507,000
1943	701,000
1944	1,233,000
Total	2,743,000

Source: Adapted from Overmans, *Deutsche militärische Verluste im Zweiten Weltkrieg,* 277, tab. 59.

Table A5.6 German Personnel Deaths by Wartime Campaign Area, 1939–1945

	Period	Deaths	Percentage of Total Deaths
Balkans	to December 1944	104,000	2.0
Italian and African campaigns	to December 1944	151,000	2.8
Homeland (including navy)	to December 1944	291,000	5.5
Western Front	to December 1944	340,000	6.4
Prisoner-of-war deaths	Postwar reconciliation	459,000	8.6
Eastern Front	June 1941–December 1944	2,743,000	51.6
Endkämpfe (all fronts/all forces)	January 1945–May 1945	1,230,000	23.1
Total		5,318,000	100.0

Source: Adapted from Overmans, *Deutsche militärische Verluste im Zweiten Weltkrieg,* 265, tab. 52.

Table A5.7 German *Wehrmacht* (and other formations) Deaths by Organization

Organization	Deaths	Percentage of Total Deaths
Heer (army)	4,202,000	79.0
Luftwaffe (air force)	433,000	8.1
Kriegsmarine (navy)	138,000	2.6
Other members	53,000	1.0
Total *Wehrmacht* (armed forces)	4,826,000	90.7
Waffen-SS	314,000	5.9
All others (including *Volkssturm* and *Polizei*)	178,000	3.4
Total all other organizations	492,000	9.3
Total	5,318,000	100.0

Source: Adapted from Overmans, *Deutsche militärische Verluste im Zweiten Weltkrieg,* 255, tab. 46.

Acronyms

AAC	Army Air Corps (United States)
AAF	Army Air Forces (United States)
AGCT	Army General Classification Test (United States)
AGF	Army Ground Forces (United States)
ASF	Army Service Forces (United States)
ASTP	Army Specialized Training Program (United States)
AWOL	absence without leave
CCC	Civilian Conservation Corps (United States)
CMTC	Civilian Military Training Camp (United States)
G-1	Staff Section for Personnel (United States)
HQ	headquarters
NCO	noncommissioned officer (United Kingdom and United States)
NFSO	National Socalist Leadership Officers (*Nationalsozialistische Führungsoffizier*) (Germany)
ROTC	Reserve Officer Training Corps (United States)
SHAEF	Supreme Headquarters Allied Expeditionary Force (United States and United Kingdom)
SS	*Schutzstaffel* (Shield Squadron) (Germany)

Glossary

Abitur, das. Graduation certificate from a *Gymnasium*-level school; equivalent to an international baccalaureate, combining the equivalent of high school graduation and college entrance.

Abteilung, die. Detachment; indicated an organization less than a battalion and more than a company, generally.

Arbeitsverwendungsfähig. Physical condition allowing labor service in the homeland, but not as a trainer or in watch units.

Arbeitsverwendungsunfähig. Physical condition requiring mustering out of the service.

Armeekorps, das. Regional army corps–level command (usually responsible for several divisions).

Auffrischung, die. Refreshing; for divisions in the field in terms of men, equipment, and training as time out of the line of battle.

British Army soldier slang terms provided by University of North Carolina State Professor Emeritus of History Charles H. Carlton.

Auftragstaktik, die. Term relating to a method of leadership demanding maximum initiative on the part of subordinate leaders, often translated as "mission orders."

Aushebung, die. Enlistment for active service.

Bataillon, das. Battalion.

Bazooka. Slang term for a US Army rocket-propelled, hand-held, anti-tank weapon.

Bewährungsbataillon, das. Probationary punishment battalion.

Bewegungskrieg, das. War of movement.

Blitzkrieg. American press–created term to describe the "lightning war" German victories between 1939 and 1941.

Chef des Heeresrüstung und Befehlshaber des Ersatzheeres. Chief of army armaments and commander in chief of the replacement army.

Deutscher Amateur-Sende und Empfangdienst, der. Radio and telecommunications school.

Einberufung, die. Enlistment for active service.

Endkämpfe, die. The final battles in the destruction and occupation of Germany, roughly from December 1944 to May 1945.

Ergänzungseinheiten, die. Members of the "white years," or the year-groups following the last-conscripted year-group of World War I until the reinstatement of the draft.

Ersatzheer, das. Replacement army, also known as the replacement and training army.

Essential men. Deferred labor for industrial or agricultural occupations in the United States.

Fechtende Truppe, die. Soldiers requiring fitness level as combat-capable, of which there were eight combat specialties.

Feldheer, das. Field army.

Feldsonderabteilung, die. Special field punishment unit.

Fizzer. Slang term in the United Kingdom for a disciplinary charge.

Freikorps, das. Nontraditional army units formed from November 1918 outside the Versailles Treaty provisions to put down unrest in the period 1918–1923.

Frontgemeinschaft, die. The community of the front; describes the notion of common support in common danger of the frontline soldier.

Führergehilfe, die. Title given to members of the Great General Staff to hide their duties in the Troop Office.

Führerheer, das. Army of leaders, from General Hans von Seeckt.

Führung und Gefecht der verbundenen Waffen, das. Combined arms leadership and battle, as the tactical precepts of the German Army, 1921–1933.

garnisonverwendungsfähig Feld. Physical condition allowing service with combat divisions but not in duties requiring combat.

garnisonverwendungsfähig Heimat. Physical condition allowing service only with homeland-stationed units.

Gebirgsdivision, die. Mountain divisions.

Gebührniskarte, die. Personal papers carried by the soldier. Similar to the *Soldbuch.*

Gefechtsstärke, die. Battle strength report only for combat divisions and below, including both combat and support troops as an indicator of unit ability to fight and sustain itself in battle.

Geheimen Polizei, die. Army security police, for areas behind the divisions and higher security duties.

Gestapo, die. Secret state police, or *Geheime Staatspolizei.*

Gesundheitsbuch, das. Personal medical record delivered to the unit upon joining, and containing all relevant health information.

Grabenstärke, die. "Trench strength," an unofficial strength report as the true measure of the actual number of men in the fighting line.

Gruppe, die. Squad.

Gruppenkommando, das. Regional army–level command with responsibility for two or more *Armeekorps,* 1920–1938.

Hauptkampflinie. Front line or line of contact between friendly and enemy forces.

Heer, das. Army.

Heeresdienstvorschrift Truppenführung, das. Army regulations for troop-leading or unit command, as the tactical precepts of the German Army, 1933–1945.

Heeresleitung, die. Army administrative headquarters, Weimar Republic.

Hilfswillige, die. Foreign troops, primarily former Russian prisoners of war, serving in an infantry division as part of its reported strength.

Infanterie, die. Infantry.

Infanteriedivision (neuer Art). The 1943 reorganization to reduce the strength of the infantry divisions by the equivalent of one regiment of infantry.

Iststärke, die. Actual strength of a unit on-hand as opposed to authorized by *Kriegsstärkenachweisungen.*.

Jäger Division, der. Form of light infantry division.

Jenkins. Slang in the United Kingdom for a minor punishment fatigue; usually given while already confined to barracks.

Kaiserheer, das. Army of the German Empire, which ended with the Versailles Treaty and became the *Reichsheer* of the Weimar Republic.

Kampfstärke, die. Frontline strength report for regiment and below including combat support soldiers and, for mechanized units, repair soldiers.

Kaserne, die. Barracks.

Kavallerie, die. Cavalry.

Korpsabteilung, die. Division-equivalent organization created ad hoc on the battlefield from disparate units of divisions surrounded or destroyed on the Eastern Front.

Kreis, der. County-level equivalent geographic and political boundary within a state (*das Land*).

Kriegsmarine, die. Navy.

Kriegsstärkenachweisung. Equivalent of the US tables of organization and equipment and the British War Establishment.

Kriegstagebuch, das. Unit war diary.

Kriegsverwendungsfähig. Doctor's examination determination of an individual as possessing the requisite physical and mental stamina for combat duty in a combat organization.

Lame-Jack. Slang in the United Kingdom for "lance corporal" (one chevron; an appointment not a rank).

Land, das. State-level equivalent geographic and political boundary.

Landeschützen, die. Home guard units for men beyond active service age.

Landser, der. Common term for a German soldier, World War II.

Landsturm, der. Nation in arms.

Landwehr, die. Militia.

Loot. Slang in the United Kingdom; "one pip loot" was equivalent of a second lieutenant; officers below the rank of captain were always referred to as "mister."

Luftwaffe, die. Air force.

Machtergreifung, die. The Nazi seizure of power through legal means.

Marineleitung, die. Navy administrative headquarters, Weimar Republic.

Nachschubtruppen, die. Supply and support services soldiers at corps and higher levels, generally.

Nationalsozialistische Deutsche Arbeiterpartei (Nazi), die. German National Socialist Workers Party.

Nationalsozialistische Fliegerkorps, das. Flying school.

Nationalsozialistische Kraftfahrkorps, das. School for driving and mechanical skills.

Oberbefehlshaber des Heeres, der. Commander in chief of the army.

Oberkommando der Wehrmacht, die. Supreme Command of the German Armed Forces.

Oberkommando des Heeres, das. Highest command of the army.

Oberste Befehlshaber der Wehrmacht, der. Commander-in-chief of the German armed forces, exercised by Adolf Hitler.

Ostfront, die. The Eastern Front.

Ostheer. The Eastern Army, 1941–1945.

Packet. Slang in the United Kingdom for a wound; "to cap a packet" meant to be wounded, often fatally.

Panzer Grenadier, der. Mechanized infantry, as divisions, or used to describe the soldier.

Panzer, der. Armored vehicle.

Panzerfaust, die. Short-range, hand-held, anti-tank rocket with a shaped charge.

Panzertruppen, die. Armored vehicle soldiers.

Professor. Slang in the United Kingdom for a senior army officer.

RAD, der. Labor office into which all conscripted men served, normally for six months, prior to enlistment for active service; *Reichsarbeitsdienst.*

Reichsheer, das. Army of the *Reich,* Weimar Republic, which ended in 1934, becoming simply the *Heer,* or army.

Reichswehr, die. Armed forces of the Weimar Republic; army and navy.

Sarnt. Slang in the United Kingdom for a sergeant ("sarge" was never used).

Schattendivision, die. "Shadow" division; formed as stationary divisions in the homeland during 1944 and converted to *Volksgrenadier* divisions.

Schnellen Truppen, die. "Fast troops": Panzer, mechanized (grenadier), and motorized.

Schwerpunkt, der. Term for the center of gravity, or key point of an enemy force or location, as the objective of the main effort in a tactical operation.

Sicherungstruppen, die. Security soldiers for duty in the rear areas at army corps and above levels, generally.

Skijägerdivision, die. Ski division.

Soldbuch, das. Personal papers always carried by the soldier to his unit and indicated all of his personal military and basic medical, information including equipment, uniforms, and weapons issued, as well as pay status; a document called a *Gebührniskarte* also served this purpose.

Sollstärke, die. Table of organization or authorized strength.

Sonderabteilungen. Special punishment and training units of the replacement army.

Squaddie. Slang in the United Kingdom for a soldier.

Stellungsdivision, die. Fortress division.

Stoßtrupps, die. Specialized assault units (1918); reconnaissance and raiding patrols (1939–1945).

Stunt. Slang in the United Kingdom for any performance of outstanding service or effectiveness.

Tagesstärke, die. Daily strength report, less foreign troops, men assigned and en route, and those in hospital for at least eight weeks.

Tauglich. Adjective to delineate fitness level for combat service.

Troop basis. Future planning document for the structure of the US Army during World War II, published as manpower availability and strategy changes occurred

Truppenamt, das. Troop Office, the location of the Great General Staff, hidden in contravention of the Versailles Treaty.

Unabkömmlichsstellung, die. Men deferred from the military for industrial or agricultural labor.

Unteroffizier, der. Junior noncommissioned officer ranks and specifically the US Army equivalent of sergeant and the British Army equivalent of corporal.

Versorgungstruppen, die. Support services soldiers from division through battalion level, generally.

Volksdeutsche, die. Ethnic Germans.

Volksgemeinschaft, die. The community of the whole people, a description of support for the larger group, friends, neighbors, nation; in Nazi political usage the term was applied racially to mean only Germans.

Volksgrenadierdivision. Newly reconstructed divisions formed after the July 20, 1944, attempt on Hitler's life.

Waffen-SS, die. Political army of Heinrich Himmler's SS.

Wehrbezirkskommando, das. Second-level subregional command beneath the *Wehrersatzinspektion* and over as many as five recruiting offices at the local level, or *Wehrmeldeamten.*

Wehrersatzinspektion, die. First-level subregional command beneath the *Wehrkreis* and over one to four *Wehrbezirkskommandos.*

Wehrkreiskommando, das. Highest or regional command level of the replacement army; it combined duty as *Armeekorps* from 1934 or as, for example, *Wehrkreiskommando* VI and *Armeekorpskommando* VI.

Wehrmacht, die. Armed forces.

Wehrmachtstrafrecht, das. Body of laws governing military service.

Wehrmeldeamt, das. Lowest level of recruiting station in which all conscripted men were mustered and examined, and the permanent repository of their service record books, the *Wehrstammbuch.*

Wehrpass, der. Supplemental document issued to new soldiers during the time between their mustering and induction, during which they were subject to military discipline; it contained essentially the same information as the *Wehrstammbuch,* and was generally discontinued for active soldiers during the war.

Wehrpflicht, die. Universal military obligation.

Wehrstammbuch, das. Soldier's service record book.

Welle, die. Waves, defining the wave system of activating infantry divisions of similar organization at specified times.

Westfeldzug, der. The Campaign in the West, 1940.

Westfront, die. The Western Front.

Windy. Slang in the United Kingdom for "afraid" or "nervous."

Zug, der. Platoon.

Bibliography

Abbott, Porter, W. *The Cambridge Introduction to Narrative.* 2nd ed. Cambridge: Cambridge University Press, 2008.

Abdulin, Mansur, and Artem Drabkin. *Red Road from Stalingrad: Recollections of a Soviet Infantryman.* Barnsley, South Yorkshire: Pen & Sword Military, 2004.

Abrams, Dominic, Michael A. Hogg, and Jose M. Marques. *Social Psychology of Inclusion and Exclusion.* New York: Psychology Press, 2005.

Absolon, Rudolf. "Das Wehrmachtsrafrecht." Kornelimünster: Bundesarchiv, Abteilung Zentralnachweisstelle, 1958.

———. *Das Wehrmachtsrafrecht im 2: Weltkrieg—Sammlung der Grundlegen Gesetze, Verordnung und Erlasse.* Kornelimünster: Bundesarchiv, Abteilung Zentralnachweisstelle, 1958.

———. *Die Wehrmacht im Dritten Reich: Band I, 30.Januar 1933 bis 2.August 1934.* Boppard am Rhein: Harald Boldt Verlag, 1969.

———. *Die Wehrmacht im Dritten Reich: Band II, 30.Januar 1933 bis 2.August 1934.* Boppard am Rhein: Harald Boldt Verlag, 1971.

———. *Die Wehrmacht im Dritten Reich: Band III, 3.August 1934 bis 4.Februar 1938.* Boppard am Rhein: Harald Boldt Verlag, 1975.

———. *Die Wehrmacht im Dritten Reich: Band V, 1.September 1939 bis 18.Dezember 1941.* Boppard am Rhein: Harald Boldt Verlag, 1988.

———. *Die Wehrmacht im Dritten Reich: Band VI, 19.Dezember 1941 bis 9.Mai 1945.* 6 Boppard am Rhein: Harald Boldt Verlag, 1995.

———. *Wehrgesetz und Wehrdienst 1935–1945: Das Personalwesen in der Wehrmacht.* Boppard am Rhein: Harald Boldt Verlag, 1960.

Adams, R. J. Q., and Philip P. Poirer. *The Conscription Controversy in Great Britain, 1900–18.* Columbus: Ohio State University Press, 1987.

Addison, Paul, and Angus Calder. *Time to Kill: The Soldier's Experience of War in the West, 1939–1945.* London: Pimlico, 1997.

Ahrenfeldt, Robert H. *Psychiatry in the British Army in World War II.* New York: Columbia University Press, 1958.

Alanbrooke. *War Diaries, 1939–1945: Field Marshal Lord Alanbrooke.* Edited by Alex Danchev and Daniel Todman. Berkeley: University of California Press, 2001.

Alexander, Joseph H. *The Final Campaign: Marines in the Victory on Okinawa.* Washington, DC: History and Museums Division, Headquarters, US Marine Corps, 1996.

Alford, C. Fred. "Hitler's Willing Executioners: What Does Willing Mean?" *Theory and Society* 26, no. 5 (1997): 719.

Alfred M. Gray Research Center. "History Division Files: Casualties." Quantico, VA: US Marine Corps University.

———. "History Division Files: Marine Rifle Squad." Quantico, VA: US Marine Corps University.

———. "History Division Files: Reference Branch." Unpublished manuscript. Chap. 9, "Recruit Training: The Mobilization Crisis, 7 December 1941–1 March 1942," and Chap. 11, "Enlisted Specialist Training." Quantico, VA: US Marine Corps University.

———. "U.S. Marine Corps Letters of Instruction: An Official Compilation, 1944." Quantico, VA: US Marine Corps University.

Allmand, Christopher, ed. *Society at War: The Experience of England and France During the Hundred Years' War.* New York: Boydell and Brewer, 1998.

Alman, Karl, Fritz-Otto Busch, Hans Holl, Karl Kollatz, and Fred Nemis. *Einzelkämpfer.* Rastatt, Germany: Druck und Verlagshaus Erich Pabel KG, 1975.

Altrichter, Friedrich. *Der Reserveoffizier: Ein Handbuch für den Offizier und Offizieranwärter des Beurlaubtenstandes aller Waffen.* Berlin: E. S. Mittler und Sohn, 1940.

Ambrose, Stephen E. *Band of Brothers: E Company, 506th Regiment, 101st Airborne—From Normandy to Hitler's Eagle's Nest.* New York: Simon and Schuster, 2001.

———. *Citizen Soldiers: The U.S. Army from the Normandy Beaches to the Bulge to the Surrender of Germany, June 7, 1944–May 7, 1945.* New York: Simon and Schuster, 1997.

———. *The Victors: Eisenhower and His Boys—The Men of World War II.* New York: Simon and Schuster, 1998.

Amphill, Colonel. "The Position of the Militia as the Constitutional Force of Great Britain." *Journal of the Royal United Services Institute* 69, no. 473 (1924): 1–23.

Anderson, Fred. *Crucible of War: The Seven Years' War and the Fate of Empire in British North America, 1754–1766.* New York: Random, 2001.

Andreski, Stanislav. *Military Organization and Society.* Berkeley: University of California Press, 1968.

Ansbacher, H. L. "German Military Psychology." *Psychological Bulletin* 38, no. 6 (1941): 370–392.

Archard, Cary, ed. *Alun Lewis: Collected Poems.* Mid Glamorgan, Wales: Seé, Poetry Wales Press, 1994.

Ardrey, Robert. *The Hunting Hypothesis: A Personal Conclusion Concerning the Evolutionary Nature of Man.* New York: Atheneum, 1976.

———. *The Social Contract: A Personal Inquiry into the Evolutionary Sources of Order and Disorder.* New York: Atheneum, 1970.

———. *The Territorial Imperative: A Personal Inquiry into the Animal Origins of Property and Nations.* New York: Atheneum, 1966.

Armstrong, William B., and Thomas G. Sticht. *Adult Literacy in the United States: A Compendium of Quantitative Data and Interpretive Comments.* San Diego, CA: San Diego Community College, 1994.

Bachrach, David S. *Religion and the Conduct of War 300–1215.* Bury St. Edmunds, Suffolk: St. Edmundsbury, 2003.

Badinski, L. "Der Einzelkämpfer mit Gewehr im modernen Gefecht." *Militär-Wochenblatt* 121, no. 37 (1937): 2221–2225.

Bähr, Walter, and Hans W. Bähr. *Die Stimmen des Menschen: Briefe und Aufzeichnungen aus der ganzen Welt, 1939–1945.* Munich: Piper, 1961.
Bald, Detlef. *Der deutsche Offizier: Sozial- und Bildungsgeschichte des deutschen Offizierkorps im 20.Jahrhundert.* Munich: Bernard und Graefe, 1982.
———. *Vom Kaiserheer zur Bundeswehr: Sozialstruktur des Militärs—Politik der Rekrutierung von Offizieren und Unteroffizieren.* Frankfurt am Main: Lang, 1981.
Baret, Joel. *I Wouldn't Want to Do It Again: D-Day in Normandy As Seen Through the Eyes of Private Fayette O. Richardson (Pathfinder) and 1st Lt Rex G. Combs, 508th Parachute Infantry Regiment, US 82nd Airborne Division.* Solihull, England: Helion, 2001.
Barkawi, Tarak. "Culture and Combat in the Colonies: The Indian Army in the Second World War." *Journal of Contemporary History* 41, no. 2 (April 2006): 325–355.
———. "Peoples, Homelands, and Wars? Ethnicity, the Military, and Battle Among British Imperial Forces in the War Against Japan." *Comparative Studies in Society and History* 46, no. 1 (January 2004): 134–163.
Barker, A. H. R., and B. Rust. *A Short History of the 50th Northumbrian Division.* Berwick-upon-Tweed: Tweedale, 1966.
Barnard, Toby. *The English Republic, 1649–1660.* New York: Addison Wesley Longman, 1997.
Barnes, B. S. *The Sign of the Double "T" (the 50th Northumbrian Division, July 1943 to December 1944).* Kingston-upon-Hull: Sentinel, 1999.
Barnhart, Michael A. *Japan Prepares for Total War: The Search for Economic Security, 1919–1941.* Ithaca: Cornell University Press, 1987.
Baron, Ulrich, and Hans-Harald Müller. "Die 'Perspektive des kleinen Mannes' in der Kriegsliteratur der Nachkriegszeiten." In *Der Krieg des kleinen Mannes: Eine Militärgeschichte von unten,* edited by Wolfram Wette, 344–360. Munich: Piper, 1992.
Bartov, Omer. *The Eastern Front, 1941–45: German Troops and the Barbarisation of Warfare.* New York: Palgrave, 1985.
———. "Film Review: Mein Krieg." *American Historical Review* 97, no. 4 (October 1992): 1155–1157.
———. *Germany's War and the Holocaust: Disputed Histories.* Ithaca: Cornell University Press, 2003.
———. *Hitler's Army: Soldiers, Nazis, and War in the Third Reich.* New York: Oxford University Press, 1992.
Bayerlein, Fritz. *Panzer Lehr Division (January–28 July 1944).* New York: Garland, 1979.
Bayley, Charles C. *The Formation of the German College of Electors in the Mid–Thirteenth Century.* Toronto: Toronto University Press, 1949.
Baynes, John Christopher Malcolm. *The History of the Cameronians (Scottish Rifles).* Aldershot: Gale and Polden, 1948.
———. *Morale: A Study of Men and Courage—The Second Scottish Rifles at the Battle of Neuve Chapelle, 1915.* New York: Praeger, 1967.
Beadon, R. H. "The Mind of an Army." *Journal of the Royal United Services Institute* 64, no. 456 (1919): 581–593.
———. "Some Strategical Theories of Captain Liddell Hart." *Journal of the Royal United Services Institute* 81, no. 524 (1936): 747–760.
Beckett, I. F. W., and Keith Simpson. *A Nation in Arms: A Social Study of the British Army in the First World War.* Manchester: Manchester University Press, 1985.
Beevor, Antony. *Ardennes 1944: Hitler's Last Gamble.* New York: Viking, 2015.
———. *Stalingrad: The Fateful Siege, 1942–1943.* New York: Penguin, 1998.
Berg, C. "Clinical Notes on the Analysis of a War Neurosis." *British Journal of Medical Psychology* 19 (1941–1943): 155–185.

Berger, Peter L., and Thomas Luckman. *The Social Construction of Reality: A Treatise in the Sociology of Knowledge.* New York: Random, 1966.
Berghahn, Volker Rolf. *Europe in the Era of Two World Wars: From Militarism and Genocide to Civil Society, 1900–1950.* Princeton: Princeton University Press, 2006.
———. *Imperial Germany, 1871–1918: Economy, Society, Culture, and Politics.* Oxford: Berghahn, 2005.
———. *Militarism: The History of an International Debate, 1861–1979.* New York: St. Martin's, 1982.
Berry, Henry. *Semper Fi, Mac: Living Memories of the U.S. Marines in World War II.* New York: Arbor, 1982.
Berthold, Will. *Das Letzte Gefecht: Roman.* Munich: Engel, 1987.
Bertua, Cristina, Neil Anderson, and Jesus F. Salgado. "The Predictive Validity of Cognitive Ability Tests: A UK Meta-Analysis." *Journal of Occupational and Organizational Psychology* 78, no. 3 (2005): 387–409.
Best, Geoffrey, and Andrew Wheatcroft. *War, Economy, and the Military Mind.* London: Rowman and Littlefield, 1976.
Bidermann, Gottlob Herbert. *In Deadly Combat: A German Soldier's Memoir of the Eastern Front.* Lawrence: University Press of Kansas, 2000.
Bidwell, Sheldon, and Dominick Graham. *Fire-Power: British Army Weapons and Theories of War, 1904–1945.* Boston: Allen and Unwin, 1985.
Bigelow, John, Jr. *Principles of Strategy.* Philadelphia: Lippincott, 1894.
Bigler, Rolf R. *Der Einsame Soldat.* Fraunenfeld, Germany: Verlag Huber AG, 1963.
Bishop, Tim, and Bruce Shand. *One Young Soldier: The Memoirs of a Cavalryman.* Norwich: Russell, 1993.
Bittner, R. H. "The Army General Classification Test." In *New Methods in Applied Psychology,* edited by George A. Kelly, 45–55. College Park: University of Maryland Press, 1947.
Black, Jeremy. "Britain as a Military Power, 1688–1815." *Journal of Military History* 64, no. 1 (2000): 159–177.
Bland, Larry I., and Sharon Ritenour Stevens, eds. *The Papers of George Catlett Marshall.* Vol. 1, *The Soldierly Spirit.* Baltimore: Johns Hopkins University Press, 1981.
———. *The Papers of George Catlett Marshall.* Vol. 4, *Aggressive and Determined Leadership.* Baltimore: Johns Hopkins University Press, 1996.
Blank, Ralf, Jörg Echternkamp, Karola Fings, Jürgen Förster, Winfried Heinemann, Tobias Jersak, Armin Nolzen, and Christoph Rass. *Die deutsche Gesellschaft, 1939 bis 1945: Politisierung, Vernichtung, Überleben.* Vol. 9/1, *Das Deutsche Reich und der Zweite Weltkrieg.* Munich: Deutsche Verlags-Anstalt, 2004.
Blumenson, Martin. *Breakout and Pursuit.* Washington, DC: Government Printing Office, 1961.
———. *Salerno to Cassino.* Washington, DC: Government Printing Office, 1965.
Blumentritt, Guenther. *Training Regular Army NCOs.* Carlisle, PA: US Army Military History Institute, 1947.
———. *Warum hat der deutsche Soldat in aussichtsloser Lager bis zum Schluß des Krieges 1939–1945?* Carlisle, PA: US Army Military History Institute, 1947.
Blunt, Roscoe C., and Roscoe C. Blunt Jr. *Foot Soldier: A Combat Infantryman's War in Europe.* Cambridge, MA: Da Capo, 2001.
Blythe, Ronald. *Private Words, Letters, and Diaries from the Second World War.* London: Viking, 1991.
———. *Writing in a War: Stories, Poems, and Essays of the Second World War.* Harmondsworth: Penguin, 1982.
Boesch, Paul. *Road to Huertgen: Forest in Hell.* Houston: Gulf, 1962.

Bogacz, Ted. "War Neurosis and Cultural Change in England, 1914–22: The Work of the War Office Committee of Enquiry into 'Shell-Shock.'" *Journal of Contemporary History* 24, no. 2 (1989): 227–256.

Böll, Heinrich. *Absent Without Leave: 2 Novellas*. Translated by Leila Vennewitz. New York: McGraw-Hill, 1965.

———. *The Casualty*. New York: Farrar, Straus, and Giroux, 1987.

Bond, Brian. *Britain, France, and Belgium, 1939–1940*. Washington, DC: Brassey's, 1990.

———. *British Military Policy Between the Two World Wars*. Oxford: Clarendon, 1980.

———. *Liddell Hart: A Study of His Military Thought*. New Brunswick, NJ: Rutgers University Press, 1977.

———. "Liddell Hart and the German Generals." *Military Affairs* 41, no. 1 (1977): 16–20.

Bonura, Michael A. *Under the Shadow of Napoleon: French Influence on the American Way of* Warfare *from the War of 1812 to the Outbreak of WWII*. New York: New York University Press, 2012.

Boog, Horst, Jürgen Förster, Joachim Hoffman, Ernst Klink, Rolf-Dieter Müller, and Gerd R. Ueberschär. *Der Angriff auf die Sowjetunion*. Vol. 4, *Das Deutsche Reich und der Zweite Weltkrieg*. Stuttgart: Deutsche Verlags-Anstalt, 1983.

Boog, Horst, Richard Lakowski, Werner Rahn, Manfred Zeidler, and John Zimmerman. *Der Zusammenbruch des deutschen Reiches: Die militärische Niederwerfung der Wehrmacht*. Vol. 10/1, *Das Deutsche Reich und der Zweite Weltkrieg*. Munich: Deutsche Verlags-Anstalt, 2008.

Boog, Horst, Werner Rahn, Reinhard Stumpf, and Bernd Wegner. *Der globale Krieg: Die Ausweitung zum Weltkrieg und der Wechsel der Initiative*. Vol. 6, *Das Deutsche Reich und der Zweite Weltkrieg*. Stuttgart: Deutsche Verlags-Anstalt, 1990.

Borthwick, James. *The 51st Highland Division in Africa and Sicily*. Glasgow: Mackenzie, 1945.

Bowman, J. E. *Three Stripes and a Gun: A Young Man's Journey Towards Maturity*. Braunton, UK: Merlin, 1987.

Boyd, Bill. *The Gentle Infantryman*. New York: St. Martin's, 1985.

Boyer, William H. *Education for Annihilation*. Honolulu: Hogarth, 1972.

Brereton, J. M. *The British Soldier: A Social History from 1661 to the Present Day*. London: Bodley Head, 1986.

Brereton, T. R. "First Lessons in Modern War: Arthur Wagner, the 1898 Santiago Campaign, and U.S. Army Lesson-Learning." *Journal of Military History* 64, no. 1 (January 2000): 79–96.

Bridgwater, Patrick. *The German Poets of the First World War*. New York: St. Martin's, 1985.

Bröckling, Ulrich. *Disziplin: Soziologie und Geschichte militärischer Gehorsamsproduktion*. Munich: Fink, 1997.

Bröckling, Ulrich, and Michael Sikora. *Armeen und ihre Deserteure: Vernachlässigte Kapital einer Militärgeschichte der Neuzeit*. Göttingen: Vandenhoeck and Ruprecht, 1998.

Brodsky, G. W. Stephen. *Gentlemen of the Blade: A Social and Literary History of the British Army Since 1660*. New York: Greenwood, 1988.

Brokaw, Tom. *The Greatest Generation*. New York: Random, 2001.

Brose, Eric Dorn. *The Kaiser's Army: The Politics of Military Technology in Germany During the Machine Age, 1870–1918*. New York: Oxford University Press, 2001.

Brown, John Sloan. *Draftee Division: The 88th Infantry Division in World War II*. Lexington: University Press of Kentucky, 1986.

Brown, R. Allen. *The Normans.* New York: St. Martin's, 1984.
Browning, Christopher. *Ordinary Men: Police Battalion 101 and the Final Solution in Poland.* New York: HarperCollins, 1992.
Bruscino, Thomas A. *A Nation Forged in War: How World War II Taught Americans to Get Along.* Knoxville: University of Tennessee Press, 2010.
Brutton, Philip. *Ensign in Italy: A Story of the Felix Factor—The Nine Lives of a Young Welsh Guards Officer Who Fought in Italy with His Regiment from the Ruins of Cassino to the Alps.* London: Cooper, 1992.
Bryant, Michael Scott. "American Military Justice from the Revolution to the UCMJ: The Hard Journey from Command Authority to Due Process." *Creighton International and Comparative Law Journal* 4 (2013): 1–14.
Buchbender, Ortwin, and Reinhold Sterz, eds. *Das andere Gesichte des Krieges: Deutsche Feldpostbriefe, 1939–1945.* Munich: Beck, 1982.
Burke, Peter. *History and Social Theory.* Ithaca: Cornell University Press, 1992.
Büsch, Otto. *Military System and Social Life in Old-Regime Prussia, 1713–1807: The Beginnings of the Social Militarization of Prusso-German Society.* Translated by John G. Gagliardo. Atlantic Highlands, NJ: Humanities Press, 1997.
Caesar, Julius. *The Battle for Gaul.* Translated by Anne Wiseman and Peter Wiseman. Boston: David R. Godine, 1980.
Caldwell, Erskine. *All Night Long: A Novel of Guerrilla Warfare in Russia.* New York: Book League of America, 1942.
Calhoun, Mark. *General Lesley J. McNair: Unsung Architect of the U.S. Army.* Lawrence: University Press of Kansas, 2015.
Callwell, C. E. "Small Wars: Their Principles and Practice." Reprint, 1896. Lincoln: University of Nebraska Press, 1996.
Cameron, Craig M. *American Samurai: Myth, Imagination, and the Conduct of Battle in the First Marine Division, 1941–1951.* New York: Cambridge University Press, 1994.
Capshew, James H. *Psychologists on the March: Science, Practice, and Professional Identity in America, 1929–1969.* New York: Cambridge University Press, 1999.
Carey, Neil, ed. *No Bended Knee: The Battle for Guadalcanal—The Memoir of General Merrill B. Twining.* Novato, CA: Presidio, 1996.
Carr, William. *Origins of the Wars of German Unification.* New York: Longman, 1991.
Carter, Ross S. *Those Devils in Baggy Pants.* New York: Appleton-Century-Crofts, 1951.
Carver, Lord Michael. *The Imperial War Museum Book of the War in Italy.* London: Sidgwick and Jackson, 2001.
Castillo, Jason J. *Endurance and War: The National Source of Military Cohesion.* Stanford: Stanford University Press, 2014.
Cato. *Guilty Men.* New York: Frederick A. Stokes, 1940.
Cervinka, Vladimir. "A Dimensional Theory of Groups." *Sociometry* 11, no. 2 (February–May 1948): 100–107.
Chandler, David G. *The Campaigns of Napoleon.* New York: Scribner, 1966.
Chapin, John C. *The 4th Marine Division in World War II.* Washington, DC: History and Museums Division, US Marine Corps, 1974.
Churchill, Winston S. *The Age of Revolution: A History of the English-Speaking Peoples.* 4 vols. New York: Dodd, Mead, 1957.
———. *The Churchill War Papers 1: At the Admiralty.* Edited by Martin Gilbert. London: Heinemann, 1993.
———. *The Churchill War Papers 2: Never Surrender.* Edited by Martin Gilbert. London: Heinemann, 1994.
———. *The Churchill War Papers 3: The Ever-widening War.* Edited by Martin Gilbert. London: Heinemann, 1993.

———. *My Early Life, 1874–1904.* New York: Simon and Schuster, 1996.
———. *Never Give In! The Best of Winston Churchill's Speeches: Selected by His Grandson.* New York: Hyperion, 2003.
———. *The Second World War.* 6 vols. London: Cassell, 1985.
Churchill, Winston, Franklin D. Roosevelt, and Warren F. Kimball. *Churchill and Roosevelt: The Complete Correspondence.* Princeton: Princeton University Press, 1984.
Citino, Robert M. "'*Die Gedanken sind frei*': The Intellectual Culture of the Interwar German Army." *Army Doctrine and Training Bulletin* 4, no. 3 (2001): 48–55.
———. *The German Way of War: From the Thirty Year's War to the Third Reich.* Lawrence: University Press of Kansas, 2005.
Clark, Christopher. *Iron Kingdom: The Rise and Downfall of Prussia, 1600–1947.* Cambridge: Belknap, 2006.
Clay, Ewart Waide. *The Path of the 50th: The Story of the 50th (Northumbrian) Division in the Second World War, 1939–1945.* Aldershot: Gale and Polden, 1950.
Claydon, John. "Interpretations of Nazi Germany." *History Review* (March 2001): 28–33.
Coffman, Edward M. *The Regulars: The American Army, 1898–1941.* Cambridge: Harvard University Press, 2004.
Cole, Eli K. "Army-Navy Joint Action." *United States Naval Institute Proceedings* 55 (November 1929): 927–937.
Cole, Hugh M. *The Ardennes: Battle of the Bulge.* Washington, DC: Government Printing Office, 1965.
Colville, John Rupert. *The Fringes of Power: 10 Downing Street Diaries, 1939–1955.* New York: Norton, 1985.
Condell, Bruce, and David T. Zabecki, eds. *On the German Art of War: Truppenführung.* Boulder: Lynne Rienner, 2001.
Congressional Research Service. *American War and Military Operations Casualties.* Report for Congress. Washington, DC: Congressional Research Service, July 29, 2020.
Conn, Stetson, ed. *American Military History.* Vol. 2. Washington, DC: Governemnt Printing Office, 1969.
Cooley, Charles Horton. *Social Organization: A Study of the Larger Mind.* New York: Charles Scribner's Sons, 1909.
Cooper, Matthew. *The German Army, 1933–1945: Its Political and Military Failure.* New York: Bonanza, 1984.
Copp, Terry, and Bill McAndrew. *Battle Exhaustion.* Toronto: McGill-Queens University Press, 1990.
———. *Psychiatry in the Canadian Army (Overseas), 1939–1945.* Waterloo, Canada: Wilfrid Laurier University, 1984.
Corum, James S. *The Roots of Blitzkrieg: Hans von Seeckt and German Military Reform.* Lawrence: University Press of Kansas, 1992.
Cox, Geoffrey. *A Tale of Two Battles: A Personal Memoir of Crete and the Western Desert, 1941.* London: Kimber, 1987.
Craig, Gordon A. *The Battle of Königgrätz: Prussia's Victory over Austria, 1866.* Philadelphia: University of Pennsylvania Press, 1964.
———. *Germany, 1866–1945.* New York: Oxford University Press, 1978.
———. *The Politics of the Prussian Army, 1640–1945.* Oxford: Oxford University Press, 1964.
Craig, Jeremy A. *The British Army and the People's War, 1939–1945.* Manchester: Manchester University Press, 2000.
Craig, Norman. *The Broken Plume: A Platoon Commander's Story, 1940–45.* London: Imperial War Museum, 1982.

Crawford, Danny J., Robert V. Aquilina, Ann A. Ferrante, and Sheila P. Gramblin. *The First Marine Division and Its Regiments.* Washington, DC: Government Printing Office, 1999.

Creighton, Colin, and Martin Shaw, eds. *The Sociology of War and Peace.* Basingstoke: Macmillan, 1987.

Crew, F. A. E. *The Army Medical Services: Campaigns.* London: HMSO, 1956.

Crisp, Robert. *Brazen Chariots: An Account of Tank Warfare in the Western Desert.* London: Muller, 1959.

Croft, W. D. "The Influence of Tanks upon Tactics." *Journal of the Royal United Services Institute* 67, no. 467 (1922): 39–53.

Cunningham, A. G. "The Training of the Army." *Journal of the Royal United Services Institute* 9 (1934): 723–732.

Currey, Cecil B. *Follow Me and Die: The Destruction of an American Division in World War II.* New York: Stein and Day, 1984.

David, Saul. *Churchill's Sacrifice of the Highland Division: France 1940.* London: Brassey's, 1994.

de Lannoy, François, and Jean-Claude Perrigault. *La Division Grossdeutschland (The Grossdeutschland Division from Regiment to Panzerkorps 1939–1945).* Bayeux: Heimdal, 1998.

Delaforce, Patrick. *Monty's Highlanders: 51st Highland Division in World War Two.* London: Tom Donovan, 1997.

Delaney, John Paul. *The Blue Devils in Italy: A History of the 88th Infantry Division in World War II.* Washington, DC: Infantry Journal Press, 1947.

Delbrück, Hans. *The Barbarian Invasions.* Translated by Walter Renfroe Jr. Reprint, 1921. Westport: Greenwood, 1990.

———. *The Dawn of Modern Warfare.* Translated by Walter Renfroe Jr. Reprint, 1920. Westport: Greenwood, 1990.

———. *Medieval Warfare.* Translated by Walter Renfroe Jr. Reprint, 1920. Westport: Greenwood, 1990.

———. *Warfare in Antiquity.* Translated by Walter Renfroe Jr. Reprint, 1920. Westport: Greenwood, 1990.

Denkert, Walter. *3rd Panzer Grendier Division (Ardennes).* Carlisle, PA: US Army Military History Institute, 1947.

Dennis, George T., ed. *Three Byzantine Military Treatises.* Washington, DC: Dumbarton Oaks, 1985.

Dennis, Peter. *Decision by Default: Peacetime Conscription and British Defence, 1919–39.* Durham, NC: Duke University Press, 1972.

Desan, Suzanne. *Crowds, Community, and Ritual in the Work of E.P. Thompson and Natalie Davis: The New Cultural History.* Berkeley: University of California Press, 1989.

d'Este, Carlo. *Decision in Normandy.* New York: Dutton, 1983.

Detwiler, Donald S., ed. *World War II German Military Studies.* Vol. 1. New York: Garland, 1979.

Dewey, P. E. "Military Recruiting and the British Labour Force During the First World War." *Historical Journal* 27, no. 1 (1984): 199–233.

Diamond, Jared. *Guns, Germs, and Steel: The Fates of Human Societies.* New York: Norton, 1997.

Dicks, Henry V. "German Personality Traits and National Socialist Ideology: A War-Time Study of German Prisoners of War." In *Propaganda in War and Crisis,* edited by Daniel Lerner, 100–161. New York: Arno, 1972.

Diehl, James M. *Paramilitary Politics in Weimar Germany.* Bloomington: Indiana University Press, 1977.

———. *The Thanks of the Fatherland: German Veterans After the Second World War.* Chapel Hill: University of North Carolina Press, 1993.

Diest, Wilhelm. "Die Aufrüstung des Heeres: Von der Einführung der Wehrpflicht zur Rheinland-Besetzung." In *Ursachen und Voraussetzungen der Deutschen Kriegspolitik,* vol. 1, *Das Deutsche Reich und der Zweite Weltkrieg,* edited by Militärgeschichtlichen Forschungsamt, 415–431. Stuttgart: Deutsche Verlags-Anstalt, 1979.

———. *Militär, Staat und Gesellschaft: Studien zur Preussisch-Deutschen Militärgeschichte.* Munich: Oldenbourg, 1991.

———. *The Wehrmacht and German Rearmament.* Toronto: University of Toronto Press, 1981.

Diest, Wilhelm, Manfred Messerschmidt, Hans-Erich Volkmann, and Wolfram Wette. *Ursachen und Voraussetzungen der Deutschen Kriegspolitik.* Vol. 1, *Das Deutsche Reich und der Zweite Weltkrieg.* Stuttgart: Deutsche Verlags-Anstalt, 1979.

Dietz, Peter. *The Last of the Regiments: Their Rise and Fall.* London: Brassey's, 1990.

Dollard, John. *Frustration and Aggression.* New Haven: Institute of Human Relations, Yale University, 1939.

———. *Victory over Fear.* New York: Reynal and Hitchcock, 1942.

Döllinger, Hans, ed. *Kain, Wo ist dein Bruder? Was der Mensch im zweiten Weltkrieg erleiden musste—Dokumentiert in Tagebüchern und Briefen.* Munich: List, 1987.

Donald, David Herbert. *Lincoln.* New York: Simon and Schuster, 1995.

Doolittle Board. *Report of the Secretary of War's Board on Officer–Enlisted Man Relationships.* Report to Robert P. Patterson, Secretary of War, May 27, 1946.

Doubler, Michael D. *Closing with the Enemy: How GIs Fought the War in Europe, 1944–45.* Lawrence: University Press of Kansas, 1994.

"Doubling the Territorial Army." *The Times* (London), March 30, 1939.

Doughty, Robert A. *The Seeds of Disaster: The Development of French Army Doctrine, 1919–1939.* Hamden, CT: Archon, 1985.

Douglas, Keith. *The Complete Poems.* 3rd ed. Edited by Desmond Graham. London: Faber and Faber, 1998.

———. *Simplify Me When I'm Dead.* Poems selected by Ted Hughes. Reprint, 1964. London: Faber and Faber, 2010.

Duffy, Christopher. *The Army of Frederick the Great.* New York: Hippocrene, 1974.

———. *The Military Experience in the Age of Reason.* London: Routledge and Kegan Paul, 1987.

———. *Red Storm on the Reich: The Soviet March on Germany, 1945.* New York: Atheneum, 1991.

Dunn, Walter S., Jr. *Hitler's Nemesis.* London: Praeger, 1994.

———. *Kursk: Hitler's Gamble, 1943.* Westport: Praeger, 1997.

Dupuy, R. Ernest, and Trevor N. Dupuy. *The Encyclopedia of Military History from 3500 B.C. to the Present.* Revised ed. New York: Harper and Row, 1977.

Dupuy, Trevor N. *A Genius for War: The German Army and General Staff, 1807–1945.* New York: Prentice-Hall, 1977.

———. *Numbers, Predictions, and War: Using History to Evaluate Combat Factors and Predict the Outcome of Battles.* Bloomington, IN: Bobbs-Merrill, 1979.

Dyakov, Yuri, and Tatyana Bushuyeva. *The Red Army and the Wehrmacht: How the Soviets Miitarized Germany, 1922–33, and Paved the Way for Fascism—From the Secret Archives of the Former Soviet Union.* Amherst, NY: Prometheus, 1995.

Echevarria, Antulio J., III. *After Clausewitz: German Military Thinkers Before the Great War.* Lawrence: University Press of Kansas, 2000.

———. *Clausewitz and Contemporary War.* New York: Oxford University Press, 2007.

Echternkamp, Jörg, ed. *Die deutsche Gesellschaft, 1939 bis 1945: Politisierung, Vernichtung, Überleben.* Vol. 9/2, *Das Deutsche Reich und der Zweite Weltkrieg.* Munich: Deutsche Verlags-Anstalt, 2005.

Eckardt, Uwe, and Stefan Koldehoff. *Deserteure in Wuppertal: Dokumentation zu den Erschiessungen, 1944–45.* Wuppertal: Stadtarchiv der Stadt Wuppertal, 1992.

Eimannsberger, L. "Panzertaktik." *Militär-Wochenblatt* 121, no. 27 (1937): 1509–1514.

Ellenberger, G. F. "The Infantry Section: French, German and British." *Journal of the Royal United Services Institute* 83 (1938): 539–551.

Ellis, Frank. *The Stalingrad Cauldron: Inside the Encirclement and Destruction of the 6th Army.* Lawrence: University Press of Kansas, 2013.

Ellis, John. *Brute Force: Allied Strategy in the Second World War.* New York: Penguin, 1990.

———. *Cassino: The Hollow Victory.* New York: McGraw-Hill, 1984.

———. *On the Front Lines: The Experience of War Through the Eyes of the Allied Soldiers in World War II.* New York: Wiley, 1990.

———. *The Sharp End: The Fighting Man in World War II.* New York: Scribner, 1980.

———. *The Stalingrad Cauldron: Inside the Encirclement and Destruction of the 6th Army.* Lawrence: University Press of Kansas, 2013.

Emirbayer, Mustafa, and Ann Mische. "What Is Agency?" *American Journal of Sociology* 103, no. 4 (1998): 962–1023.

English, John A., and Bruce I. Gudmundsson. *On Infantry.* Revised ed. Westport: Praeger, 1994.

Evans, Michael, and Alan Ryan. *The Human Face of Warfare: Killing, Fear, and Chaos in Battle.* London: Allen and Unwin, 2000.

Evans, Richard J. *Rituals of Retribution: Capital Punishment in Germany, 1600–1987.* New York: Oxford University Press, 1996.

Evans-Jones, Robin. "Roman Law in Britain." In *Festschrift für Joseph Georg Wolf zum 70.Geburtstag.* Translated by U. Manthe and C. Krampe, 83–110. Berlin: Duncker and Humboldt, 2000.

Eyck, Erich. *A History of the Weimar Republic.* Translated by Harlan P. Hanson and Robert G. L. Waite. Cambridge: Harvard University Press, 1962.

Fahey, John Edward. *A History of the Marine Corps Recruit Depot, San Diego, California, 1911–1974.* San Diego, CA: University of San Diego Press, 1974.

Fann, Willerd R. "On the Infantryman's Age in Eighteenth Century Prussia." *Military Affairs* 41, no. 4 (December 1977): 165–170.

Farago, Ladislas. *Patton: Ordeal and Triumph.* New York: Ivan Obolensky, 1963.

Farrar, L. L. *Arrogance and Anxiety: The Ambivalence of German Power, 1848–1914.* Iowa City: University of Iowa Press, 1981.

———. *The Short-War Illusion: German Policy, Strategy & Domestic Affairs, August–December 1914.* Santa Barbara, CA: ABC-Clio, 1973.

Farrar, L. L. ed. *War: A Historical, Political, and Social Study.* Santa Barbara, CA: ABC-Clio, 1978.

Farrar-Hockley, Anthony. *Infantry Tactics.* London: Squadron/Signal, 1976.

Farrell, Theo. "The Dynamics of British Military Transformation." *International Affairs* 84, no. 4 (2008): 777–807.

Fawcett, Gabriel. "The Wehrmacht Exhibition." *History Today* 52, no. 4 (April 2002): 2–4.

Feifer, George. *Tennozan: The Battle of Okinawa and the Atomic Bomb.* New York: Ticknor and Fields, 1992.

Felber, Abraham. *The Old Breed of Marine: A World War II Diary.* Jefferson, NC: McFarland, 2003.

Feld, Maury D. *The Structure of Violence: Armed Forces as Social Systems.* Beverly Hills: Sage, 1977.

Fergusson, Sir Bernard. *The Black Watch and the King's Enemies*. London: Collins, 1950.

Fetchenhauer, Detlef, ed. *Solidarity and Prosocial Behavior: An Integration of Sociological and Psychological Perspectives*. New York: Springer, 2006.

Fine, Gary Alan. "Group Culture and Interaction Order: Local Sociology on the Meso-Level." *Annual Review of Sociology* 38 (2012): 159–179.

Finer, Samuel E. "State- and Nation-Building in Europe: The Role of the Military." In *The Formation of National States in Western Europe,* edited by Charles Tilly, 84–163. Princeton: Princeton University Press, 1975.

Finkelstein, Norman G., and Ruth Bettina Birn. *A Nation on Trial: The Goldhagen Thesis and Historical Truth*. New York: Metropolitan, 1998.

Fischer, David Hackett. *Washington's Crossing*. Oxford: Oxford University Press, 2004.

Fisher, B. D. "The Training of the Regimental Officer." *Journal of the Royal United Services Institute* 74 (1929): 241–261.

Fisher, Earnest F., Jr. *Cassino to the Alps*. Washington, DC: Government Printing Office, 1977.

———. *Guardians of the Republic: A History of the Noncommissioned Officer Corps of the U.S. Army*. Mechanicsburg, PA: Stackpole, 2001.

Fitton, Robert A., ed. *Leadership: Quotations from the Military Tradition*. Boulder: Westview, 1994.

Fitzharris, Joseph C. "Field Officer Courts and U.S. Civil War Military Justice." *Journal of Military History* 68, no. 1 (January 2004): 47–72.

Flynn, George Q. *Conscription and Democracy: The Draft in France, Great Britain, and the United States*. Westport: Greenwood, 2002.

Foerster, Roland G. *Die Wehrpflicht: Entstehung, Erscheinungsformen und politisch-militärische Wirkung*. Munich: R. Oldenbourg, 1994.

———. *"Unternehmen Barbarossa": Zum historischen Ort der deutsch-sowjetischen Beziehungen von 1933 bis Herbst 1941*. Munich: R. Oldenbourg, 1993.

Foerster, Roland G., Karl-Heinz Frieser, and Günter Roth. *Operational Thinking in Clausewitz, Moltke, Schlieffen, and Manstein*. Bonn: Verlag E. S. Mittler und Sohn, 1989.

Foertsch, H. *Die Wehrmacht im Nationalsozialistischen Staat*. Hamburg: Verlagsbuchhandlung Broschek, 1935.

Foley, William. *Visions from a Foxhole: A Rifleman in Patton's Ghost Corps*. New York: Presidio, 2003.

Ford, Guy Stanton. *Stein and the Era of Reform in Prussia, 1807–1815*. Princeton: Princeton University Press, 1922.

Ford, Ken. *D-Day 1944: Sword Beach and the British Airborne Landings*. Westport: Praeger, 2004.

Förster, Jürgen. *Die Wehrmacht im NS-Staat*. Osnabrück: Oldenbourg Verlag, 2009.

———. "The Dynamics of Volksgemeinschaft: The Effectiveness of the German Military Establishment in the Second World War." In *Military Effectiveness,* edited by Allan R. Millett and Williamson Murray, 180–220. New York: Cambridge University Press, 1988.

———. "Motivation and Indoctrination in the Wehrmacht, 1939–1945." In *Time to Kill: The Soldiers Experience of War in the West 1939–1945,* edited by Paul Addison and Angus Calder, 263–271. London: Pimlico, 1997.

Förster, Stig. "Militär und staatsbürgerlicher Partizipation." In *Die Wehr-pflicht: Entstehung, Erscheinungsformen und politisch-militärische Wirkung,* edited by Roland G. Foerster, 55–70. Munich: Oldenbourg Verlag, 1994.

Forster, William B., et al. *Physical Standards in World War II*. Edited by Robert S. Anderson. Washington, DC: Office of the Surgeon General, 1967.

Forty, George. *British Army Handbook, 1939–1945.* Sparkford, UK: Haynes, 2002.

Fox, F. *The Royal Inniskilling Fusiliers in the Second World War.* London: Gale and Polden, 1951.

Fox, Robin. "Fatal Attraction: War and Human Nature." *The National Interest* (Winter 1992–1993): 11–20.

Fox, Robin Lane. *The Search for Alexander.* Boston: Little, Brown, 1980.

Frank, Benis M., and Henry I. Shaw. *Victory and Occupation.* Washington, DC: Government Printing Office, 1968.

Frank, Joseph Allan. *"Seeing the Elephant": Raw Recruits at the Battle of Shiloh.* New York: Greenwood, 1989.

Frank, Richard B. *Guadalcanal.* New York: Random, 1990.

Fraser, David. *And We Shall Shock Them: The British Army in the Second World War.* London: Hodder and Stoughton, 1983.

Fraser, George MacDonald. *Quartered Safe out Here: A Recollection of the War in Burma.* London: Harvill, 1992.

Frederickson, George M. "From Exceptionalism to Variability: Recent Developments in Cross-National Comparative History." *Journal of American History* 82, no. 2 (1995): 587–604.

Frei, Norbert, Hermann Kling, and Margit Brandt, eds. *Der nationalsozialistische Krieg.* Frankfurt: Campus Verlag, 1990.

French, David. *The British Way in Warfare, 1688–2000.* London: Unwin Hyman, 1990.

———. "Discipline and the Death Penalty in the British Army in the War Against Germany During the Second World War." *Journal of Contemporary History* 33, no. 4 (October 1998): 531–545.

———. "Doctrine and Organization in the British Army, 1919–1932." *Historical Journal* 44, no. 2 (June 2001): 497–515.

———. *Military Identities: The Regimental System, the British Army, and the British People.* Oxford: Oxford University Press, 2005.

———. *Raising Churchill's Army: The British Army and the War Against Germany, 1919–1945.* Oxford: Oxford University Press, 2000.

Fretter-Pico, Maximilian. *Battle Sector IX (Wehrkreis IX) Situation (2 Apr 45).* Washington, DC: Government Printing Office, 1947.

———. *Mißbrauchte Infanterie: Deutsche Infanteriedivisionen im osteuropäischen Grossraum, 1941 bis 1944—Erlebnisskizzen, Erfahrungen und Erkenntnisse.* Frankfurt am Main: Verlag für Wehrwesen Bernard und Graefe, 1957.

Friedkin, Noah E. "Social Cohesion." *Annual Review of Sociology* 30 (2004): 409–425.

Friedl, Karl E. "Body Composition and Military Performance: Origins of the Army Standards." In *Body Composition and Physical Performance: Applications for the Military Services,* edited by Bernadotte M. Marriott and Judith Grumstrup-Scott. Washington, DC: Committee on Military Nutrition Research, Institute of Medicine, 1992.

Frieser, Karl-Heinz. *The Blitzkrieg Legend.* Annapolis, MD: Naval Institute Press, 2005.

Frieser, Karl-Heinz, Klaus Schmider, Klaus Schönherr, Gerhard Schreiber, Kristián Ungváry, and Bernd Wegner. *Die Ostfront 1943/44: Der Krieg im Osten und an den Nebenfronten.* Vol. 8, *Das Deutsche Reich und der Zweite Weltkrieg.* Munich: Deutsche Verlags-Anstalt, 2007.

Fritscher, Werner. *Dokumente zur Deutschen Wehrpsychologie, 1914–1945.* Munich: Verlag für Wehrwissenschaften, 1990.

Fritz, Stephen G. *Frontsoldaten: The German Soldier in World War II.* Lexington: University Press of Kentucky, 1995.

Frost, Robert I. *The Northern Wars: War, State, and Society in Northeastern Europe, 1558–1721.* Essex: Pearson Education, 2000.

Fuchs, Helmut. *Wer spricht von Siegen: Der Bericht über unfreiwillige Jahre in Rußland.* Munich: A. Knaus, 1987.
Fuchs, Karl. *Sieg Heil! War Letters of Tank Gunner Karl Fuchs, 1937–1941.* Translated by Horst Fuchs Richardson and Dennis Showalter. Hamden, CT: Archon, 1987.
Fugate, Bryan I., and L. S. Dvoretskii. *Thunder on the Dnepr: Zhukov-Stalin and the Defeat of Hitler's Blitzkrieg.* Novato, CA: Presidio, 1997.
Fuller, J. F. C. "The Application of Recent Developments in Mechanics and Other Scientific Knowledge to Preparation and Training for Future War on Land." *Journal of the Royal United Services Institute* 65, no. 458 (1920): 239–274.
———. *Armored Warfare: An Annotated Edition of Lectures on F.S.R. III (Operations Between Mechanized Forces).* Harrisburg, PA: Military Service, 1943.
———. *Generalship: Its Diseases and Their Cure—A Study of the Personal Factor in Command.* Harrisburg, PA: Military Service, 1936.
———. *Memoirs of an Unconventional Soldier.* London: Nicholson and Watson, 1936.
———. "Moral, Instruction, and Leadership." *Journal of the Royal United Services Institute* 65, no. 459 (1920): 656–674.
———. "Progress in the Mechanicalization of Modern Armies." *Journal of the Royal United Services Institute* 70, no. 477 (1925): 73–89.
———. "Tactics and Mechanization." *Infantry Journal* 30, no. 5. (May 1927): 457–465.
Fuller, William C. *Strategy and Power in Russia, 1600–1914.* New York: Free Press, 1992.
Furer, Julius Augustus. *Administration of the Navy Department in World War II.* Washington, DC: Historical Division, US Navy, 1959.
Fussell, Paul. *The Boys' Crusade.* New York: Modern Library, 2003.
———. *Doing Battle: The Making of a Skeptic.* New York: Little, Brown, 1996.
———. *Wartime: Understanding and Behavior in the Second World War.* New York: Oxford University Press, 1989.
Gabriel, Richard A. *The Culture of War: Invention and Early Development.* New York: Greenwood, 1990.
———. *No More Heroes: Madness and Psychiatry in War.* New York: Hill and Wang, 1987.
———. *The Painful Field: The Psychiatric Dimension of Modern War.* New York: Greenwood, 1988.
Gabriel, Richard A., and Karen S. Metz. *A History of Military Medicine.* New York: Greenwood, 1992.
Gaddis, John Lewis. *The Landscape of History.* New York: Oxford University Press, 2002.
Gagliardo, John G. "Translator's Preface." In *Military System and Social Life in Old Regime Prussia, 1713–1807,* i–iii. Boston: Humanities Press International, 1997.
Gailey, Harry A. *Peleliu 1944.* Annapolis, MD: Nautical and Aviation Publication Company of America, 1983.
Garforth, F. I. de la P. "WOSB (OCTU)." *Occupational Psychology* 19 (1945): 97–106.
Garland, Albert N., and Howard M. Smyth. *Sicily and the Surrender of Italy.* Washington, DC: Government Printing Office, 1965.
Gat, Azar. *A History of Military Thought.* New York: Oxford University Press, 2001.
Gayle, Gordon. *Bloody Beaches: The Marines at Peleliu.* Washington, DC: Government Printing Office, 1996.
Geanakoplos, Deno John. *Byzantium: Church, Society, and Civilization Seen Through Contemporary Eyes.* Chicago: University of Chicago Press, 1984.
———. *Interaction of the "Sibling" Byzantine and Western Cultures in the Middle Ages and Italian Renaissance (330–1600).* New Haven: Yale University Press, 1976.

Geary, Dick. "Working-Class Identities in the Third Reich." In *Nazism, War, and Genocide: Essays in Honour of Jeremy Noakes,* edited by Neil Gregor, 42–55. Exeter: University of Exeter Press, 2005.

Geertz, Clifford. *The Interpretation of Cultures: Selected Essays.* New York: Basic Books, 1973.

Gelhorn, Martha. *The Face of War.* London: Hart Davies, 1959.

Gellately, Robert. *Backing Hitler: Consent and Coercion in Nazi Germany.* Oxford: Oxford University Press, 2001.

General Board, European Theater. *Military Justice in Theater of Operations: Study no. 83.* Headquarters, European Theater of Operations, November 1945.

———. *Organization, Equipment, and Tactical Employment of the Airborne Division.* Ft. Leavenworth, KS: US Army Command and General Staff College, Combined Arms Research Library.

General Staff, War Office. *Comrades in Arms: Three Talks to Junior Officers or Officer Cadets to Assist Them in Handling Their Men.* London: HMSO, 1942.

———. *Field Service Pocket Book (1914).* London: HMSO, 1914.

———. *Field Service Pocket Book (1939).* London: HMSO, 1939.

———. *Field Service Regulations,* pt. 1, *Operations (1909).* London: HMSO, 1909.

———. *Field Service Regulations,* vol. 1, *Organization and Administration, 1930: Reprinted with Amendments (Numbers 1–11).* London: HMSO, 1939.

———. *Field Service Regulations,* vol. 2, *Operations (1920).* London: HMSO, 1920.

———. *Field Service Regulations,* vol. 2, *Operations (1924).* London: HMSO, 1924.

———. *Field Service Regulations,* vol. 2, *Operations (1929).* London: HMSO, 1929.

———. *Field Service Regulations,* vol. 2, *Operations, General, 1935 (with Amendments 1 and 2 Annotated 1937).* London: HMSO, 1935.

———. *The Handbook of the German Army, 1940.* Reprint. Nashville, TN: Battery, 1996.

———. *Handbook of the German Army in War, April, 1918.* Reprint. London: Frontline, 2008.

———. *Infantry Section Leading.* London: HMSO, 1938.

———. *Infantry Training,* pt. 8, *Fieldcraft, Battle Drill, Section, and Platoon Tactics.* London: HMSO, 1944.

———. *Infantry Training: Training and War, 1937.* London: HMSO, 1937.

———. *Infantry Training,* vol. 1, *Training.* London: HMSO, 1932.

———. *Infantry Training,* vol. 2, *Provisional.* London: HMSO, 1921.

———. *Infantry Training,* vol. 2, *War.* London: HMSO, 1926.

———. *The Soldier's Welfare: Notes for Officers.* 2nd ed. London: HMSO, 1943.

George, Alexander L. "Primary Groups, Organization, and Military Performance." In *Handbook of Military Institutions,* edited by Roger W. Little, 34–66. Beverly Hills: Sage, 1971.

George, Clay E. *Pilot Studies of Team Effectiveness.* Ft. Benning, GA: US Army Infantry Human Research Unit, 1963.

———. *Some Determinants of Small-Group Effectiveness.* Fort Benning, GA: US Army Infantry Human Research Unit, 1962.

Gerald, Astor. *Operation Iceberg: The Invasion and Conquest of Okinawa in World War II.* New York: D. I. Fine, 1995.

Gerns, Ditte. *Hitlers Wehrmacht in der Sowjetunion: Legenden, Wahrheit, Tradition, Dokumente.* Frankfurt am Main: Verlag Marxistische Blätter, 1985.

Gibb, J. W. *Training in the Army: The Second World War, 1939–1945.* London: HMSO, 1950.

Gibbon, Edward. *The Decline and Fall of the Roman Empire.* Reprint, 1776. Chicago: Encyclopeadia Britannica, 1984.

Gilbert, Felix. "Machiavelli: The Renaissance of the Art of War." In *Makers of Modern Strategy: From Machiavelli to the Nuclear Age,* edited by Peter Paret, 11–31. Princeton: Princeton University Press, 1986.

Gilbert, Martin. *The Second World War: A Complete History.* New York: Holt, 1989.

Ginzburg, Eli. *The Lost Divisions.* New York: Columbia University Press, 1959.

Glantz, David M. *The Battle of Kursk.* Lawrence: University Press of Kansas, 1999.

———. *Before Stalingrad: Barbarossa—Hitler's Invasion of Russia, 1941.* Gloucestershire: Tempus, 2001.

———. *Endgame at Stalingrad.* Lawrence: University Press of Kansas, 2014.

———, ed. *The Initial Period of the War on the Eastern Front, 22 June–August 1941: Proceedings of the Fourth "Art of War Symposium."* 3rd ed. Reprint, 1993. London: Cass, 2001.

———. *To the Gates of Stalingrad: Soviet-German Combat Operations, April–August 1942.* Lawrence: University Press of Kansas, 2009.

———. *When Titans Clashed: How the Red Army Stopped Hitler.* Lawrence: University Press of Kansas, 1995.

———. *Zhukov's Greatest Defeat: The Red Army's Epic Disaster in Operation Mars, 1942.* Lawrence: University Press of Kansas, 1999.

Glantz, David M., and Jonathan M. House. *Armageddon in Stalingrad, September–November 1942.* Lawrence: University Press of Kansas, 2009.

Glass, Charles. *The Deserters: A Hidden History of World War II.* New York: Penguin, 2013.

Glick, Edward Bernard. *Between Israel and Death.* Harrisburg, PA: Stackpole, 1974.

———. *Peaceful Conflict: The Non-Military Use of the Military.* Harrisburg, PA: Stackpole, 1967.

———. *Soldiers, Scholars, and Society: The Social Impact of the American Military.* Pacific Palisades, CA: Goodyear, 1971.

Godfrey, Ernest Gordon, and Robert Frederick Kinglake Goldsmith. *The History of the Duke of Cornwall's Light Infantry, 1939–45.* Aldershot: Duke of Cornwall's Light Infantry (Regimental History Committee), 1966.

Goldhagen, Daniel Jonah. *Hitler's Willing Executioners: Ordinary Germans and the Holocaust.* New York: Knopf, 1996.

Goldhagen, Daniel Jonah, Christopher R. Browning, Leon Wieseltier, Michael Berenbaum, and US Holocaust Research Institute. *The "Willing Executioners"/"Ordinary Men" Debate: Selections from the Symposium, April 8, 1996.* Washington, DC: US Holocaust Research Institute, 1996.

Goldman, Nancy L., and David R. Segal. *The Social Psychology of Military Service.* Beverly Hills: Sage, 1976.

Gole, Henry G. *The Road to Rainbow.* Annapolis, MD: Naval Institute, 2003.

Golovschanski, Anatoly. *"Ich will raus aus diesem Wahnsinn": Deutsche Briefe von der Ostfront, 1941–1945.* Wuppertal: P. Hammer, 1991.

Goodacre, Daniel M., III. "The Use of a Sociometric Test As a Predictor of Combat Unit Effectiveness." *Sociometry* 2, no. 3 (May–June (1951): 148–152.

Goodman, Mitchell. *The End of It: A Novel.* New York: Horizon, 1961.

Gordon, W. "Roman Law in Scotland." In *The Civil Law Tradition in Scotland,* edited by Robin Evans-Jones, 13–40. Edinburgh: Stair Society, 1995.

Graham, Dominick. *Against Odds: Reflections on the Experiences of the British Army, 1914–45.* New York: St. Martin's, 1999.

Graham, Dominick, and Shelford Bidwell. *Tug of War: The Battle for Italy, 1943–45.* Barnsley, South Yorkshire: Pen and Sword, 2004.

Grant, Roderick. *The 51st Highland Division at War.* London: Ian Allen, 1977.

Granzow, Klaus. *Tagebuch eines Hitlerjungen: Kriegsjugend im Pommern, 1943–45.* Bremen: Limes Verlag, 1989.
Gray, J. Glenn. *The Warriors: Reflections on Men in Battle.* New York: Harcourt, Brace, 1959.
Gray, Jeffrey Alan. *The Psychology of Fear and Stress.* Cambridge: Cambridge University Press, 1987.
Green, Peter J. *Dad's War with the United States Marines.* Rockville, MD: Seaboard, 2005.
Greene, Joshua, Shiva Kumar, Steven Gorn, Gerhart Friedlander, Diemut Majer, and William F. Meinecke. *Hitler's Courts: Betrayal of the Rule of Law in Nazi Germany.* Old Westbury, NY: Touro College, Jacob Fuchsberg Law Center, 2006.
Greenlaw, Elizabeth. "'Parade Ground Soldiers': French Army Assessments of the British on the Somme in 1916." *Journal of Military History* 63, no. 2 (1999): 283–312.
Gregor, Manfred. *Die Brücke: Roman.* Wien: K. Desch, 1958.
Gregor, Neil. *Nazism, War, and Genocide: Essays in Honour of Jeremy Noakes.* Exeter: University of Exeter Press, 2005.
Griffith, James. "Multilevel Analysis of Cohesion's Relation to Stress, Well-Being, Identification, Disintegration, and Perceived Combat Readiness." *Military Psychology* 14, no. 3 (2002): 217–239.
Griffith, James, and Mark Vaitkus. "Relating Cohesion to Stress, Strain, Disintegration, and Performance: An Organizing Framework." *Military Psychology* 11, no. 1 (2000): 27–55.
Griffith, Paddy. *Battle Tactics of the Western Front: The British Army's Art of Attack, 1916–18.* New Haven: Yale University Press, 1994.
———. *Forward into Battle: Fighting Tactics from Waterloo to the near Future.* Reprint, 1981. Novato, CA: Presidio, 1991.
Griffith, Robert K. *Men Wanted for the U.S. Army: America's Experience with an All-Volunteer Army Between the World Wars.* Westport: Greenwood, 1982.
Griffith, Samuel B. *The Battle for Guadalcanal.* Philadelphia: Lippincott, 1963.
Griffiths, Bill. *An Introduction to Early English Law.* Norfolk: Anglo-Saxon, 1995.
Grinker, Roy R. *Men Under Stress.* Reprint, 1945. New York: McGraw-Hill, 1963.
Groehler, Olaf. *Selbstmörderische Allianz: Deutsche-Russische Militärbeziehungen, 1920–1941.* Berlin: Vision Verlag, 1992.
Groß, Gerhard P. *Mythos und Wirklichkeit: Geschichte des operativen Denkens im Deutschen Heer von Moltke d.A. bis Heusinger.* Paderborn: Ferdinand Schoeningh, 2012.
Grossjohan, Georg. *Five Years, Four Fronts: A German Officer's World War II Combat Memoir.* New York: Random, 2005.
Grossman, Dave. *On Killing: The Psychological Cost of Learning to Kill in War and Society.* Boston: Little, Brown, 1995.
Grossman, Vasily. *A Writer at War: A Soviet Journalist with the Red Army, 1941–1945.* Edited by Antony Beevor. New York: Random, 2005.
Grossmann, Horst. *Geschichte der Rheinisch-Westfalischen 6: Infanterie-Division, 1939–1945.* Bad Nauheim: H. H. Podzun, 1958.
Grupe, Friedrich. *Jahrgang 1916: Die Fahne war mehr als der Tod.* Munich: Universitas, 1989.
Guderian, Hans. "Die Panzertruppen und ihr Zusammenwirken mit den anderen Waffen." *Militärwissenschaftliche, Rundschau* 1, no. 5 (1936): 607–627.
———. "Kraftfahrkampftruppen." *Militärwissenschaftliche, Rundschau* 1, no. 1 (1936): 52–78.
Guderian, Heinz. *Achtung-Panzer! Die Entwicklung der Panzerwaffe, Ihre Kampftaktik und Ihre operativen Möglichkeiten.* Stuttgart: Union Deutsche Verlagsgesellschaft, 1937.

———. *Die Panzertruppen und ihr Zusammenwirken mit den anderen Waffen.* Berlin: Mittler, 1938.
———. *Panzer Leader.* New York: Dutton, 1952.
Gurfein, M. I., and Morris Janowitz. "Trends in Wehrmacht Morale." In *Propaganda in War and Crisis,* edited by Daniel Lerner, 200–208. New York: Arno, 1972.
Guy, Alan J., and Peter B. Boyden. *Soldiers of the Raj: The Indian Army, 1600–1947.* London: National Army Museum Chelsea, 1997.
Haase, Norbert, and Otl Aicher. *Deutsche Deserteure.* Berlin: Rotbuch Verlag, 1987.
Habeck, Mary R. *Storm of Steel: The Development of Armor Doctrine in Germany and the Soviet Union, 1919–1939.* Ithaca: Cornell University Press, 2003.
Hackett, John. *The Profession of Arms.* London: Sedgwick and Jackson, 1983.
Hackett, John Winthrop. "The Profession of Arms: The 1962 Lees Knowles Lectures Given at Trinity College, Cambridge." Washington, DC: US Army Center for Military History, 1986.
Halder, Franz. *The Halder Diaries: The Private War Journals of Colonel General Franz Halder.* 2 vols. Reprint, 1946. Boulder: Westview, 1976.
Haldon, John. *Byzantium at War, 600–1453.* Oxford: Osprey, 2002.
Halleck, H. Wager. *Military Art and Science.* New York: Appleton, 1862.
Hamburger Institut für Sozialforschung. *The German Army and Genocide: Crimes Against War Prisoners, Jews, and Other Civilians in the East, 1939–1944.* New York: New Press, 1999.
Hamerow, Theodore S. *The Social Foundations of German Unification, 1858–1871.* Princeton: Princeton University Press, 1969.
Hamilton, Nigel. *Master of the Battlefield: Monty's War Years, 1942–1944.* New York: McGraw-Hill, 1983.
———. *Monty: The Field Marshal, 1944–1976.* London: Hamilton, 1986.
———. *Monty: The Making of a General, 1887–1942.* New York: McGraw-Hill, 1981.
Hamilton, Richard, and Herwig, Holger H., eds. *War Planning 1914.* Cambridge: Cambridge University Press, 2010.
Hancock, M. Donald. *The Bundeswehr and the National People's Army: A Comparative Study of German Civil-Military Polity.* Denver: University of Denver Press, 1973.
Hanson, Victor Davis. *Carnage and Culture: Landmark Battles in the Rise of Western Power.* New York: Random, 2002.
———, ed. *Makers of Ancient Strategy: From the Persian Wars to the Fall of Rome.* New York: Oxford University Press, 2010.
———. *Ripples of Battle: How the Wars of the Past Still Determine How We Fight, How We Live, and How We Think.* New York: Anchor, 2003.
———. *A War Like No Other: How the Athenians and the Spartans Fought the Peloponnesian War.* New York: Random, 2005.
———. *Western Way of War: Infantry Battle in Classical Greece.* Berkeley: University of California Press, 1989.
Hardach, Gerd. *The First World War, 1914–1918.* Berkeley: University of California Press, 1977.
Harman, Nicholas. *Dunkirk: The Patriotic Myth.* New York: Simon and Schuster, 1980.
Harnack, Adolf. *Militia Christi: The Christian Religion and the Military in the First Three Centuries.* Translated by David McGinnes Gracie. Reprint, 1905. Philadelphia: Fortress, 1981.
Harries-Jenkins, Gwyn, and Jacques van Doorn. *The Military and the Problem of Legitimacy.* Beverly Hills: Sage, 1976.
Harrison, Gordon A. *Cross-Channel Attack.* Washington, DC: Government Printing Office, 1951.

Harrison, Mark. *Medicine and Victory: British Military Medicine in the Second World War.* New York: Oxford University Press, 2004.

Harrison, T. "The British Soldier: Changing Attitudes." *British Journal of Psychology* 19 (1944–1945): 34–39.

Hart, Stephen A. *Montgomery and "Colossal Cracks": The 21st Army Group in Northwest Europe, 1944–45.* London: Praeger, 2000.

———. "Montgomery, Morale, Casualty Conservation and 'Colossal Cracks': 21st Army Group's Operational Technique in North West Europe, 1944–1945." *Journal of Strategic Studies* 19, no. 4 (1996): 132–153.

Hastings, Max. *Armageddon: The Battle for Germany, 1944–45.* New York: Knopf, 2004.

———. *Bomber Command.* New York: Dial Press and J. Wade, 1979.

———. *Overlord: D-Day and the Battle for Normandy.* New York: Simon and Schuster, 1984.

Hattendorf, John B. "Henry Lloyd and the Military Enlightenment of Eighteenth-Century Europe." *Albion: A Quarterly Journal Concerned with British Studies* 36, no. 1 (2004): 141–142.

Haupt, W. *Elite German Divisions in World War II: Waffen-SS, Fallschirmjäger, Mountain Troops, 1995.* Translated by David Johnston. Atglen, PA: Schiffer, 2001.

———. *Sturmfahrt nach Riga: Juni 1941 (das Infanterieregiment 43 der 1.Infanteriedivision); deutsche Vorausabteilungen im Baltikum; das erbitterte Ringen um die lettische Stadt.* Rastatt: Pabel, 1976.

Hauptmann im Generalstab. "Mittel und Wege Der Schwerpunktbildung." *Militär-Wochenblatt* 116, no. 6 (1931): 207–211.

Hayes, Denis. *Conscription Conflict: The Conflict of Ideas in the Struggle for and Against Military Conscription in Britain Between 1901 and 1939.* London: Sheppard, 1949.

Heal, Felicity. *Reformation in Britain and Ireland.* New York: Oxford University Press, 2003.

Heather, Peter J. "Holding the Line: Frontier Defense and the Later Roman Empire." In *Makers of Ancient Strategy: From the Persian Wars to the Fall of Rome,* edited by Victor Davis Hanson, 227–246. Princeton: Princeton University Press, 2010.

Heer, Hannes, and Klaus Naumann. *War of Extermination: The German Military in World War II, 1941–1944.* New York: Berghahn, 1999.

Heeres, Oberkommando des Heeresdienstvorschrift. *Ausbildungsdiens-tvorschrift für die Infanterie.* Berlin: Mittler und Sohn, 1938.

———. *Ausbildungsvorschrift für die Infanterie Heft V.: A. Das Infanteriebataillon; B. Das Infanterieregiment.* Berlin: Mittler und Sohn, 1935.

———. "Der Sturmangriff: Kriegserfahrungen eines Frontoffiziers on 1917." Reprint. Berlin: Mittler und Sohn, 1945.

———. "Merkblatt 5a/17 Prüfungsordnung für Unteroffizier-Beweber des Heeres." Berlin: Oberkommando des Heeres, 1943.

———. *Taschenbuch für den Winterkrieg: Nachdruck mit eingearbeiteten Ergänzungen nach dem Stande vom 1.November 1942.* Berlin: Zander, 1942.

———. *Truppenführung.* Berlin: Verlag von E. S. Mittler und Sohn, 1936.

Heeresleitung, Dienst Vorschrift. *Führung und Gefecht der verbundenen Waffen (F.u.G.) vom 1.September 1920.* Berlin: Verlag: "Offene Worte" Charlottenburg, 1930.

Heinl, Robert Debs, Jr. *Dictionary of Military and Naval Quotations.* Annapolis, MD: Naval Institute, 1966.

Heinrich, Willi. *The Cross of Iron.* Translated by Richard Winston and Clara Winston. Indianapolis: Bobbs-Merrill, 1956.

———. *The Willing Flesh: A Novel.* Translated by Richard Winston and Clara Winston. London: Weidenfeld and Nicolson, 1969.
Heller, Joseph. *Catch-22: A Novel.* New York: Simon and Schuster, 1961.
Hemingway, Ernest. *Across the River and into the Trees.* New York: Scribner, 1950.
———, ed. *Men at War: The Best War Stories of All Time.* Updated from 1942 and 1955 editions. New York: Bramhall, 1979.
Henderson, Diana. *Highland Soldier: A Social Study of the Highland Regiments, 1820–1920.* Edinburgh: J. Donald, 1989.
Henderson, William Darryl. *Cohesion: The Human Element in Combat—Leadership and Societal Influence in the Armies of the Soviet Union, the United States, North Vietnam, and Israel.* Washington, DC: National Defense University Press, 1985.
Herodotus. *The History.* Translated by David Grene. Chicago: University of Chicago Press, 1987.
Hersey, John. *A Bell for Adano.* New York: Knopf, 1944.
———. *The War Lover.* New York: Knopf, 1959.
Hershey, Lewis B. *Special Monograph no. 5: The Classification Process.* Washington, DC: Government Printing Office, 1950.
———. *Special Monograph no. 15: Physical Examination of Selective Ser-vice Registrants.* Washington, DC: Government Printing Office, 1948.
———. *Special Monograph no. 16: Problems of Selective Service.* Washington, DC: Government Printing Office, 1952.
Hillgruber, Andreas. *Deutschlands Rolle in der Vorgeschichte der beiden Weltkriege.* Gottlingen: Vandenhoek and Ruprecht, 1967.
———. *Germany and the Two World Wars.* Cambridge: Harvard University Press, 1981.
———. *Probleme des zweiten Weltkrieges.* Berlin: Kiepenheuer and Witsch, 1967.
Hitler, Adolf. *Mein Kampf.* Translated by Ralph Manheim. Reprint, 1925. New York: Houghton Mifflin, 1999.
Hoehne, Gustav. *In Snow and Mud: 31 Days of Attack Under Seydlitz During Early Spring of 1942.* New York: Garland, 1979.
Hoffmeier, Heinrich. *Ich habe keine Hoffnung mehr: Soldatenbriefe aus Russland, 1942–1943.* Berlin: Quintessenz Verlags, 2004.
Hogg, Michael A., ed. *The Sage Handbook of Social Psychology.* London: Sage, 2003.
Hogg, Michael A., and Dominic Abrams. *Social Identifications: A Social Psychology of Intergroup Relations and Group Processes.* London: Routledge, 1988.
Holborn, Hajo. *A History of Modern Germany.* New York: Knopf, 1969.
———. "The Prusso-German School: Moltke and the Rise of the General Staff." In *Makers of Modern Strategy from Machiavelli to the Nuclear Age,* edited by Peter Paret, Gordon A. Craig, and Felix Gilbert, 281–295. Princeton: Princeton University Press, 1986.
Horne, Alistair. *To Lose a Battle: France 1940.* Boston: Little, Brown, 1969.
Horne, Christine. "The Enforcement of Norms: Group Cohesion and Meta-Norms." *Social Psychology Quarterly* 64, no. 3 (2001): 253–266.
Hossbach, Friedrich. *Infanterie im Ostfeldzug, 1941/42.* Osterode: Giebel and Oehlschlägel, 1951.
———. *Von der militärischen Verantwortlichkeit in der Zeit vor dem zweiten Weltkriege.* Göttingen: Vandenhoeck and Ruprecht, 1948.
———. *Zwischen Wehrmacht und Hitler, 1934–1938.* Göttingen: Vandenhoeck and Ruprecht, 1965.
Houle, Cyril O., Elbert W. Burr, Thomas H. Hamilton, and John R. Yale. *The Armed Services and Adult Education.* Washington, DC: American Council on Education, 1947.

Houlihan, Thomas L., III, ed. *Kriegsprache.* Lake Orion, MI: Maps at War, 2009.
Houston, Donald Eugene. *Hell on Wheels: The 2d Armored Division.* San Rafael, CA: Presidio, 1977.
Howard, Michael. *The Franco-Prussian War: The German Invasion of France, 1870–71.* Reprint, 1961. London: Rupert Hart-Davis, 2000.
———. "World War One: The Crisis in European History—The Role of the Military Historian." *Journal of Military History* 57, no. 2 (1993): 127–238.
Howard, Michael, and Basil Henry Liddell Hart. *The Theory and Practice of War.* Bloomington: Indiana University Press, 1975.
Hoyt, Edwin Palmer. *Stalin's War: Tragedy and Triumph, 1941–1945.* New York: Cooper Square, 2003.
———. *To the Marianas: War in the Central Pacific, 1944.* New York: Van Nostrand Reinhold, 1980.
Hudson, C. E. "Infantry Require Cover from Flanking Machine-Gun Fire." *Journal of the Royal United Services Institute* 67, no. 466 (1922): 366–371.
Hughes, Daniel J. *Moltke on the Art of War: Selected Writings.* Novato, CA: Presidio, 1993.
"The Human Element in War." *Infantry Journal* 30, no. 5 (May 1927): 463.
Ikenberry, G. John. *After Victory: Institutions, Strategic Restraint, and the Rebuilding of Order After Major Wars.* Princeton: Princeton University Press, 2001.
Industrial College of the Armed Forces. *Cohesion in the US Military/Defense Management Study Group on Military Cohesion.* Washington, DC: National Defense University Press, 1984.
Ingraham, Harry H., and Frederick J. Manning. "Cohesion: Who Needs It, What Is It, and How Do We Get It to Them?" *Military Review* (June 1981): 3–12.
International Institute for Strategic Studies. *Force in Modern Societies: The Military Profession.* London, 1973.
International Military Tribunal. *Nazi Conspiracy and Aggression: Supplement A.* Washington, DC: Government Printing Office, 1947.
Irgang, Frank J. *Etched in Purple: One Soldier's War in Europe.* Caldwell, ID: Caxton, 1949.
Iriye, Akire. *The Globalizing of America.* Vol. 3. New York: Cambridge University Press, 1993.
Irrgang, Astrid Julia. *Leutnant der Wehrmacht Peter Stölten in seinen Feldpostbriefen: Vom richtigen Leben im Falschen.* Freiburg im Breisgau: Rombach Verlag, 2007.
Isely, Jeter A., and Philip A. Crowl. *The U.S. Marines and Amphibious War.* Reprint, 1951. Princeton: Princeton University Press, 1979.
Isserson, G. S. "The Evolution of Operational Art." Translated by Bruce W. Menning. Ft. Leavenworth, KS: SAMS Theoretical, 2005.
Jackman, Steven D. "Shoulder to Shoulder: Close Control and 'Old Prussian Drill' in German Offensive Infantry Tactics, 1871–1914." *Journal of Military History* 68, no. 1 (January 2004): 73–104.
Jacobsen, Hans Adolf, and Oleg Prudkow. *Deutsch-Russische Zeitenwende: Krieg und Frieden, 1941–1995.* Baden-Baden: Nomos Verlagsgesellschaft, 1995.
Jahoda, Gustav. "'Ordinary Germans' Before Hitler: A Critique of the Goldhagen Thesis." *Journal of Interdisciplinary History* 29, no. 1 (1998): 69–88.
Janowitz, Morris. *The New Military: Changing Patterns of Organization.* New York: Sage, 1964.
———. *The Professional Soldier: A Social and Political Portrait.* Glencoe, IL: Free Press, 1960.
———. *Sociology and the Military Establishment.* London: Kegan Paul Routledge, 1959.

Janowitz, Morris, and Stephen D. Westbrook. *The Political Education of Soldiers.* Beverly Hills: Sage, 1983.

Jenkins, John G. "New Opportunities and New Responsibilities for the Psychologist." In *New Methods in Applied Psychology,* edited by George A. Kelly, 1–13. College Park: University of Maryland Press, 1947.

Johns, John H. *Cohesion in the U.S. Military.* Washington, DC: National Defense University Press, 1984.

Johnston, David. "The General Influence of Roman Institutions of State and Public Law." In *The Civilian Tradition and Scots Law,* translated by Carey Miller and R. Zimmermann, 87–101. Berlin: Duncker and Humboldt, 1997.

Joint Board. *Joint Action of the Army and the Navy.* Washington, DC: Government Printing Office, 1936.

Jomini, Henri Antoine. *The Art of War.* Translated by G. H. Mendell and W. P. Craighill. Mechanicsburg, PA: Stackpole, 1996.

Jones, Edgar. "'LMF': The Use of Psychiatric Stigma in the Royal Air Force During the Second World War." *Journal of Military History* 70, no. 2 (April 2006): 439–58.

———. "The Psychology of Killing: The Combat Experience of British Soldiers During the First World War." *Journal of Contemporary History* 41, no. 2 (April 2006): 229–246.

Jones, Franklin D., et al. *War Psychiatry.* Washington, DC: Government Printing Office, 1995.

Jones, James. *From Here to Eternity.* New York: Scribner, 1951.

———. *The Thin Red Line.* New York: Scribner, 1962.

———. *Whistle.* New York: Delacorte, 1978.

Joslen, H. F. *Orders of Battle.* Vol. 1, *United Kingdom and Colonial Formations and Units in the Second World War, 1939–1945.* London: HMSO, 1960.

Jung, Hans Joachim. *The History of Panzerregiment "Grossdeutschland."* Winnipeg: J. J. Fedorowicz, 1990.

Kalmbach, Peter. *Wehrmachtjustiz.* Berlin: Metropol Verlag, 2012.

Kann, Robert A. "Reflections on the Causes of Eighteenth-Century Warfare in Europe." In *East Central European Society and War in the Pre-Revolutionary Eighteenth Century,* edited by Gunther E. Rothenberg, Bèla K. Király, and Peter F. Sugar, 29–48. Boulder: Social Science Monograph, 1982.

Kannapin, Norbert. *Die deutsche Feldpost, 1939–1945: Organisation und Lokalisation der Feldpostämter und Feldpostdienstellen.* Osnabrück: Biblio Verlag, 1987.

Kappeler, Andreas. *The Russian Empire.* Essex: Pearson Education, 2001.

Karschkes, Helmut. *Menschlichkeit im Krieg: Soldaten der deutschen Wehrmacht und ihre gegner Berichten.* Stuttgart: Leopold Stocker Verlag, 2000.

Karsten, Peter, ed. *Civil-Military Relations.* Portland, OR: Taylor and Francis, 1998.

———. *Law, Soldiers, and Combat: The Causes of War Crimes.* Westport: Greenwood, 1978.

———, ed. *The Military in America: From the Colonial Era to the Present.* New York: Free Press, 1986.

———, ed. *Military, State, Society Symbiosis.* Portland, OR: Taylor and Francis, 1998.

———, ed. *Motivating Soldiers: Morale or Mutiny.* Portland, OR: Taylor and Francis, 1998.

———. *Patriot-Heroes in England and America.* Madison: University of Wisconsin Press, 1978.

———. *Recruiting, Drafting, and Enlisting: Two Sides of the Raising of Military Forces.* New York: Garland, 1998.

———. *Soldiers and Society: The Effects of Military Service and War on American Life.*Westport: Greenwood, 1978.

———. *Theory, Method, and Practice in Social and Cultural History.* New York: New York University Press, 1993.
———, ed. *Training and Socializing of Military Personnel.* Portland, OR: Taylor and Francis, 1998.
Kautz, Fred. *The German Historians: Hitler's Willing Executioners and Daniel Goldhagen.* Montreal: Black Rose, 2003.
Keast, William R. *Study no. 6: The Procurement and Branch Distribution of Officers.* Washington, DC: Historical Section, US Army Ground Forces, 1946.
———. *Study no. 7: Provision of Enlisted Replacements.* Washington, DC: Historical Section, US Army Ground Forces, 1946.
———. *Study no. 32: Major Developments in the Training of Enlisted Replacements.* Washington, DC: Historical Section, US Army Ground Forces, 1946.
Keefer, Louis E. *Scholars in Foxholes: The Story of the Army Specialized Training Program in World War II.* Jefferson, NC: McFarland, 1988.
Keegan, John. *The Battle for History: Re-Fighting World War II.* New York: Vintage, 1996.
———. *The Book of War.* New York: Penguin Putnam, 1999.
———. *The Face of Battle.* New York: Random, 1976.
———. *The First World War.* New York: Vintage, 1998.
———. "Regimental Ideology." In *War, Economy, and the Military Mind,* edited by G. Best and A. Wheatcroft, 10–24. London: Croom Helm, 1976.
———. *The Second World War.* London: Penguin, 1990.
———. *Six Armies in Normandy: From D-Day to the Liberation of Paris, June 6th–August 25th, 1944.* London: Jonathon Cape, 1982.
Keegan, John, and Joseph Darracott. *The Nature of War.* New York: Holt, Rinehart, and Winston, 1981.
Keegan, John, Richard Holmes, and John Gau. *Soldiers: A History of Men in Battle.* London: Hamilton, 1985.
Keene, Jennifer Diane. "Intelligence and Morale in the Army of a Democracy: The Genesis of Military Psychology During the First World War." *Military Psychology* 6, no. 4 (1994): 235–253.
Kegley, Charles W., Jr., and Gregory A. Raymond. "The Long Cycle of Global War and the Transformation of Alliance Norms." *Journal of Peace Research* 26, no. 3 (1989): 265–284.
Kellett, Anthony. *Combat Motivation: The Behavior of Soldiers in Battle.* Boston: Kluwer-Nijhoff, 1982.
Kelly, George A. "New Methods in Applied Psychology." Paper presented at the Maryland Conference on Military Contributions to Methodology in Applied Psychology, College Park, 1947.
Kennedy, Paul M. *Engineers of Victory: The Problem Solvers Who Turned the Tide in the Second World War.* New York: Random, 2013.
———. *The Rise and Fall of British Naval Mastery.* Amherst, NY: Humanity, 1998.
———. *The Rise and Fall of the Great Powers: Economic Change and Military Conflict from 1500–2000.* New York: Random, 1987.
Kennedy, Robert M. *The German Campaign in Poland (1939).* Reprint. Washington, DC: Government Printing Office, 1956.
Kennett, Lee B. *G.I.: The American Soldier in World War II.* New York: Scribner, 1987.
Kershaw, Ian. *The End: The Defiance and Destruction of Hitler's Germany, 1944–45.* New York: Penguin, 2011.
Kersting, Franz-Werner. "Book Review: Wehrmacht zwischen Tradition und Ideologie—Der NS-Führungsoffizier im zweiten Weltkrieg." *Historische Zeitschrift* 255, no. 3 (1992): 791–792.

Kier, Elizabeth. *Imagining War: French and British Military Doctrine Between the Wars.* Princeton: Princeton University Press, 1997.

Kilian, Katrin A. "Kriegsstimmungen: Emotionen einfacher Soldaten in Feldpostbriefen." In *Die deutsche Gesellschaft, 1939 bis 1945: Politisierung, Vernichtung, Überleben,* vol. 9/2, *Das Deutsche Reich und der Zweite Weltkrieg,* edited by Jörg Echternkamp, 251–288. Munich: Deutsche Verlags-Anstalt, 2005.

Kimball, Warren F. "Stalingrad: A Chance for Choices." *Journal of Military History* 60, no. 1 (January 1996): 89–114.

Klee, Ernst, and Willi Dressen. *"Gott Mit Uns": Der Deutsche Vernichtungskrieg im Osten, 1939–1945.* Frankfurt: S. Fisher, 1989.

Knappe, Siegfried, and Ted Brusaw. *Soldat: Reflections of a German Soldier.* New York: Orion, 1992.

Knoch, Peter, ed. *Kriegsalltag: Die Rekonstruktion des Kriegsalltags als Aufgabe der historischen Forschung und der Friedenserziehung.* Stuttgart: Metzler, 1989.

Koch, H. W. *A Constitutional History of Germany in the Nineteenth and Twentieth Centuries.* New York: Longman, 1984.

Korn, James H. *Illusions of Reality: A History of Deception in Social Psychology.* Albany: State University of New York Press, 1997.

Korol, V. E. "The Price of Victory: Myths and Reality." *Journal of Slavic Military Studies* 9, no. 2 (1996): 417–426.

Kreidberg, Marvin A., and Merton G. Henry. *History of Military Mobilization in the United States Army, 1775–1945.* Washington, DC: Government Printing Office, 1955.

Kretschmer, Volcker, and Detlef Vogel. "Feldpostbriefe im zweiten Weltkrieg: Propagandainstrument und Spiegelbild von Kriegswirkungen." *Sozialwissenschaftliche Informationen* 2 (1990): 103–110.

Kriegsministerium. *Exerzier-Reglement für die Infanterie.* Berlin: Mittler und Sohn, 1888.

———. *Exerzier-Reglement für die Infanterie.* Berlin: Mittler und Sohn, 1906.

———. *Grundzuege der hoeheren Truppenfuehrung.* Berlin: Reichsdruckerei, January 1910.

Kroener, Bernhard. *Generaloberst Friedrich Fromm, eine Biographie: "Der Starke Mann in Heimatskriegsgebiet."* Paderborn: Ferdinand Schöningh, 2005.

Kroener, Bernhard, Rolf-Dieter Müller, and Hans Umbreit. *"Auf Dem Weg Zu Einer 'Nationalsozialistischen Volksarmee': Die Soziale Öffnung des Heeresoffizierkorps im Zweiten Weltkrieg."* In *Von Stalingrad Zur Währungsreform: Zur Sozialgeschichte des Umbruchs in Deutschland,* edited by Martin Broszat, Klaus-Dietmar, Henke and Hans Woller. Munich: Oldenbourg, 1988.

———. "Generationserfahrungen und Elitenwandel: Strukturveränderungen im Deutschen Offizierkorps, 1933–1945." In *Eliten in Deutschland und Frankreich im 19. und 20.Jahrhundert: Strukturen und Beziehungen,* edited by Rainer Hudemann and Georges-Henri Soutu. Darmstadt: Wissenschaftliche Buchgesellschaft, 1991.

———. *Organisation und Mobilisierung des deutschen Machtbereichs: Kriegsverwaltung, Wirtschaft und personelle Ressourcen, 1939–1941.* Vol. 5/1, *Das Deutsche Reich und der Zweite Weltkrieg.* Stuttgart: Deutsche Verlags-Anstalt, 1988.

———. *Organisation und Mobilisierung des deutschen Machtbereichs: Kriegsverwaltung, Wirtschaft und personelle Ressourcen, 1942–1944/45.* Vol. 5/2, *Das Deutsche Reich und der Zweite Weltkrieg.* Stuttgart: Deutsche Verlags-Anstalt, 1999.

———. *"'Störer' und 'Versager': Die Sonderabteilungen Der Wehrmacht— Soziale Disziplinierung Aus Dem Geist des Ersten Weltkrieges."* In *Adel—Geistlichkeit—Militär,* edited by Michael Busch and Jörg Hillman. Bochum: Winkler, 1999.

Krüger, Horst. *Das zerbrochene Haus: Eine Jugend in Deutschland.* Munich: Rütten and Loening, 1966.

Kuby, Erich. *Mein Krieg: Aufzeichnung aus 2129 Tagen.* Munich: Nymphenburger Verlagshandlung, 1975.
Kuhn, Thomas. *The Structure of Scientific Revolutions.* Chicago: University of Chicago Press, 1996.
Kühne, Thomas. "Kameradschaft: 'Das Beste im Leben des Mannes'—Die deutschen Soldaten des zweiten Weltkriegs in Erfahrungs- und geschlechtlicher Perspektive." *Geschichte und Gesellschaft* 22, no. 4 (1996): 504–529.
———. *Kameradschaft: Die Soldaten des nationalsozialistischen Krieges und das 20.Jahrhundert.* Göttingen: Vandenhoeck and Rupprecht, 2006.
———. "Gruppenkohäsion und Kameradschaftsmythos in der Wehrmacht." In *Die Wehrmacht Mythos und Realität,* edited by Rolf-Dieter Müller and Hans-Erich Volkmann. Munich: Oldenbourg Verlag, 2012.
———. "Zwischen Männerbund und Volksgemeinschaft: Hitlers Soldaten und der Mythos der Kameradschaft." *Archiv für Sozialgeschichte* 22 (1998): 504–529.
Laffin, John. *British VC's of World War 2: A Study in Heroism.* Gloucestershire: Sutton, 1997.
———. *Jackboot: The Story of the German Soldier.* New York: Barnes and Noble, 1995.
Lambert, A. F. "Army Recruiting." *Journal of the Royal United Services Institute* 80, no. 519 (1935): 572–679.
Lang, Kurt. *Military Institutions and the Sociology of War: A Review of the Literature with Annotated Bibliography.* Beverly Hills: Sage, 1972.
Lange, Horst. *Tagebücher aus dem zweiten Weltkrieg.* Mainz: V. Hase und Koehler, 1979.
Lau, Thomas. *Normenwandel der deutschen militärischen Elite seit 1918.* New York: P. Lang, 1988.
Lawrence, T. E. "Evolution of a Revolt." *Army Quarterly and Defence Journal* (October 1920). Ft. Leavenworth, KS: CSI, 1990.
Ledig, Gert. *The Stalin Front: A Novel of World War II.* Translated by Michael Hofmann. New York: New York Review of Books, 2004.
———. *The Stalin Organ.* London: Granta, 2004.
———. *The Tortured Earth.* Translated by Mervyn Savill. Chicago: H. Regnery, 1956.
Lee, David. *Up Close and Personal: The Reality of Close-Quarter Fighting in World War II.* London: Greenhill, 2006.
Leed, Eric J. *No Man's Land: Combat and Identity in World War I.* Cambridge: Cambridge University Press, 1979.
Leistenschneider, Stephan. *Auftragstaktik im preußen-deutschen Heer, 1871 bis 1914.* Berlin: Mittler und Sohn, 2002.
Lejeune, John A. *The Reminiscences of a Marine.* Reprint, 1930. Philadelphia: Dorrance, 1990.
Lenz, Siegfried. *Ein Kriegsende.* Hamburg: Hoffmann und Campe, 1984.
———. *Ein Kriegsende Siegfried Lenz, seine Roman-Klassiker und sein Leben.* Hamburg: Studio Hamburg, 2008.
Lerner, Paul. "Psychiatry and Casualties of War in Germany, 1914–18." *Journal of Contemporary History* 35, no. 1 (January 2000): 13–28.
Lerwill, Leonard. *The Personnel Replacement System in the United States Army.* Washington, DC: Government Printing Office, 1954.
L'Etang, Hugh. *Fit to Lead?* London: Heinemann Medical, 1980.
———. *The Pathology of Leadership.* New York: Hawthorn, 1970.
Levi, Margaret. "The Institution of Conscription." *Social Science History* 20, no. 1 (Spring 1996): 133–167.
Lewis, Adrian R. *Omaha Beach: A Flawed Victory.* Chapel Hill: University of North Carolina Press, 2001.
Lewis, S. J. *Forgotten Legions.* New York: Praeger, 1985.

Lichtenstern, Karl. *Schießausbildung und Feuer der Infanterie im Gefecht.* Berlin: Mittler und Sohn, 1897.
Liddell Hart, Basil Henry. "Armoured Forces in 1928." *Journal of the Royal United Services Institute* 73, no. 492 (1928): 720–729.
———. "Army Manoeuvres, 1925." *Journal of the Royal United Services Institute* 70 (1925): 647–655.
———. "Army Training, 1927: Conversion by Demonstration." *Journal of the Royal United Services Institute* 72, no. 487 (1927): 746–754.
———. "The 'Man-in-the-Dark' Theory of Infantry Tactics and the 'Expanding Torrent' System of Attack." *Journal of the Royal United Services Institute* 66, no. 461 (1921): 1–22.
———. *The Other Side of the Hill: Germany's Generals, Their Rise and Fall, with Their Own Account of Military Events, 1939–1945.* London: Cassell, 1948.
———. "Suggestions on the Future Development of the Combat Unit: The Tank as a Weapon of Infantry." *Journal of the Royal United Services Institute* 64, no. 456 (1919): 666–669.
———. "The 'Ten Commandments' of the Combat Unit: Suggestions on Its Theory and Training." *Journal of the Royal United Services Institute* 64, no. 454 (1919): 289–293.
Lince, George. *Too Young the Heroes: A World War II Marine's Account of Facing a Veteran Enemy at Guadalcanal, the Solomons, and Okinawa.* Jefferson, NC: McFarland, 1997.
Lincoln, John. *Thank God and the Infantry: From D-Day to VE-Day with the 1st Battalion, the Royal Norfolk Regiment.* Stroud, Gloucestershire: Sutton, 1994.
Linderman, Gerald F. *Embattled Courage: The Experience of Combat in the American Civil War.* New York: Free Press, 1989.
———. *The World Within War: America's Combat Experience in World War II.* Cambridge: Harvard University Press, 1999.
Linklater, Eric. *The Highland Division.* London: HMSO, 1942.
Linn, Brian McAllister. *The Echo of Battle: The Army's Way of War.* Cambridge: Harvard University Press, 2007.
Little, Roger W. "Buddy Relations and Combat Performance." In *The New Military: Changing Patterns of Organization,* edited by Morris Janowitz, 195–224. New York: Sage, 1964.
———. "Handbook of Military Institutions." Paper presented at the Inter-University Seminar on Armed Forces and Society, Beverly Hills, 1971.
———. *Selective Service and American Society.* New York: Sage, 1969.
Löffler, Klara. *Aufgehoben: Soldatenbriefe aus dem zweiten Weltkrieg—Eine Studie zur subjectiven Wirklichkeit des Krieges.* Bamburg: WVB, 1992.
Lorenz, Konrad. *On Aggression.* New York: Bantam, 1967.
Lucas, James. *Last Days of the Reich: The Collapse of Nazi Germany, May 1945.* London: Arms and Armour, 1986.
———. *War in the Desert: The Eighth Army at El Alamein.* New York: Beaufort, 1982.
———. *War on the Eastern Front, 1941–1945: The German Soldier in Russia.* New York: Stein and Day, 1980.
———. *World War Two Through German Eyes.* New York: Arms and Armour, 1988.
Lupfer, Timothy T. *The Dynamics of Doctrine: The Changes in German Tactical Doctrine During the First World War.* Ft. Leavenworth, KS: Combat Studies Institute, 1981.
Luvaas, Jay, ed. *Frederick the Great on the Art of War.* Reprint, 1966. New York: Da Capo, 1999.
Luvaas, Jay, and Thomas E. Griess, eds. *The Regulations of the Prussian Infantry.* Translated by William Faucitt. Reprint, 1759. New York: Greenwood, 1968.

Maaß, Winfried. *Die Funfzigjährigen: Porträt einer verratenen Generation.* Hamburg: Hoffmann und Campe, 1980.
MacDonald, Charles B. *Company Commander.* Reprint, 1949. Short Hills, NJ: Burford, 1999.
———. *The Last Offensive.* Washington, DC: Government Printing Office, 1973.
———. *The Siegfried Line Campaign.* Washington, DC: Government Printing Office, 1963.
Machiavelli, Niccolò. *The Art of War.* Translated and edited by Christopher Lynch. Chicago: University of Chicago Press, 2005.
MacIsaac, David. "Voices from the Central Blue." In *Makers of Modern Strategy: From Machiavelli to the Nuclear Age,* edited by Peter Paret, 624–647. Princeton: Princeton University Press, 1986.
Madden, Brian John George. *A History of the 6th Battalion, the Black Watch: (Royal Highland Regiment) 1939–1945.* Perth: D. Leslie, 1948.
Madden, Stuart. "The Graeco-Roman Antecedents of Modern Tort Law." Bepress Legal Series, Paper 823 (2005): 1–59.
Madej, Victor. "Effectiveness and Cohesion of the German Ground Forces in World War II." *Journal of Political and Military Sociology* 6 (Fall 1978): 233–248.
Maehl, William Harvey. *Germany in Western Civilization.* Tuscaloosa: University of Alabama Press, 1979.
Mael, Fred A., and Cathie E. Alderks. "Leadership Team Cohesion and Subordinate Work Unit Morale." *Military Psychology* 5, no. 3 (1993): 141–158.
Magenheimer, Heinz. *Hitler's War: Germany's Key Strategic Decisions.* Translated by Helmut Bogler. London: Cassell, 1998.
Mahon, John K., and Romana Danysh. *Infantry Part I: Regular Army.* Washington, DC: Government Printing Office, 1972.
Maier, Klaus A., Horst Rohde, Bernd Stegemann, and Hans Umbreit. *Die Errichtung der Hegemonie auf dem europäischen Kontinent.* Vol. 2, *Das Deutsche Reich und der Zweite Weltkrieg.* Stuttgart: Deutsche Verlags-Anstalt, 1979.
Mailer, Norman. *The Naked and the Dead.* New York: Rinehart, 1948.
Majer, Diemut. *Grundlagen des nationalsozialistischen Rechtssystems: Führerprinzip, Sonderrecht, Einheitspartei.* Stuttgart: W. Kohlhammer, 1987.
Malik, Keenan. "In Defense of Human Agency in Consciousness, Genetics and Society." In *Engelsberg Seminar on "Consciousness, Genetics and Society,"* edited by Kurt Almqvist and Eric Wallrup. Avesta, Sweden: Axsson Johnson, 2002.
Manchester, William. *Goodbye, Darkness: A Memoir of the Pacific War.* Boston: Little, Brown, 1979.
———. *A World Lit Only by Fire: The Medieval Mind and the Renaissance—Portrait of an Age.* New York: Little, Brown, 1992.
Mangelsdorff, A. David. "Preface to Special Issue." *Military Psychology* 11, no. 1 (2000): 1–3.
Mann, Thomas. *Diaries, 1918–1933.* Translated by Richard Winston and Hermann Kesten. London: Robin Clark, 1984.
Manning, F. J. "Morale, Cohesion, and Espirit de Corps." In *Handbook of Military Psychology,* edited by R. Gal and A. David Mangelsdorff, 520–541. Chichester: Wiley, 1991.
Manoschek, Walter. *Opfer der NS-Militärjustiz: Urteilspraxis, Strafvollzug, Entschüldigungspolitik in Österreich.* Wien: Mandelbaum, 2003.
Manoschek, Walter, and Reinhold Gärtner. *Die Wehrmacht im Rassenkrieg: Der Vernichtungskrieg hinter der Front.* Wien: Picus, 1996.
Mansoor, Peter R. *The GI Offensive in Europe: The Triumph of American Infantry Divisions, 1941–1945.* Lawrence: University Press of Kansas, 1999.
Manstein, Erich von. *Lost Victories.* Novato, CA: Presidio, 1982.

Marshall, George C., ed. *Infantry in Battle*. Reprint, 1939. Richmond, VA: Garrett and Massie, 1986.
Marshall, S. L. A. *Men Against Fire: The Problem of Battle Command in Future War*. New York: William Morrow, 1947.
Marshall, S. L. A., John G. Westover, and A. Joseph Webber. *Bastogne: The Story of the First Eight Days in Which the 101st Airborne Division Was Closed Within the Ring of German Forces*. Washington, DC: Government Printing Office, 1988.
Martin, Michel Louis, and Ellen Stern McCrate, eds. *The Military, Militarism, and the Polity: Essays in Honor of Morris Janowitz*. London: Collier Macmillan, 1984.
Massie, Robert K. *Dreadnought: Britain, Germany, and the Coming of the Great War*. New York: Random, 1991.
Matheny, Michael R. *Carrying the War to the Enemy: American Operational Art to 1945*. Norman: University of Oklahoma Press, 2011.
Matthies, Kurt. *Ich hörte die Lerchen singen: Ein Tagebuch aus dem Osten, 1941/45*. Munich: Kösel-Verlag, 1956.
Maurice, Emperor. *Strategikon*. Translated by George T. Dennis. Philadelphia: University of Pennsylvania Press, 1984.
McHugh, John. "The Labour Party and the Parliamentary Campaign to Abolish the Military Death Penalty, 1919–1930." *Historical Journal* 42, no. 1 (1999): 233–249.
McKee, Alexander. *Caen, Anvil of Victory: Formerly "Last Round Against Rommel."* Reprint, 1964. New York: St. Martin's, 1984.
McMilling, J. "Individual Training." *Journal of the Royal United Services Institute* 75 (1930): 794–780.
———. "The Training of the Infantry Soldier." *Journal of the Royal United Services Institute* 74 (1929): 515–519.
McPherson, A. B. *Discipline*. London: War Office, 1950.
McPherson, James. *Battle Cry of Freedom: The Civil War Era*. New York: Oxford University Press, 1988.
McRandle, James, and James Quirk. "The Blood Test Revisited: A New Look at German Casualty Counts in World War I." *Journal of Military History*, no. 70 (July 2006): 667–702.
Mearsheimer, John J. *The Tragedy of Great Power Politics*. New York: Norton, 2001.
Mendelsohn, John, and Donald S. Detwiler. *The Einsatzgruppen or Murder Commandos*. Clark, NJ: Lawbook Exchange, 2010.
Mercur, James. *The Art of War*. New York: Wiley and Sons, 1894.
Merkatz, Friedrich. *Das Zusammenwerken der Infanterie mit ihren schweren Waffen*. Berlin: R. Eisenschmidt, 1925.
Merridale, Catherine. "Culture, Ideology and Combat in the Red Army, 1939–45." *Journal of Contemporary History* 41, no. 2 (2006): 305–324.
Messenger, Charles. *Call to Arms: The British Army, 1914–18*. London: Weidenfeld and Nicolson, 2005.
Messerschmidt, Manfred. *Die Wehrmachtjustiz, 1933–1945*. Munich: Ferdinand Schöningh, 2008.
———. "The Wehrmacht and the Volksgemeinschaft." *Journal of Contemporary History* 18, no. 4 (1983): 719–744.
Messerschmidt, Manfred, and Fritz Wüllner. *Die Wehrmachtjustiz im Dienste des Nationalsozialismus*. Baden-Baden: Nomos Verlagsgesellschaft, 1987.
Michalka, Wolfgang. *Der zweite Weltkrieg: Analysen, Grundzüge, Forschungs-bilanz*. Munich: Piper, 1989.
Mielert-Pflugradt, Gerda, ed. *Russiche Erde: Kriegsbriefe aus Russland*. Stuttgart: Deutsche Verlags-Anstalt, 1950.

Milgram, Norman A., Ruth Orenstein, and Ezer Zafrir. "Stressors, Personal Resources, and Social Supports in Military Performance During Wartime." *Military Psychology* 1, no. 4 (1989): 185–199.

Miller, Lynn H. *Global Order: Values and Power in International Politics* Boulder: Westview, 1985.

Miller, Timothy S., and John Nesbitt, eds. *Peace and War in Byzantium.* Washington, DC: Catholic University of America Press, 1995.

Millet, John D. *The Organization and Role of Army Services Forces.* Reprint, 1954. Washington, DC: Government Printing Office, 1998.

Millett, Allan R. *Semper Fidelis: The History of the United States Marine Corps.* New York: Macmillan, 1980.

Millett, Allan R., and Williamson Murray, eds. *Military Effectiveness.* Vol. 1, *The First World War.* 2nd ed. New York: Cambridge University Press, 2010.

———. *Military Effectiveness.* Vol. 2, *The Interwar Years.* 2nd ed. New York: Cambridge University Press, 2010.

———. *Military Effectiveness.* Vol. 3, *The Second World War.* 2nd ed. New York: Cambridge University Press, 2010.

Model, Hansgeorg. *Der deutsche Generalstabsoffizier: Seine Auswahl und Ausbildung in Reichswehr, Wehrmacht und Bundeswehr.* Frankfurt am Main: Bernard und Graefe, 1968.

Mohrmann, Wolf-Dieter. *Der Krieg ist Hart und Grausam! Feldpostbriefe an den Osnabrücker Regierungspräsidenten, 1941–44.* Osnabrück: Wenner, 1984.

Monateri, P. G. "Black Gaius: A Quest for the Multicultural Origins of the 'Western Legal Tradition.'" *Hastings Law Journal* 51, no. 3 (March 2000): 479–556.

Moran, Charles McMoran Wilson. *The Anatomy of Courage.* Boston: Houghton Mifflin, 1967.

Moreno, J. L. "Note on Cohesion in Social Groups." *Sociometry* 13, no. 2 (May 1950): 176.

Morris, B. S. "Officer Selection in the British Army, 1942–45." *Occupational Psychology* 23 (1949): 219–234.

Moskos, Charles C. *Soldiers and Sociology.* Alexandria, VA: US Army Research Institute for the Behavioral and Social Sciences, 1989.

Mosse, George L. "Shell-Shock as a Social Disease." *Journal of Contemporary History* 35, no. 1 (2000): 101–108.

Müller, Klaus-Dieter, Konstantin Nikischkin, and Günther Wagenlehner. *Die Tragödie der Gefangenschaft in Deutschland und in Der Sowjetunion, 1941–1956.* Köln: Böhlau, 1998.

Müller, Klaus-Jürgen. *General Ludwig Beck: Studien und Dokumente zur Politisch-Militärischen Vorstellungswelt und Tätigkeit des Generalstabchefs des deutschen Heeres, 1933–1938.* Boppard am Rhein: Harald Boldt Verlag, 1980.

———. *Generaloberst Ludwig Beck: Eine Biographie.* Paderborn: Ferdinand Schöningh, 2009.

Müller, Rolf-Dieter. *Das Tor Zur Weltmacht.* Boppard am Rhein: Harald Boldt Verlag, 1984.

———. "Die Konzequenzen der 'Volksgemeinschaft': Ernährung, Ausbeutung und Vernichtung." In *Der zweite Weltkrieg: Analysen, Grundzüge, Forschungsbilanz,* edited by Wolfgang Michalka, 240–248. Munich: Piper Verlag, 1989.

Müller, Rolf-Dieter, and Gerd R. Ueberschär. *Hitler's War in the East, 1941–1945: A Critical Assessment.* New York: Berghan, 2002.

Müller, Rolf-Dieter, Gerd Ueberschär, and Wolfram Wette. *Wer züruckweicht wird Erschossen! Kriegsalltag und Kriegsende in Südwestdeutschland, 1944/45.* Freiburg im Breisgau: Dreisam-Verlag, 1985.

Müller, Rolf-Dieter, and Hans-Erich Volkmann, eds. *Die Wehrmacht: Mythos und Realität.* Munich: Oldenbourg Verlag, 2012.

Müller-Hillebrand, Burkhart. *Das Heer bis zum Kriegsbeginn.* Vol. 1, *Das Heer, 1933–1945.* Darmstadt: Mittler und Sohn, 1954.

———. *Der Zweifrontenkrieg: Das Heer vom Beginn des Feldzuges gegen die Sowjetunion bis zum Kriegsende.* Vol. 3, *Das Heer, 1933–1945.* Frankfurt am Main: Mittler und Sohn, 1969.

———. *Die Blitzfeldzüge, 1939–1941: Das Heer im Kriege bis zum Beginn des Feldzuges gegen die Sowjetunion im Juni 1941.* Vol. 2, *Das Heer, 1939–1945.* Frankfurt: Mittler und Sohn, 1956.

Murphy, Audie. *To Hell and Back.* New York: Holt, 1949.

Murray, Williamson. *German Military Effectiveness.* Baltimore: Nautical and Aviation, 1992.

Murray, Williamson, MacGregor Knox, and Alvin Bernstein, eds. *The Making of Strategy: Rulers, States, and War.* New York: Cambridge University Press, 1994.

Myers, C. S. "The Selection of Army Personnel." *Occupational Psychology* 18 (1943): 1–5.

Myrer, Anton. *The Big War.* New York: Appleton-Century-Crofts, 1957.

Nalty, Bernard C. *Cape Gloucester: The Green Inferno.* Washington, DC: Government Printing Office, 1994.

Nalty, Bernard C., John F. Shiner, George M. Watson, and Alfred M. Beck. *With Courage: The U.S. Army Air Forces in World War II.* Washington, DC: Government Printing Office, 1994.

Natzio, G. "British Army Servicemen and Women, 1939–1945." *Journal of the Royal United Services Institute* 138 (1993): 36–43.

Neill, Donald A. "The Influence of the Ancients on the Military Thought of the Seventeenth and Eighteenth Centuries." *Journal of Military History* 62, no. 3 (1998): 487–520.

Neillands, Robin. *The Battle for the Rhine: The Battle of the Bulge and the Ardennes Campaign, 1944.* Woodstock, NY: Overlook, 2007.

Nenninger, Timothy K. "Tactical Dysfunction in the AEF, 1917–18." *Military Affairs* 51, no. 4 (October 1987): 177–181.

———. "'Unsystematic as a Mode of Command': Commanders and the Process of Command in the American Expeditionary Forces, 1917–18." *Journal of Military History* 64, no. 3 (2000): 739–768.

Neufeldt, Hans-Joachim, Jürgen Huck, and Georg Tessin. *Zur Geschichte der Ordnungspolizei, 1936–1945.* Koblenz: Bundesarchiv, 1957.

Newman, Aubrey. *Follow Me: The Human Element in Leadership.* Novato, CA: Presidio, 1981.

Newton, Steven H. *Kursk, The German View: Eyewitness Reports of Operation Citadel by the German Commanders.* Cambridge, MA: Da Capo, 2002.

Nicholson, W. N. "The Training of an Infantry Battalion." *Journal of the Royal United Services Institute* 74 (1929): 49–54.

Nikolaieff, A. M. "The Red Army in the Second World War." *Russian Review* 7, no. 1 (1947): 49–60.

Nordyke, Phil. *All American, All the Way: The Combat History of the 82nd Airborne Division in World War II.* St. Paul, MN: Zenith, 2005.

Nosworthy, Brent. *The Bloody Crucible of Courage: Fighting Methods and Combat Experience of the Civil War.* New York: Carroll and Graf, 2003.

Novotny, Fred. *The Good Soldier: From Austrian Social Democracy to Communist Captivity with a Soldier of Panzer-Grenadier Division "Grossdeutschland."* Bedford, PA: Aberjona, 2003.

Nye, Robert. *The Origins of Crowd Psychology: Gustav Lebon and the Crisis of Mass Democracy in the Third Republic.* London: Sage, 1975.
Oliver, Laurel W., Joan Harman, Elizabeth Hoover, Stephanie M. Hayes, and Nancy A. Pandhi. "A Quantitative Integration of the Military Cohesion Literature." *Military Psychology* 11, no. 1 (2000): 57–83.
O'Neill, Robert. "Doctrine and Training in the German Army." In *The Theory and Practice of War,* edited by Michael Howard, 73–85. New York: Praeger, 1965.
———. *The German Army and the Nazi Party, 1933–1939.* London: Cassell, 1966.
Otterbein, Keith. "Cross-Cultural Studies of Armed Conflict." In *Buffalo Studies in International Conflict,* edited by Glenn Snyder, 91–109. Buffalo: State University of New York Press, 1968.
Otway, T. B. H. *Airborne Forces.* London: War Office, 1951.
Overmans, Rüdiger. *Deutsche militärische Verluste im Zweiten Weltkrieg.* Munich: R. Oldenburg Verlag, 2004.
Overton, Richard E. *God Isn't Here.* Clearfield, UT: American Legacy Historical Press, 2007.
Overy, R. J. "Mobilization for Total War in Germany, 1939–1941." *English Historical Review* 103, no. 408 (1988): 613–639.
Overy, Richard. *Russia's War: Blood upon the Snow.* New York: Penguin Putnam, 1997.
———. *Why the Allies Won.* New York: Norton, 1996.
Pabst, Helmut. *Der Ruf der äußersten Grenze: Tagebuch eines Frontsoldaten.* Tübingen: Fritz Schlichtenmayer, 1953.
Pakenham, Thomas. *The Boer War.* New York: Random, 1979.
Palmer, Robert R. "Frederick the Great, Guibert, Bulöw: From Dynastic to National War." In *Makers of Modern Strategy from Machiavelli to the Nuclear Age,* edited by Peter Paret, Gordon A. Craig, and Felix Gilbert, 91–119. Princeton: Princeton University Press, 1986.
———. *Study no. 3: Ground Forces in the War Army—A Statistical Table.* Washington, DC: Historical Section, US Army Ground Forces, 1946.
———. *Study no. 4: Mobilization of the Ground Army.* Washington, DC: Historical Section, US Army Ground Forces, 1946.
———. *Study no. 5: Procurement of Enlisted Personnel for the AGF—The Problem of Quality.* Washington, DC: Historical Section, US Army Ground Forces, 1946.
Palmer, Robert R., Bell I. Wiley, and William R. Keast. *The Procurement and Training of Ground Combat Troops.* Washington, DC: Government Printing Office, 1948.
Panayi, Panikos. *Weimar and Nazi Germany: Continuities and Discontinuities.* New York: Longman, 2001.
Paret, Peter. *The Cognitive Challenge of War: Prussia 1806.* Princeton: Princeton University Press, 2009.
———. "Napoleon and the Revolution in War." In *Makers of Modern Strategy from Machiavelli to the Nuclear Age,* edited by Peter Paret, Gordon A. Craig, and Felix Gilbert, 123–142. Princeton: Princeton University Press, 1986.
———. *Yorck and the Era of Prussian Reform.* Princeton: Princeton University Press, 1966.
Paret, Peter, Gordon A. Craig, and Felix Gilbert, eds. *Makers of Modern Strategy from Machiavelli to the Nuclear Age.* Princeton: Princeton University Press, 1986.
Parker, Geoffrey. *The Thirty Years' War.* Reprint, 1984. New York: Routledge, 1997.
Parker, Ian, and John Shotter, eds. *Deconstructing Social Psychology.* London: Routledge, 1990.
Parker, Matthew. *Monte Cassino: The Hardest Fought Battle of World War II.* New York: Doubleday, 2004.
Parkinson, C. N. *Always a Fusilier.* London: Sampson Low, 1949.
Parks, Craig D. *Group Performance and Interaction.* Boulder: Westview, 1999.

Parth, Wolfgang W. *Vorwärts Kameraden, wir müssen Zurück: Roman.* Munich: Kindler, 1958.

Patton, George S. *War As I Knew It.* Edited by Beatrice Ayer Patton and Paul D. Harkins. New York: Bantam, 1980.

Paul, Gerhard. *Ungehorsame Soldaten: Dissens, Verweigerung und Widerstand deutscher Soldaten (1939–1945).* St. Ingbert: Röhrig Universitätsverlag, 1994.

Paulus, Bertold. *Es Grüßt euch alle, Bertold von Koblenz nach Stalingrad: Die Feldpostbriefe des Pioniers Bertold Paulus aus Kastel.* Nonnweiler-Otzenhausen Burr Verlag, 1993.

Paulus, Paul B., ed. *Psychology of Group Influence.* Hillsdale, NJ: Erlbaum, 1989.

Pemberton, A. L.. "The New Psychology and the Services." *Journal of the Royal United Services Institute* 73, no. 490 (1928): 321–328.

Perret, Geoffrey. *There's a War to Be Won: The United States Army in World War II.* New York: Random, 1991.

———. *Winged Victory: The Army Air Forces in World War II.* New York: Random, 1993.

Petrov, Vladimir. *"June 22, 1941": Soviet Historians and the German Invasion.* Columbia: University of South Carolina Press, 1968.

Pfeifer, Jochen. *Der deutsche Kriegsroman, 1945–1960: Ein Versuch zur Vermittlung von Literatur und Sozialgeschichte.* Königstein: Scriptor, 1981.

Phillips, Gervase. "Military Morality Transformed: Weapons and Soldiers on the Nineteenth-Century Battlefield." *Journal of Interdisciplinary History* 41, no. 4 (2011): 565–590.

Pick, Daniel. *War Machine: The Rationalisation of Slaughter in the Modern Age.* New Haven: Yale University Press, 1993.

Picq, Ardant du. "Battle Studies." In *Roots of Strategy: Book 2,* 9–300. Harrisburg, PA: Stackpole, 1987.

Pigott, A. J. K. *Manpower Problems.* London: War Office, 1949.

Pitts, Jesse Richard. *Return to Base: Memoirs of a B-17 Copilot, Kimbolton, England, 1943–1944.* Charlottesville, VA: Howell, 2004.

Place, T. Harrison. "British Perceptions of the Tactics of the German Army, 1938–40." *Intelligence and National Security* 9, no. 3 (1994): 495–519.

Place, Timothy H. *Military Training in the British Army, 1940–1944: From Dunkirk to D-Day.* London: Cass, 2000.

Po-chia Hsia, R. *The World of Catholic Renewal, 1540–1770.* Reprint, 1998. New York: Cambridge University Press, 2005.

Podzun, Hans-Henning. *Weg und Schicksal Der 21: Infanterie-Division.* Kiel: Podzun, 1951.

Pöppel, Martin. *Heaven and Hell: The War Diary of a German Paratrooper.* Translated by Louise Willmot. Tunbridge Wells, UK: Spellmount, 1988.

Pownall, Henry. *Chief of Staff: The Diaries of Lieutenant General Sir Henry Pownall.* Vol. 1, *1933–1940.* Edited by Brian Bond. London: Leo Cooper, 1972.

Price, A. Preston. *The Last Kilometer: Marching to Victory in Europe with the Big Red One, 1944–1945.* Annapolis, MD: Naval Institute, 2002.

Proskouriakov, Alexander. *Feldpost aus Stalingrad.* Berlin: Dr. Köster, 2004.

Prüller, Wilhelm. *Diary of a German Soldier.* Translated by H. C. Robbins Landon and Sebastian Leitner. New York: Coward-McCann, 1963.

Purkiss, Diane. *The English Civil War: Papists, Gentlewomen, Soldiers, and Witchfinders in the Birth of Modern Britain.* New York: Basic, 2006.

Quarrie, Bruce. *Panzer-Grenadier Division Grossdeutschland.* London: Osprey, 1977.

Queckbörner, Ludwig. *Die Schützenkompanie: Ein Handbuch für den Dienstunterricht.* Berlin: Mittler und Sohn, 1939.

Quinnett, Robert L. "The German Army Confronts the NSFO." *Journal of Contemporary History* 13, no. 1 (January 1978): 53–64.
Radine, Lawrence. *The Taming of the Troops: Social Control in the United States Army.* Westport: Greenwood, 1977.
Rapport, Leonard. "Das Sozialprofil von Kampfverbänden des deutschen Heeres, 1939 bis 1945." In *Die Deutsche Kriegsgesellschaft, 1939 bis 1945,* edited by Jörg Echternkamp, 641–741. Munich: Deutsche Verlags-Anstalt, 2004.
Rapport, Leonard, Arthur Northwood, and Samuel Lyman Atwood Marshall. *Rendezvous with Destiny: A History of the 101st Airborne Division.* Washington, DC: Infantry Journal, 1948.
Rass, Christoph. *"Menschenmaterial": Deutsche Soldaten an der Ostfront—Innenansichten einer Infanteriedivision, 1939–1945.* Paderborn: Ferdinand Schöningh, 2003.
Rass, Christoph, and René Rohrkamp. *Deutsche Soldaten, 1939–1945: Handbuch einer biographischen Datenbank zu Mannschaften und Unteroffizieren von Heer, Luftwaffe und Waffen SS.* Freiburg im Breisgau: Deutsche Bundesarchiv-Militärarchiv, 2007.
Ratzmer, V. "Die Ausbildung des Gruppenführers." *Militär-Wochenblatt* 119, no. 44 (1935): 1775–1777.
Raynsford, R. M. "Army Estimates." *Fighting Forces* 14, no. 1 (April 1937): 4–6.
———. "Recruiting for the Army." *Fighting Forces* 13, no. 5 (December 1936): 386–388.
Rees, John R. "The Development of Psychiatry in the British Army." In *Military Neuro-Psychiatry: Proceedings of the Association of Nervous and Mental Diseases, December 15–16, 1944, New York,* 48–56. Baltimore: Association of Nervous and Mental Diseases, 1946.
———. *The Shaping of Psychiatry by War.* New York: Norton, 1945.
Reese, Roger R. *Stalin's Reluctant Soldiers: A Social History of the Red Army, 1925–1941.* Lawrence: University Press of Kansas, 1996.
Reinhardt, Klaus. *Moscow: The Turning Point: The Failure of Hitler's Strategy in the Winter of 1941–42.* Translated by M. A. Karl B. Keenan. Oxford: Berg, 1992.
Reinhardt, Klaus, and Gerhard L. Weinberg. "Moscow: The Turning Point—The Failure of Hitler's Strategy in the Winter of 1941–42." *Central European History* 27, no. 2 (1995): 251.
Reinicke, Adolf. *Das Reichsheer, 1921–1934.* Osnabrück: Biblio Verlag, 1986.
Reisner, Karl. *Schiessausbildung und Feuer der Infanterie im Gefecht.* Berlin: Mittler und Sohn, 1897.
Rejai, Mostafa. *World Military Leaders: A Collective and Comparative Analysis.* Westport: Praeger, 1996.
Rempel, Gerhard. *Hitler's Children: The Hitler Youth and the SS.* Chapel Hill: University of North Carolina Press, 1989.
Rendulic, Lothar. "Die Einheitsgruppe." *Militär-Wochenblatt* 116, no. 46 (1932): 1613–1616.
———. "Die letzten 200 meter." *Militär-Wochenblatt* 116, no. 26 (1932): 929–933.
Renouf, Tom. *Black Watch.* London: Little, Brown, 2011.
Reoch, Ernest. *The St. Valery Story: A New Account of the Last Stand of the 51st Highland Division in France in 1940.* Inverness, UK: Highland, 1965.
Reynolds, B. T. "Recruiting and Unemployment." *Journal of the Royal United Services Institute* 81, no. 524 (1936): 739–746.
Richter, Werner. *Die 1: Ostpreußische Infanterie-Division.* Munich: Eigenverlag, 1975.
Riefrens, R. "Die letzten 200 m. in Angriff." *Militär-Wochenblatt* 119, no. 9 (1934): 331–333.

Rodgers, Thomas E. "Billy Yank and G.I. Joe: An Exploratory Essay on the Sociopolitical Dimensions of Soldier Motivation." *Journal of Military History* 69, no. 1 (2005): 93–121.
Rohrbeck, A. *Befehlstechnik und Kampverfahren: Ein Hilfsbuch für junge Offiziere und Unteroffiziere.* Berlin: Mittler und Sohn, 1925.
Rohrkamp, René. "Weltanschaulich gefestigte Kämpfer: Die Soldaten der Waffen-SS, 1933–1945: Organisationsgeschichte—Personalwesen—Sozialstruktur." PhD dissertation, Aachen, 2009.
Romeiser, John B., ed. *Combat Reporter: Don Whitehead's World War II Diary and Memoirs.* New York: Fordham University Press, 2006.
Römer, Felix. *Der Kommissarbefehl.* Munich: Schoeningh Ferdinand GmbH, 2008.
———. *Kameraden: Die Wehrmacht von Innen.* Munich: Piper, 2012.
———. "Milieus in the Military: Soldierly Ethos, Nationalism, and Conformism Among Workers in the Wehrmacht." *Journal of Contemporary History* 48, no. 1 (2013): 125–149.
Rommel, Erwin. *Attacks.* Translated by J. R. Driscoll. Reprint, 1937. Vienna, VA: Athena, 1979.
Ropp, Theodore. *War in the Modern World.* Revised ed. New York: Collier, 1959.
Rose, Arnold M. "The Social Psychology of Desertion from Combat." *American Sociological Review* 16, no. 5 (1951): 614–629.
Rosen, William. *Justinian's Flea.* New York: Penguin, 2007.
Ross, Bill D. *Iwo Jima: Legacy of Valor.* New York: Vanguard, 1985.
———. *Peleliu: Tragic Triumph—The Untold Story of the Pacific War's Forgotten Battle.* New York: Random, 1991.
Rostow, W. W. *The World Economy: History and Prospect.* Austin: University of Texas Press, 1980.
Rottman, Gordon L. *U.S. Airborne Units in the Mediterranean Theater, 1942–1945.* New York: Osprey, 2006.
———. *U.S. Marine Corps, Pacific Theater of Operations, 1943–44.* New York: Osprey, 2004.
Royster, Charles. *A Revolutionary People at War.* Chapel Hill: University of North Carolina Press, 1979.
Ruggero, Ed. *Combat Jump: The Young Men Who Led the Assault into Fortress Europe, July 1943.* New York: HarperCollins, 2004.
Rundquist, E. A. "Development of an Interview for Selection Purposes." In *New Methods in Applied Psychology,* edited by George A. Kelly, 85–95. College Park: University of Maryland Press, 1947.
Rush, Robert S. "A Different Perspective: Cohesion, Morale, and Operational Effectiveness in the German Army, Fall 1944." *Armed Forces and Society: An Interdisciplinary Journal* 25, no. 3 (Spring 1999): 477–500.
———. *Hell in Hürtgen Forest.* Lawrence: University Press of Kansas, 2001.
Rush, Robert S., Elizabeth Sharp, and Ian Palmer. *GI: The US Infantryman in World War II.* Oxford: Osprey, 2003.
RUSI. "Infantry Tactics, 1914–1918." *Journal of the Royal United Services Institute* 64, no. 455 (1919): 460–469.
———, ed. "Military Notes: The Peace Conference (Notes Summarizing the End of the War, German Disarmament, Present Dispositions, and the Anti-Bolshevik Deployments)." *Journal of the Royal United Services Institute* 64, no. 454 (1919): 330–353.
Ryan, Garry D., and Timothy K. Nenninger, eds. *Soldiers and Civilians: The U.S. Army and the American People.* Washington, DC: National Archives and Records Administration, 1987.

Rybach, Timothy. "Stalingrad: Letters from the Dead." *New Yorker,* February 1, 1993.
Sajer, Guy. *The Forgotten Soldier.* New York: Harper and Row 1971.
Sakurai, Melvin M. "Small Group Cohesiveness and Detrimental Conformity." *Sociometry* 38, no. 3 (1975): 340–357.
Salmond, J. B. *The History of the 51st Highland Division, 1939–1945.* Edinburgh: Blackwood, 1953.
Sarkesian, Sam Charles. *Soldiers, Society, and National Security.* Boulder: Lynne Rienner, 1995.
Sassoon, Siegfried. *The War Poems.* London: Heinemann, 2014.
Scales, Len. "Late-Medieval Germany: An Under-Stated Nation?" In *Power and the Nation in European History,* edited by Len Scales and Oliver Zimmer, 166–191. New York: Cambridge University Press, 2005.
Scheff, Thomas J. *Bloody Revenge: Emotions, Nationalism, and War.* Boulder: Westview, 1994.
Scheffel, Charles, and Barry Basden. *Crack! And Thump: With a Combat Infantry Officer in World War II.* Llano, TX: Camroc, 2007.
Schifferle, Peter J. *America's School for War: Fort Leavenworth, Officer Education and Victory in World War II.* Lawrence: University Press of Kansas, 2010.
Schneider, Heinrich. *Bunte Bilder aus dem Friedens- und Kriegserleben der Nachrichten-Abteilung 1: Ehemals im Verband der 1. (Ostpr.) Infanterie-Division.* Kameraden: Verlag Ostpreuss, 1962.
Schnitter, Helmut. "Die überlieferte Defensionspflicht." In *Die Wehrpflicht: Entstehung, Erscheinungsformen und politisch-militärische Wirkung,* edited by Roland G. Foerster, 29–38. Munich: Oldenbourg Verlag, 1994.
Schoenbaum, David. *Hitler's Social Revolution: Class and Status in Nazi Germany, 1933–1939.* Garden City, NY: Doubleday, 1966.
Schrijvers, Peter. *The Crash of Ruin: American Combat Soldiers in Europe During World War II.* New York: New York University Press, 1998.
Schröder, Hans Joachim. *Die gestohlene Jahre. Erzählungsgeschichten und Geschichtungserzählung im Interview: Der zweiten Weltkrieg aus der Sicht ehemaliger Mannschaftssoldaten.* Tübingen: Niemeyer, 1992.
———. *Kasernenzeit: Arbeiter Erzählen von der Militärausbildung im Dritten Reich.* Frankfurt am Main: Campus-Verlag, 1985.
Schröder, Willy. *Nur ein Kriegstagebuch: Rein menschliche Reflektionen.* Baden-Baden: Verlag Presse Informations Agentur, 1982.
Schulte, Theo J. *The German Army and Nazi Policies in Occupied Russia.* New York: Berg, 1989.
———. "The German Soldier in Occupied Russia." In *Time to Kill: The Soldier's Experience of War in the West, 1939–1945,* edited by Paul Addison and Angus Calder, 274–283. London: Pimlico, 1997.
Schumann, Willy. *Being Present: Growing Up in Hitler's Germany.* Kent, OH: Kent State University Press, 1991.
Schweling, Otto Peter, and Erich Schwinge. *Die deutsche Militärjustiz in der Zeit des Nationalsozialismus.* Marburg: Elwert, 1977.
Scott, Winfield. "Vera Cruz and Its Castle." In *Battles of America by Sea and Land.* Vol. 2, *War of 1812 and Mexican Campaigns,* edited by Robert Tomes. New York: James S. Virtue, 1878.
Seidler, Franz Wilhelm. *Die Militärgerichtsbarkeit der deutschen Wehr-macht, 1939–1945: Rechtsprechung und Strafvollzug.* Munich: Herbig, 1991.
Selz, Barbara. *Das Grüne Regiment: Der Weg der 256. Infanterie-Division aus der Sicht des Regiments 481.* Freiburg im Breisgau: Kehrer, 1970.
Semin, G. R., and Barbara Krahe. *Issues in Contemporary German Social Psychology: History, Theories, and Application.* London: Sage, 1987.

Semsch, Philip L. "Elihu Root and the General Staff." *Military Affairs* 27, no. 1 (Spring 1963): 16–27.

Seton-Watson, Hugh. *Nations and States: An Enquiry into the Origins of Nations and the Politics of Nationalism.* Boulder: Westview, 1977.

———. *The Russian Empire, 1801–1917.* New York: Oxford University Press, 1967.

Sewell, William H. "A Theory of Structure: Duality, Agency, and Transformation." *American Journal of Sociology* 98, no. 1 (July 1992): 1–29.

Shanahan, William O. "Enlightenment and War: Austro-Prussian Military Practice, 1760–1690." In *East Central European Society and War in the Pre-Revolutionary Eighteenth Century,* by Gunther E. Rothenberg, Bèla K. Király, and Peter F. Sugar, 83–105. Boulder: Social Science Monograph, 1982.

Shapiro, Milton J. *The Screaming Eagles: The 101st Airborne Division in World War II.* New York: Messner, 1976.

Sharpe, Mike, and Brian Leigh Davis. *Grossdeutschland: Guderian's Eastern Front Elite.* Surrey: Ian Allan, 2001.

Shaw, Irwin. *The Young Lions.* New York: Random, 1948.

Sheffield, G. D. "Montgomery and 'Colossal Cracks': The 21st Army Group in Northwest Europe, 1944–45, by Stephen Ashley Hart." *Journal of Strategic Studies* 24 (2001): 257.

Shepard, Ben. *A War of Nerves: Soldiers and Psychiatrists in the Twentieth Century.* Cambridge: Harvard University Press, 2001.

———. *War in the Wild East: The German Army and Soviet Partisans.* Cambrige: Harvard University Press, 2004.

Sherbrooke, R. L. "The New Infantry Weapons." *Journal of the Royal United Services Institute* 83 (1938): 136–144.

———. "Regimental Depots." *Journal of the Royal United Services Institute* 77 (1932): 571–517.

Shils, Edward A., and Morris Janowitz. "Cohesion and Disintegration in the Wehrmacht in World War II." *Public Opinion Quarterly* 12, no. 2 (1948): 280–315.

Shirer, William L. *Berlin Diary: The Journal of a Foreign Correspondent, 1934–1941.* New York: Knopf, 1941.

———. *The Rise and Fall of the Third Reich: A History of Nazi Germany.* New York: Simon and Schuster, 1960.

Showalter, Dennis E. "Army and Society in Imperial Germany: The Pains of Modernization." *Journal of Contemporary History* 18, no. 4 (October 1983): 583–618.

———. "From Deterrence to Doomsday Machine: The German Way of War, 1890–1914." *Journal of Military History* 64, no. 3 (2000): 679–710.

Siebold, Guy L. "The Evolution of the Measurement of Cohesion." *Military Psychology* 11, no. 1 (2000): 5–26.

Siebold, Guy L., and Twila J. Lindsay. "The Relation Between Demographic Descriptors and Soldier-Perceived Cohesion and Motivation." *Military Psychology* 11, no. 1 (2000): 109–128.

Simkins, Peter. *Kitchener's Army: The Raising of New Armies, 1914–1916.* Manchester: Manchester University Press, 1988.

Simon, Walter M. *The Failure of the Prussian Reform Movement, 1807–1819.* New York: Howard Fertig, 1971.

Simonov, Konstantin M. *Man wird nicht als Soldat geboren.* Berlin: Verlag Volk und Welt, 1973.

Simonov, Konstantin M., and Corinna Wojtek. *Die Lebenden und die Toten: Roman.* Berlin: Verlag Volk und Welt, 1973.

Sixsmith, E. K. G. "The British Army in May 1940: A Comparison with the BEF 1914." *RUSI Journal* 127, no. 3 (1982): 8–10.

Skelley, Alan Ramsay. *The Victorian Army at Home: The Recruitment and Terms and Conditions of the British Regular, 1859–1899.* London: Croom Helm, 1977.

Skinner, H. Allen, Jr. "Transformation of the German Reichsheer." Master's thesis, US Army Command and General Staff College, Fort Leavenworth, KS, 2006.

Slack, G. Granville. *Being the National Service Acts, 1939–1941: The Emergency Powers (Defence) Acts, 1939–1940, and Regulations and Orders Thereunder, and All Other Relevant Acts and Regulations.* Edited by J. W. Whitlock. London: Butterworth, 1942.

Sledge, E. B. *With the Old Breed at Peleliu and Okinawa.* New York: Oxford University Press, 1981.

Sloan, Bill. *Brotherhood of Heroes: The Marines at Peleliu, 1944: The Bloodiest Battle of the Pacific War.* New York: Simon and Schuster, 2005.

Smith, E. D. *Even the Brave Falter.* London: Hale, 1978.

Smurthwaite, David, Mark Nicholls, and Linda Washington. *Against All Odds: The British Army of 1939–40.* London: National Army Museum, 1990.

Snyder, Louis L., ed. *Documents of German History.* New Brunswick, NJ: Rutgers University Press, 1958.

Sokolov, Boris V. "The Cost of War: Human Losses for the USSR and Germany, 1939–1945." *Journal of Slavic Military Studies* 9, no. 1 (1996): 152–193.

Solas, Eduardo, Renée E. DeRouin, and Paul A. Gade. "The Military's Contribution to Our Science and Practice: Peoples, Places, and Findings." In *Historical Perspectives in Industrial and Organizational Psychology,* edited by Laura Koppes, 211–246. London: Routledge, 2007.

Solomon, Zahava. *Combat Stress Reaction: The Enduring Toll of War.* New York: Plenum, 1993.

Sonnenberger, Martin. "Mission Command During the War of Movement in World War I: Initiative and Synchronization of the German Right Wing in August and Early September 1914." Master's thesis, US Army Command and General Staff College, Fort Leavenworth, KS, 2014.

Spannenkrebs, V. Major. "Motorizierte Schützen." *Militär-Wochenblatt* 120, no. 37 (1936): 1637–1642.

Sparrow, J. H. A. *Morale.* London: War Office, 1949.

Spiers, Edward M. *The Army and Society, 1815–1914.* London: Longman, 1980.

Spires, David N. *Image and Reality: The Making of the German Officer, 1921–1933.* Westport: Greenwood, 1984.

Spohn. *Ratschläge und Winke für die praktische Ausbildung und Schulung der Infanterie-Kompagnie.* Berlin: Verlag von R. Eisenschmidt, 1907.

Stearns, Peter N. *Encyclopedia of Social History.* New York: Garland, 1994.

Stearns, Stephen J. "Conscription and English Society in the 1620s." *Journal of British Studies* 11, no. 2 (May 1972): 1–23.

Steele, Matthew. "The Conduct of War." *Journal of the Military Service Institution of the United States* 42 (1908): 22–31.

Stein, Peter. *Roman Law in European History.* New York: Cambridge University Press, 1999.

Steinzer, Franz. *Die 2: Panzer-Division—Bewaffnung, Einsätze, Männer (gebundene Ausgabe).* Eggolsheim: Dörfler im Nebel Verlag, 2003.

Stewart, Nora Kinzer. *Mates and Muchachos: Unit Cohesion in the Falklands/Malvinas.* Washington, DC: Brassey's, 1991.

Stewart, Richard W. *The United States Army in a Global Era, 1917–2003.* Vol. 2. Washington, DC: Center of Military History, US Army, 2005.

Stoler, Mark A. *Allies and Adversaries: The Joint Chiefs of Staff, the Grand Alliance, and U.S. Strategy in World War II.* Chapel Hill: University of North Carolina Press, 2000.

Stolfi, R. H. S. *Barbarossa Revisited: A Critical Reappraisal of the Opening Stages of the Russo-German Campaign, June–December 1941.* Chicago: University of Chicago Press, 1982.
Stone, John. "Technology, Society, and the Infantry Revolution of the Fourteenth Century." *Journal of Military History* 68, no. 2 (April 2004): 361–380.
Story, Joanna. *Charlemagne: Empire and Society.* New York: Manchester University Press, 2005.
Stouffer, Samuel A. *The American Soldier: Combat and Its Aftermath.* Vol. 2. Princeton: Princeton University Press, 1949.
Stoves, Rolf. *Die Gepanzerten und Motorisierten deutschen Grossverbände: Divisionen und selbständige Brigaden, 1935–1945.* Friedberg: Podzun-Pallas-Verlag, 1986.
———. *1.Panzer-Division, 1935–1945: Chronik einer der drei Stamm-Divisionen der deutschen Panzerwaffe.* Bad Nauheim: H.-H. Podzun, 1961.
Strachan, Hew. "Training, Morale, and Modern War." *Journal of Contemporary History* 41, no. 2 (April 2006): 211–227.
Strassler, Robert B., ed. *The Landmark Thucydides: A Comprehensive Guide to the Peloponnesian War.* New York: Touchstone, 1996.
Strauß, Franz J. *Die Geschichte Der 2.(Wiener) Panzer-Division.* Friedeberg: Podzun-Pallas, 1977.
Streit, Christian. "Soviet Prisoners of War in the Hands of the Wehrmacht." In *War of Extermination: The German Military in World War II, 1941–1944,* edited by Hannes Heer and Klaus Naumannn, 80–92. New York: Bergahn, 2000.
Stremlow, Mary V. *Free a Marine to Fight: Women Marines in World War II.* Quantico, VA: United States Marine Corps, 1979.
Stubbs, Mary Lee, and Stanley Russell Conner. *Armor-Cavalry Part I: Regular Army and Army Reserve.* Washington, DC: Government Printing Office, 1969.
Stübig, Heinz. "Die Verfassung Preußens in der Reformzeit." In *Die Wehrpflicht: Entstehung, Erscheinungsformen und politisch-militärische Wirkung,* edited by Roland G. Foerster, 39–54. Munich: Oldenbourg Verlag, 1994.
Stumpf, Reinhard. *Die Wehrmacht-Elite: Rang- und Herkunftsstruktur der deutschen Generale und Admirale, 1933–1945.* Boppard am Rhein: H. Boldt, 1982.
Sturdy, David J. *Fractured Europe.* Oxford: Blackwell, 2002.
Swinton, E. D. "The Defense of Duffer's Drift." *Infantry Journal.* Reprint, April 1905. Ft. Leavenworth, KS: US Army Command and General Staff College, 1989.
Tallett, Frank. *War and Society in Early-Modern Europe, 1495–1715.* London: Routledge, 1992.
Tapert, Annette. *Lines of Battle: Letters from American Servicemen, 1941–1945.* New York: Times, 1987.
Tessin, Georg. *Formationsgeschichte der Wehrmacht, 1933–1939.* Boppard am Rhein: Harald Boldt Verlag, 1959.
———. *Verbände und Truppen der deutschen Wehrmacht und Waffen-SS im Zweiten Weltkrieg, 1939–1945.* Vol. 15. Frankfurt/Main: Mittler und Sohn, 1966.
Thompson, James Westfall. "German Feudalism." *American Historical Review* 28, no. 3 (April 1923): 440–474.
Thoß, Bruno. "Vom Führerheer zur Wehrmacht." *Militärgeschichte: Zeitschrift für historische Bildung,* no. 2 (2009): 4–7.
Tiger, Lionel. *Men in Groups.* New York: Random, 1969.
Tilly, Charles. *Coercion, Capital, and European States, AD 990–1992.* Oxford: Blackwell, 1992.
———, ed. *The Formation of National States in Western Europe.* Princeton: Princeton University Press, 1975.
Toland, John. *No Man's Land: 1918, the Last Year of the Great War.* Garden City, NY: Doubleday, 1980.

Tolhurst, Michael. *Saint Vith: US 106th Infantry Division.* London: Leo Cooper, 1999.
Towell, Pat. *Forging the Sword: Unit-Manning in the U.S. Army.* Washington, DC: Center for Strategic and Budgetary Assessments, 2004.
Travers, T. H. E. "Command and Leadership Styles in the British Army: The 1915 Gallipoli Model." *Journal of Contemporary History* 29, no. 3 (1994): 402–433.
Tregaskis, Richard. *Guadalcanal Diary.* New York: Random, 1943.
Ueberschär, Gerd R. *Hitlers militärische Elite: 68 Lebensläufe.* Darmstadt: Der Theiss Verlag, 2015.
Ueberschär, Gerd R., and Wolfram Wette, eds. *"Unternehmen Barbarossa": Der deutsche Überfall auf die Sowjetunion, 1941.* Paderborn: Fischer Taschenbuch, 1984.
UK Secretary of State for War. *Report of the Army and Air Force Courts-Martial Committee, 1946.* London: HMSO, 1949.
UK War Office. *King's Regulations for the Army and Army Reserve 1928.* London, 1928.
———. *Manual of Military Law 1929.* Reprint. London, 1939.
———. *Notes on Administration of Discipline.* London, 1941.
———. *Regulations for the Territorial Army (Including the Territorial Army Reserve) and for County Associations, 1936.* London, 1936.
———. *Report of the War Office Committee on Enquiry into "Shell-Shock."* London, 1922.
———. *Training Regulations 1934.* London, 1934.
Unger, Roberto Mangabeira. *Plasticity into Power: Comparative-Historical Studies of the Institutional Conditions of Economic and Military Success.* Cambridge: Cambridge University Press, 1987.
Ungerson, B. *Personnel Selection.* London: War Office, 1953.
US Army. Army Regulation 40-105. *Standards of Physical Examination for Entrance into the Regular Army, National Guard, and Organized Reserves.* Washington, DC: Department of War, May 29, 1923.
———. Army Regulation 40-105. *Standards of Physical Examination for Entrance into the Regular Army, National Guard, and Organized Reserves.* Washington, DC: Department of War, November 15, 1932.
———. Army Regulation 40-105. *Standards of Physical Examination for Entrance into the Regular Army, National Guard, and Organized Reserves.* Washington, DC: Department of War, August 17, 1940.
———. Army Regulation 600-8-22 (Military Awards). Washington, DC: Department of the Army, 1995.
———. Army Regulation 600-8-22 (Military Awards). Washington, DC: Department of the Army, 2006.
———. Army Regulation 615-25. *Enlisted Men: Classification.* Washington, DC: Department of War, September 3, 1940.
———. Army Regulation 615-25. *Enlisted Men: Initial Classification.* Washington, DC: Department of War, July 31, 1942.
———. Army Regulation 615-28. *Enlisted Men: Classification, Reclassification, Assignment, and Reassignment.* Washington, DC: Department of War, May 28, 1942.
———. Command and General Staff College. *Truppenführung.* Translated by US Army. Reprint, 1936. Ft. Leavenworth, KS: Department of Army, 1989.
———. *4th Infantry Division.* Baton Rouge, LA: US Army and US Navy, 1946.
———. Infantry Field Manual 7-10. *Rifle Company, Rifle Regiment.* Washington, DC: Department of War, June 2, 1942.
———. Infantry Field Manual 7-10. *Rifle Company, Infantry Regiment.* Washington, DC: Department of War, March 18, 1944.

———. Mobilization Regulation 1-7. *Reception of Selective Service Men.* Washington, DC: Department of War, September 1940.
———. Mobilization Regulation 1-9. *Reception of Selective Service Men.* Washington, DC: Department of War, January 15, 1934.
———. Mobilization Regulation 1-9. *Standards of Physical Examination During Mobilization.* Washington, DC: Department of War, August 31, 1940.
———. Table of Organization and Equipment 7-37T. *Infantry, Rifle Company, Parachute.* Washington DC: Government Printing Office, December 16, 1944.
———. Table of Organization 7-37. *Infantry, Rifle Company, Parachute.* Washington DC: Government Printing Office, February 17, 1942.
———. War Department, Division Historical, and History Center of Military. *Small Unit Actions: France—2d Ranger Battalion at Pointe Du Hoc; Saipan—27th Division on Tanapag Plain; Italy—351st Infantry at Santa Maria Infante; France—4th Armored Division at Singling.* Washington, DC: Government Printing Office, 1986.
———. War Department Field Manual 7-20. *Infantry Battalion.* Washington, DC: Government Printing Office, 1944.
US Army Air Forces. *The History of the US Army Air Forces in World War II.* Edited by W. F. Craven and J. L. Cate. Washington, DC: Government Printing Office, 1947.
US Army Ground Forces. *Study no. 3: A Statistical Table (The Troop Basis).* Washington, DC: Historical Section, 1946.
———. *Study no. 8: Reorganization of Ground Troops for Combat.* Washington, DC: Historical Section, 1946.
———. *Study no. 9: Organization and Training of New Ground Combat Elements.* Washington, DC: Historical Section, 1946.
———. *Study no. 11: Training in the Ground Army, 1942–1945.* Washington, DC: Historical Section, 1946.
———. *Study no. 12: The Building and Training of Infantry Divisions.* Washington, DC: Historical Section, 1946.
———. *Study no. 36: The Training of Negro Troops.* Washington, DC: Historical Section, 1946.
US Chief of Staff. *Field Service Regulations: FM 100-5.* Edited by US Department of War. Washington, DC: Government Printing Office, 1944.
———. *Field Service Regulations 1923: United States Army.* Edited by US Department of War. Washington, DC: Government Printing Office, 1924.
———. *Field Service Regulations: Operations, FM 100-5.* Edited by US Department of War. Washington, DC: Government Printing Office, 1941.
———. *Tentative Field Service Regulations, FM 100-5.* Edited by US Department of War. Washington, DC: Government Printing Office, 1939.
US Department of Army. *Army Almanac: A Book of Facts Concerning the Army of the United States.* Washington, DC: Government Printing Office, 1950.
———. *Combat in Russian Forests and Swamps.* New York: Garland, 1979.
———. *Effects of Climate on Combat in European Russia.* New York: Garland, 1979.
———. *Military Improvisations During the Russian Campaign.* New York: Garland, 1979.
———. *Operations of Encircled Forces: German Experiences in Russia.* New York: Garland, 1979.
———. *Small Unit Actions During the German Campaign in Russia.* New York: Garland, 1979.
———. *Terrain Factors in the Russian Campaign.* New York: Garland, 1979.
US Department of Navy. *Articles for the Government of the Navy, 1930.* Washington, DC: Bureau of Navigation, 1932.
———. *History of the Medical Department of the U.S. Navy in World War II.* Washington, DC: Government Printing Office, 1953.

US Department of War. *Annual Reports, 1920–1941*. Washington, DC: Army Adjutant General's Office, 1920–1942.
———. *The Articles of War*. Washington, DC: Government Printing Office, 1920.
———. Courts-Martial Records Branch. *Statistics*. Washington, DC: Government Printing Office, 1959.
US Marine Corps. *Marine Corps Manual, 1940*. Washington, DC: Government Printing Office, 1940.
———. *Warfighting*. Washington, DC: US Government Printing Office, 1997.
US Marine Corps Schools. *Tactics and Technique of Infantry*. Quantico, VA, 1928.
US Office of Naval Operations, Division of Fleet Training. *Fleet Training Publication (F.T.P.) 167, Landing Operations Doctrine, United States Navy*. Washington, DC: Government Printing Office, 1938. http://www.ibiblio.org/hyperwar/USN/ref/Amphibious/index.html.
US Secretary of Navy. *Manual of the Medical Department*. Washington, DC: Government Printing Office, 1942.
Vaitkus, Mark, and James Griffith. "An Evaluation of Unit Replacement on Unit Cohesion and Individual Morale in the U.S. Army All-Volunteer Force." *Military Psychology* 2, no. 4 (2008): 221–239.
van Creveld, Martin. *Fighting Power: German and U.S. Army Performance, 1939–1945*. Westport: Greenwood, 1982.
Vandegrift, A. A., and Robert B. Asprey. *Once a Marine: The Memoirs of General A. A. Vandegrift, United States Marine Corps*. New York: Norton, 1964.
Vannoy, Allyn R. *Against the Panzers: United States Infantry Versus German Tanks, 1944–45: A History of Eight Battles Told Through Diaries, Unit Histories, and Interviews*. Jefferson, NC: McFarland, 1996.
Vansittart, Peter. *Voices from the Great War*. New York: Watts, 1984.
Vernon, Philip E. *Personnel Selection in the British Forces*. London: University of London Press, 1949.
von Altrock, Constantin. "Charakterbildung und Notzeit." *Militär-Wochenblatt* 116, no. 36 (1932): 1265–1270.
———, ed. "Das Ethos des Soldatenberufs." *Militär-Wochenblatt: Unabhängig Zeitschrift für die deutsche Wehrmacht* 116, no. 27 (January 1932): 970.
———. "Der Soldat und die nationale Revolution." *Militär-Wochenblatt* 118, no. 7 (August 1933): 209–211.
———. "Hitler." *Militär-Wochenblatt* 117, no. 40 (April 1933): 1301.
———. "Politizierung der Wehrmacht." *Militär-Wochenblatt* 116, no. 1 (July 1931): 2–3.
von Clausewitz, Carl. *On War*. Translated by Peter Paret and Michael Howard. Princeton: Princeton University Press, 1984.
von Rabenau, Friedrich. *Seeckt: Aus seinem Leben, 1918–1936*. Vol. 2. Leipzig: Hafe und Kohler Verlag, 1940.
von Schiller, Friedrich. *Wallenstein: A Historical Drama in Three Parts*. New York: Frederick Ungar, 1958.
von Seeckt, Johannes Friedrich "Hans." *Gedanken Eines Soldaten*. Berlin: Verlag für Kulturpolitik, 1929.
———. *Thoughts of a Soldier*. Translated by Gilbert Waterhouse. London: Ernest Benn, 1930.
Wagner, Margaret, David M. Kennedy, Linda Burnett Osborne, and Susan Rayburn. *The Library of Congress World War II Companion*. Illustrated ed. New York: Simon and Schuster, 2007.
Waites, B. A. "The Effect of the First World War on Class and Status in England, 1910–1920." *Journal of Contemporary History* no. 1 (1976): 27–48.
Wallach, Jehuda. "Feldmarschall Erich von Manstein und die deutsche Judenausrottung in Russland." *Jahrbuch des Instituts für deutsche Geschichte* 4 (1975): 457–472.

Ward, Candace, ed. *World War One British Poets: Brooke, Owen, Sassoon, Rosenberg, and Others.* Mineola, NY: Dover, 1997.
Ward, John, Colonel. "Labour and Its Relation to the Army." *Journal of the Royal United Services Institute* 66, no. 461 (1921): 22–31.
Watson, Alexander. "Culture and Combat in the Western World, 1900–1945." *Historical Journal* 51, no. 2 (June 2008): 529–546.
———. "Self-Deception and Survival: Mental Coping Strategies on the Western Front, 1914–18." *Journal of Contemporary History* 41, no. 2 (April 2006): 247–268.
Watson, Bruce. *Desert Battle: Comparative Perspectives.* Westport: Praeger, 1995.
———. *When Soldiers Quit: Studies in Military Disintegration.* Westport: Praeger, 1997.
Watson, Mark Skinner. *Chief of Staff: Prewar Plans and Preparations.* Washington, DC: Government Printing Office, 1950.
Wavell, A. P. "The Army and the Prophets." *Journal of the Royal United Services Institute* 75, no. 500 (1930): 665–675.
Wawro, Geoffrey. *The Franco-Prussian War: The German Conquest of France in 1870–1871.* New York: Cambridge University Press, 2003.
Weaver, Michael E. *Guard Wars: The 28th Infantry Division in World War II.* Bloomington: Indiana University Press, 2010.
Wegner, Bernd. *From Peace to War: Germany, Soviet Russia, and the World, 1939–1941.* Providence, RI: Berghahn, 1997.
———. *Hitlers politische Soldaten, Die Waffen-SS 1933–1945: Studien zu Leitbild, Struktur and Funktion einer nationalsozialistischen Elite.* Paderborn: Schöningh, 1982.
———. *Zwei Wege nach Moskau: Vom Hitler-Stalin-Pakt bis zum "Unternehmen Barbarossa."* Munich: Piper, 1991.
Weigley, Russell F. "American Strategy from Its Beginnings Through the First World War." In *Makers of Modern Strategy from Machiavelli to the Nuclear Age,* edited by Peter Paret, 408–443. Princeton: Princeton University Press, 1986.
———. *The American Way of War.* New York: Macmillan, 1973.
———. *Eisenhower's Lieutenants: The Campaign of France and Germany, 1944–1945.* Bloomington: Indiana University Press, 1981.
———. *History of the United States Army.* New York: Macmillan, 1967.
———. *New Dimensions in Military History.* San Rafael, CA: Presidio, 1975.
Weinberg, Gerhard. "Dokumentation: Adolf Hitler und der NS-Führungsoffizier (NSFO)." *Vierteljahrsjefte für Zeitsgeschichte* 12 (1964): 443–456.
———. "Unexplored Questions About the German Military During World War II." *Journal of Military History* 62, no. 2 (1998): 371–380.
Weinert, Erich. *Stalingrad Diary.* Translated by Egon Larsen. London: ING, 1944.
Welcker, Ingrid. *"Qualifikation zum Offizier?"* Frankfurt am Main: Peter Lang, 1982.
Wette, Wolfram, ed. *Der Krieg des kleinen Mannes: Eine Militärgeschichte von Unten.* Munich: Piper, 1992.
———. "'Es Roch nach ungeheuerlichem': Zeitzeugenbericht eines Panzerschützen über die Stimmung in einer Einheit des deutschen Ostheeres am Vorabend des Überfalls auf die Sowjetunion 1941." *Zeitschrift für Sozialgeschichte des 20. und 21.Jahrhunderts* 4, no. 4 (1989): 62–73.
———. *The Wehrmacht: History, Myth, Reality.* Translated by Deborah Lucas Schneider. London: Harvard University Press, 2006.
Wette, Wolfram, and Gerd R. Ueberschär, eds. *Stalingrad: Mythos und Wirklichkeit einer Schlacht.* Frankfurt am Main: Fischer Taschenbuch Verlag, 1992.
Wheeler, William. *The Letters of Private Wheeler.* Boston: Houghton Mifflin, 1952.
Whitaker, Leighton C., ed. *Understanding and Preventing Violence: The Psychology of Human Destructiveness.* Boca Raton, FL: CRC, 2000.

White, William Richard. *The Marine: A Guadalcanal Survivor's Final Battle.* Annapolis, MD: Naval Institute, 2002.
Whitehouse, Stanley, and George B. Bennett. *Fear Is the Foe: A Footslogger from Normandy to the Rhine.* London: Hale, 1995.
Whiting, Charles. *Death of a Division.* New York: Stein and Day, 1981.
———. *Poor Bloody Infantry.* London: Stanley Paul, 1987.
Whitlock, Flint. *The Fighting First: The Untold Story of the Big Red One on D-Day.* Boulder: Westview, 2004.
———. *The Rock of Anzio: From Sicily to Dachau—A History of the 45th Infantry Division.* Boulder: Westview, 1998.
Whitmarsh, Patrick J., and Robert H. Sulzen. "Prediction of Simulated Infantry-Combat Performance from a General Measure of Individual Aptitude." *Military Psychology* 1, no. 2 (1989): 111–116.
Whittow, Mark. *The Making of Byzantium, 600–1025.* Berkeley: University of California Press, 1996.
Whyte, William Hollingsworth. *A Time of War: Remembering Guadalcanal—A Battle Without Maps.* New York: Fordham University Press, 2000.
Wigmore, John H. "Modern Penal Methods in Our Army." *Journal of the American Institute of Criminal Law and Criminology* 9, no. 2 (August 1918): 163–172.
Wiley, Bell I., and William P. Govan. *Study no. 16: History of the Second Army.* Washington, DC: Historical Section, US Army Ground Forces, 1946.
Willems, Emilio. *A Way of Life and Death: Three Centuries of Prussian-German Militarism: An Anthropological Approach.* Nashville, TN: Vanderbilt University Press, 1986.
Willer, Robb. "Groups Reward Individual Sacrifice: The Status Solution to the Collective Action Problem." *American Sociological Review* 74, no. 1 (February 2009): 23–43.
Williams, Robin M., Jr. "Field Observations and Surveys in Combat Zones." *Social Psychology Quarterly* 47, no. 2 (June 1984): 186–192.
Williamson, Gordon. *Infantry Aces of the Reich.* London: Arms and Armour, 1991.
Willis, John T. "The United States Court of Military Appeals: Its Origin, Operation, and Future." *Military Law Review* no. 55 (January 1972): 39–94.
Wilmot, Chester. *The Struggle for Europe.* Old Saybrook, CT: Konecky and Konecky, 1952.
Wilson, George. *If You Survive.* New York: Ballantine, 1987.
Wilson, John B. *Maneuver and Firepower: Evolution of Divisions and Separate Brigades.* Washington, DC: Government Printing Office, 1999.
Wilson, Sloan. *The Man in the Gray Flannel Suit.* New York: Simon and Schuster, 1955.
Winter, Denis. *Death's Men: Soldiers of the Great War.* London: Penguin, 1979.
Winter, J. "Shell-Shock and the Cultural History of the Great War." *Journal of Contemporary History* 35, no. 1 (2000): 7–11.
Winter, J. M. "Britain's 'Lost Generation' of the First World War." *Population Studies* 31, no. 3 (November 1977): 449–466.
Woltersdorf, Hans Werner. *Gods of War: A Memoir of a German Soldier.* Novato, CA: Presidio, 1990.
Wool, Harold. *The Military Specialist: Skilled Manpower for the Armed Forces.* Baltimore: Johns Hopkins University Press, 1968.
Wormald, Patrick. "Germanic Power Structures: The Early English Experience." In *Power and the Nation in European History,* edited by Len Scales and Oliver Zimmer, 105–124. New York: Cambridge University Press, 2005.
Wright, Stephen L. *The Last Drop: Operation Varsity, March 24–25, 1945.* Mechanicsburg, PA: Stackpole, 2008.

Wüllner, Fritz. *Die NS-Militärjustiz und das Elend der Geschichtsschreibung: Ein grundlengender Forschungsbericht.* Baden-Baden: Nomos Verlagsgesellschaft, 1997.
Wyatt, Thomas C., and Reuven Gal, eds. *Legitimacy and Commitment in the Military.* New York: Greenwood, 1990.
Zamulin, Valerii Nikolaevich, and Stuart Britton. *Demolishing the Myth: The Tank Battle at Prokhorovka, Kursk, July 1943—An Operational Narrative.* Solihull, UK: Helion, 2010.
Zeidler, Manfred. *Reichswehr und Rote Armee, 1920–1933: Wege und Stationen einer ungewöhnlichen Zusammenarbeit.* Munich: R. Oldenbourg, 1993.
Zeidler, Manfred, and Ute Schmidt. *Gefangene in deutschem und sowjetischem Gewahrsam, 1941–1956: Dimensionen und Definitionen.* Dresden: Hannah-Arendt-Institut für Totalitarismusforschung, 1999.
Zetterling, Niklas, and Anders Frankson. *Kursk 1943: A Statistical Analysis.* Portland, OR: Cass, 2000.
Ziemke, Earl F. "Franz Halder at Orsha: The German General Staff Seeks a Consensus." *Military Affairs* 39, no. 4 (1975): 173–176.
Zimmermann, Bodo. *Die (neue) Gruppe.* Berlin: Verlag "Offene Worte," 1933.
Zoepf, Arne W. G. *Wehrmacht zwischen Tradition und Ideologie: Der NS-Führungsoffizier im zweiten Weltkrieg.* Frankfurt am Main: P. Lang, 1988.
Zydowitz, Kurt von. *Die Geschichte Der 58. Infanterie-Division, 1939–1945.* Kiel: H. H. Podzun, 1952.

Index

About the Book

IT HAS LONG BEEN ACCEPTED WISDOM THAT GERMANY'S infantrymen possessed superior tactical ability relative to their Anglo-American adversaries in World War II. Now, drawing on newly available information, Stephen Lauer unpacks that assumption, exploring the conscription, classification, and training methods of the US, British, and German infantries from 1919 through 1945.

How did conscripted citizens become foot soldiers willing to fight, and even die, for each other in the face of brutal physical and mental demands? How was it decided which men to assign to combat units? How did each nation engender the social bonds that were essential if soldiers were to succeed—and survive—in their small unit milieus? Addressing these questions of manpower quality, *Forging the Anvil* is a landmark study of the key factors that influenced the creation of World War II infantries and sustained them in the crucible of close combat.

The late **G. Stephen Lauer** was associate professor of history and theory in the School of Advanced Military Studies, US Army Command and General Staff College, at Fort Leavenworth and a retired lieutenant colonel in the US Marine Corps.